Advances in Experimental Medicine and Biology

Volume 937

More information about this series at http://www.springer.com/series/5584

Ondrej Slaby • George A. Calin
Editors

Non-coding RNAs in Colorectal Cancer

Springer

Editors
Ondrej Slaby
Central European Institute
of Technology
Masaryk University
Brno, Czech Republic

George A. Calin
Department of Experimental
Therapeutics Division of Cancer
Medicine
The University of Texas MD Anderson
Cancer Center
Houston, TX, USA

ISSN 0065-2598 ISSN 2214-8019 (electronic)
Advances in Experimental Medicine and Biology
ISBN 978-3-319-42057-8 ISBN 978-3-319-42059-2 (eBook)
DOI 10.1007/978-3-319-42059-2

Library of Congress Control Number: 2016949077

Printed on acid-free paper

This Springer imprint is published by Springer Nature
The registered company is Springer International Publishing AG Switzerland

Preface

It was assumed that humans, being highly complex organisms, would have many more genes than less complex organisms. However, the completion of the Human Genome Project estimated the number of human genes to be between 20,000 and 25,000, which is similar to genome of *Caenorhabditis elegans* (roundworm), estimated to have around 20,000 genes, and the number of mice genes. This revelation meant that organism complexity could not be mainly the result of a higher number of protein-coding genes. Although there was no correlation between complexity and the number of genes, there was a clear correlation with the relative amount of noncoding sequences in the genome. In humans, only around 3 % of the genome is protein coding, while the rest consists of introns, regulatory sequences, and noncoding RNA. These days, 13 years after the completion of the Human Genome Project, research has rapidly progressed, and we are now beginning to understand the importance of noncoding sequences in cellular regulatory processes. In cancer, noncoding RNAs function as regulatory molecules acting as oncogenes and tumor suppressors with very important roles in cancer biology.

This edited volume reflects the current state of knowledge about the roles of noncoding RNAs in the formation and progression of colorectal cancer and the potential translation of this knowledge to diagnosis and therapy of the disease. The main focus lies on involvement of noncoding RNAs in molecular pathology of colorectal cancer, together with cutting-edge translational research performed to transfer noncoding RNAs from bench to the bedside. We are sure that the emergence of noncoding RNAs represents a new dimension of colorectal cancer pathogenesis and it will be absolutely necessary to consider that in future translational studies. This book will be a state-of-the-art resource for scientists or physicians starting out with noncoding RNA research in colorectal cancer but is also intended for the experienced researchers who want to incorporate noncoding RNA concepts into their colorectal cancer research.

Brno, Czech Republic — Ondrej Slaby
Houston, TX, USA — George A. Calin

Contents

Part I Non-coding RNAs: Biology and Implications in Colorectal Cancer Pathogenesis

1 Non-coding RNAs: Classification, Biology and Functioning 3
Sonja Hombach and Markus Kretz

2 Involvement of Non-coding RNAs in the Signaling Pathways of Colorectal Cancer 19
Yinxue Yang, Yong Du, Xiaoming Liu, and William C. Cho

3 MicroRNAs and Inflammation in Colorectal Cancer 53
Claire Josse and Vincent Bours

4 Interplay Between Transcription Factors and MicroRNAs Regulating Epithelial-Mesenchymal Transitions in Colorectal Cancer 71
Markus Kaller and Heiko Hermeking

5 Non-coding RNAs Functioning in Colorectal Cancer Stem Cells 93
Daniele Fanale, Nadia Barraco, Angela Listì, Viviana Bazan, and Antonio Russo

6 MicroRNA Methylation in Colorectal Cancer 109
Sippy Kaur, Johanna E. Lotsari-Salomaa, Riitta Seppänen-Kaijansinkko, and Päivi Peltomäki

7 Polymorphisms in Non-coding RNA Genes and Their Targets Sites as Risk Factors of Sporadic Colorectal Cancer 123
Pavel Vodicka, Barbara Pardini, Veronika Vymetalkova, and Alessio Naccarati

Part II Non-coding RNAs: New Class of Biomarkers in Colorectal Cancer

8 Non-coding RNAs as Biomarkers for Colorectal Cancer Screening and Early Detection 153
Ondrej Slaby

9 Circulating Non-coding RNA as Biomarkers in Colorectal Cancer ... 171
Manuela Ferracin, Laura Lupini, Alessandra Mangolini, and Massimo Negrini

10 Non-coding RNAs Enabling Prognostic Stratification and Prediction of Therapeutic Response in Colorectal Cancer Patients ... 183
Samantha O. Perakis, Joseph E. Thomas, and Martin Pichler

Part III Non-coding RNAs: Therapeutic Targets and Colorectal Cancer Therapeutics

11 Involvement of Non-coding RNAs in Chemo- and Radioresistance of Colorectal Cancer ... 207
Daniele Fanale, Marta Castiglia, Viviana Bazan, and Antonio Russo

12 Non-coding RNAs: Therapeutic Strategies and Delivery Systems ... 229
Hui Ling

13 MicroRNAs as Therapeutic Targets and Colorectal Cancer Therapeutics ... 239
Hirofumi Yamamoto and Masaki Mori

Index ... 249

Part I

Non-coding RNAs: Biology and Implications in Colorectal Cancer Pathogenesis

Non-coding RNAs: Classification, Biology and Functioning

1

Sonja Hombach and Markus Kretz

Abstract

One of the long-standing principles of molecular biology is that DNA acts as a template for transcription of messenger RNAs, which serve as blueprints for protein translation. A rapidly growing number of exceptions to this rule have been reported over the past decades: they include long known classes of RNAs involved in translation such as transfer RNAs and ribosomal RNAs, small nuclear RNAs involved in splicing events, and small nucleolar RNAs mainly involved in the modification of other small RNAs, such as ribosomal RNAs and transfer RNAs. More recently, several classes of short regulatory non-coding RNAs, including piwi-associated RNAs, endogenous short-interfering RNAs and microRNAs have been discovered in mammals, which act as key regulators of gene expression in many different cellular pathways and systems. Additionally, the human genome encodes several thousand long non-protein coding RNAs >200 nucleotides in length, some of which play crucial roles in a variety of biological processes such as epigenetic control of chromatin, promoter-specific gene regulation, mRNA stability, X-chromosome inactivation and imprinting. In this chapter, we will introduce several classes of short and long non-coding RNAs, describe their diverse roles in mammalian gene regulation and give examples for known modes of action.

Keywords

Non-coding RNA • miRNA • piRNA • snoRNA • snRNA • tRNA • rRNA • lncRNA • Classification • Biogenesis • Function

1.1 Introduction

Gene expression is a central process required for all aspects of life, and its regulation defines development and homeostasis of all cells and

S. Hombach (✉) • M. Kretz
Institute of Biochemistry, Genetics and Microbiology, University of Regensburg, Regensburg, Germany
e-mail: sonja.hombach@ur.de; markus.kretz@vkl.uni-regensburg.de

O. Slaby, G.A. Calin (eds.), *Non-coding RNAs in Colorectal Cancer*, Advances in Experimental Medicine and Biology 937, DOI 10.1007/978-3-319-42059-2_1

tissues. A central component of this process is the genomic DNA localized in the cell nucleus, serving as a template for the transcription of messenger RNAs, which in turn translocate into the cytoplasm and act as blueprints for the translation of proteins. Several classes of non-protein coding RNAs (ncRNAs) are needed for these processes to function: Small nuclear RNAs (snRNAs) are mainly involved in splicing events of mRNAs. Transfer RNAs (tRNAs) decode the mRNA sequence into peptide or protein by specifically recognizing three-nucleotide sequences of mRNAs and recruiting amino acids to the ribosome in the right order. Ribosomal RNAs (rRNAs) are thought to represent the most abundant RNA molecules in the cell and form the framework of ribosomes, macromolecular structures essential for protein translation. These housekeeping RNAs are constitutively expressed and essential for normal function of the cell. A significant portion of these housekeeping RNAs may carry chemical modifications which are added by a class of small nucleolar RNAs (snoRNAs) [1].

The discovery of small regulatory ncRNAs in the 1990s completely changed our understanding of ncRNAs as regulatory molecules. Andrew Fire, Craig Mello et al. could show that small double-stranded RNAs (dsRNAs) were able to mediate post-transcriptional gene silencing of complementary mRNAs in the nematode Caenorhabditis elegans by a process called RNA interference [2, 3]. Quickly, endogenous dsRNAs such as small interfering RNAs (siRNAs) and microRNAs (miRNAs) were found in a multitude of organisms, such as plants, flies and mammals. A growing number of new regulatory small RNA classes has been discovered in recent years [4], with piwi-associated RNAs, miRNAs and siRNAs belonging to the best investigated classes to date [5–12].

With the rapidly increasing development of high throughput, in depth transcriptome sequencing techniques, our understanding of the protein-coding and non-coding portion of the mammalian transcriptome increased exponentially. Recent studies suggest that while about two thirds of the mammalian genome is actively transcribed, only approximately 1,9 % encodes for proteins [13–15]. A significant portion of this transcriptional activity appears to be represented by another class of regulatory ncRNAs, the long non-coding RNAs (lncRNAs). These RNA molecules are characterized by a length of at least 200 nucleotides, a lack of protein-coding potential and often harbor a poly-A tail and can be spliced, similar to mRNAs. While an estimated abundance of 5400 to more than 10,000 lncRNA transcripts has been reported in humans [15–17], exact genomic annotations and functional significance are still unknown for many lncRNAs to date. Nevertheless, a rapidly growing number of lncRNAs have been shown to play crucial roles in a variety of biological processes such as epigenetic control of gene expression, promoter-specific gene regulation [18–20], X-chromosome inactivation [21–23], imprinting [24–28], maintenance of nuclear architecture [29–31].

Both small and long regulatory non-coding RNAs have been implicated in many different diseases and many types of cancer [32–34]. In the following paragraphs, several functional characteristics of short and long ncRNAs will be elucidated and examples for known modes of actions in gene regulatory processes will be discussed.

1.2 Characteristics and Modes of Action of Short Regulatory Non-coding RNAs

Small ncRNAs have emerged as key regulators of gene expression in many different cellular pathways and systems. A multitude of small regulatory RNA classes have been identified in recent years, with miRNAs, piRNAs and siRNAs being the most thoroughly investigated classes.

1.2.1 MicroRNAs

MiRNAs directly interact with partially complementary target sites located in the 3′ untranslated region of target mRNAs and repress their expression [35]. They play essential roles during differentiation and development. More than 60 % of all mRNAs are estimated to contain miRNA target sites at their 3′UTR region, suggesting a tight regulation as well as their involvement in normal cellular homeostasis and in diseased states [36]. In addition, it has been shown that many miRNAs are able to target up to several hundred mRNAs, suggesting a complex and combinatorial mode of miRNA action in mRNA regulation [37]. In recent years, a growing number of studies could reveal the involvement of miRNAs in the development of a multitude of diseases [33, 38], among them different types of cancers [38], heart diseases such as hypertrophy and ischemia [36, 39, 40] as well as associations to mental disorders such as schizophrenia or major depression disorders [41].

In general, mammalian miRNAs are genomically encoded and transcribed by RNA Polymerase II as primary miRNA transcripts (pri-miRNAs), which get processed by the microprocessor complex consisting of the RNAse III enzyme Drosha and the pri-miRNA binding protein DGCR8 (Fig. 1.1). The resulting pre-miRNA gets translocated to the cytosol by exportin5, where it is further processed into an approximately 21nt long dsRNA by the RNAse III enzyme Dicer and one of its two mammalian cofactors TRBP or PACT. Subsequently, either arm of this dsRNA gets incorporated into the RNA-induced silencing complex (RISC). Within this complex, the now mature single-stranded miRNA directly binds a member of the Argonaute protein family and acts as a guide to partially complementary regions predominantly located within the 3′-UTR of target mRNAs. Subsequent binding of TNRC6 proteins play a pivotal role for all downstream events leading to translational repression and degradation of the target mRNA in animals. Interaction of TNRC6 with the poly(A)-binding protein (PABP) appears to interfere with PABP function in protein translation, likely by interrupting the interaction between the 5′-cap structure and the 3′-poly(A) tail of the mRNA. Subsequently, degradation of the target mRNA is initiated by deadenylation and decapping, making the mRNAs accessible for exoribonucleases [12, 32, 42, 43]. Some miRNAs are encoded by a gene cluster, such as the miR-17~92 gene cluster. A recent study showed that processing of the pri-miR-17~92a transcript containing all six miRNAs from this cluster, includes formation of a processing intermediate called progenitor-miRNA (pro-miRNA) [44]. While the pro-miRNA is efficiently processed by the microprocessor complex (Fig. 1.1), the pri-miR-17~92a on the other hand can adopt an RNA conformation blocking the DROSHA/DGCR8 complex, thus acting as a posttranscriptional regulator of miR-17~92a processing in embryonic stem cells. Biogenesis of the pro-miRNA is mediated by the ribonuclease CPSF3 and the splicing factor ISY1 as well as other U2 snRNP components. An increase of ISY1 expression during embryonic stem cell differentiation appears to induce processing of all miRNAs within the cluster except for miR-92, which seems to be independently processed at the pri-miR-17~92a stage. Thus, the developmentally regulated formation of pro-miRNA in differentiating cells as a processing intermediate for miR-17~92 adds an additional layer to post-transcriptional control of miR-17~92 expression. Whether this mechanism is present in other miRNA clusters or represents a unique feature of miR-17~92, remains to be determined.

1.2.2 piRNAs

Unlike siRNAs and miRNAs, piRNAs are not processed by the RNAse III enzyme Dicer and are incorporated into the PIWI subfamily of AGO proteins. While miRNAs are widely expressed in most mammalian cells and tissues, piRNAs in mammals appear to mainly function in the germline where they target and repress expression of transposable and repetitive elements to maintain genomic stability [10, 45]. PiRNAs are located in gene clusters enriched in mobile repetitive elements and are transcribed as long primary RNAs that are further processed to primary piRNAs, which in turn act as guides for the generation of secondary piRNAs. Biogenesis of mature piRNAs varies between mammals, flies and nematodes, but their crucial role in maintaining

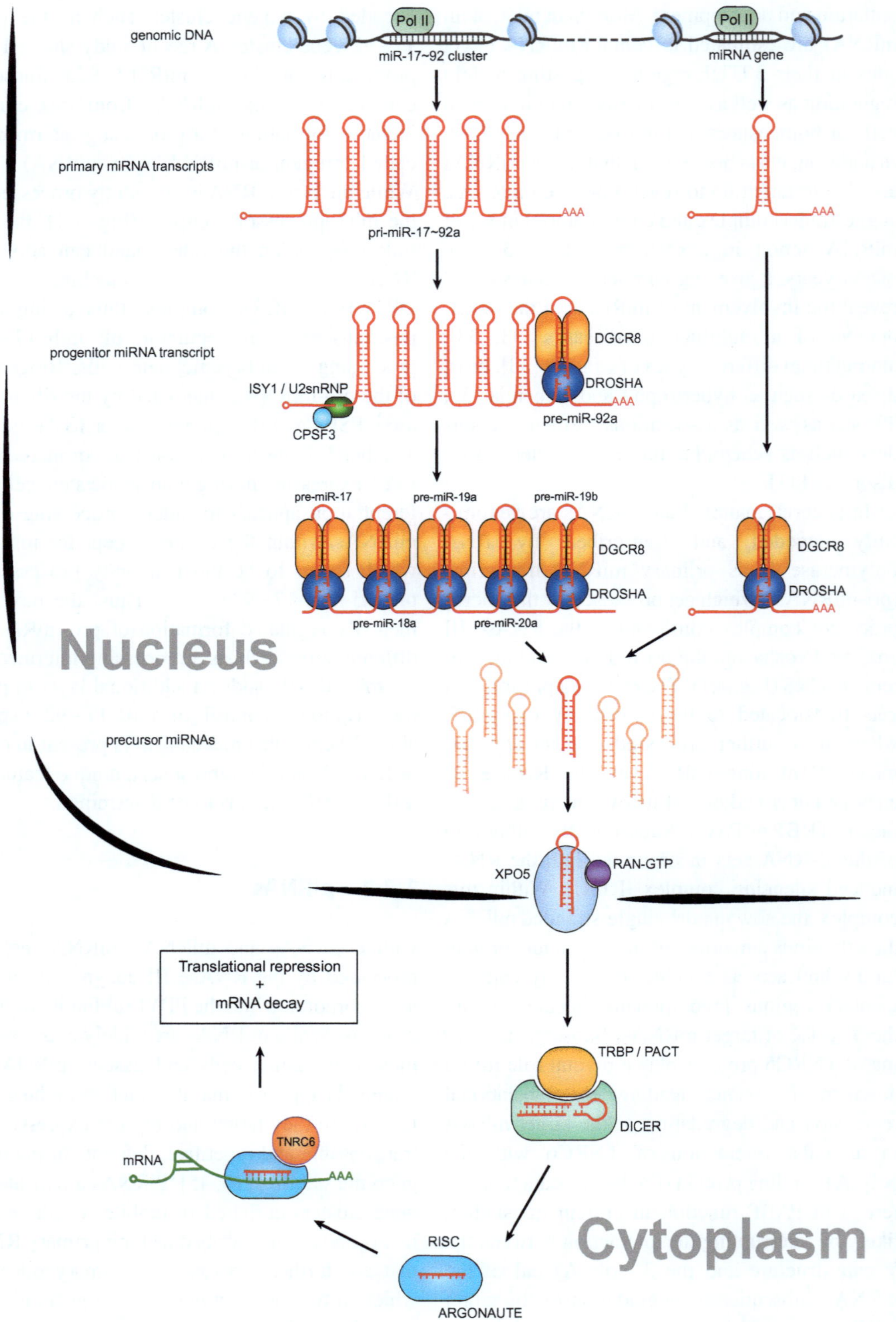
genomic DNA
Pol II
miR-17~92 cluster
Pol II
miRNA gene
primary miRNA transcripts
AAA
AAA
pri-miR-17~92a
progenitor miRNA transcript
DGCR8
DROSHA
ISY1 / U2snRNP
AAA
CPSF3
pre-miR-92a
pre-miR-17
pre-miR-19a
pre-miR-19b
DGCR8
DROSHA
DGCR8
DROSHA
AAA
pre-miR-18a
pre-miR-20a
Nucleus
precursor miRNAs
XPO5
RAN-GTP
Translational repression
+
mRNA decay
TRBP / PACT
DICER
TNRC6
mRNA
AAA
RISC
ARGONAUTE
Cytoplasm

genomic stability by targeting and repression of repetitive transposable elements was verified in multiple organisms using transgenic model systems. Deficiency of two proteins of the PIWI family in mice for example, resulted in activation of retrotransposons in the male germ line, arrest of gametogenesis and male sterility [10]. This result strongly indicates that piRNAs seem to be functionally important for maintaining genomic stability specifically in the male germline. Another group of small regulatory RNAs, called endogenous siRNAs (endo-siRNAs), seem to control silencing of repetitive transposable elements in the female germline of mammals.

1.2.3 siRNAs

Besides miRNAs, a number of additional RNAi pathways utilize dsRNA to generate mature regulatory small RNAs through cleavage by the RNAse III enzyme Dicer. In the case of endogenous siRNAs (endo-siRNAs), dsRNA structures were shown to originate from extended hairpin structures or from base-pairing of sense and antisense transcripts originating either from bidirectional transcription or complementary transcripts derived from separate loci [7, 46]. Exogenous siRNAs (exo-siRNAs) on the other hand, are processed from dsRNAs taken up from the cellular environment. This mechanism has been shown in nematodes, and also appears to play a role in antiviral defense mechanisms in plants and drosophila, where upon infection, viral dsRNA gets processed to siRNAs targeting the viral mRNAs they originate from [47, 48]. Recent work suggests a similar mechanism in mammalian stem- and germ cells. Most differentiated cells possess a protein-mediated interferon response system, which mediates anti-viral responses upon infiltration of viral long dsRNAs [49]. Several recent reports suggest, that in pluripotent cells this mechanism appears to be at least partially replaced by the RNAi pathway as a cellular anti-viral response system [50–53].

Interestingly, endo-siRNAs seem to also play a role in silencing transposable elements in mammals. While the piRNA pathway acts as major defense mechanism against repetitive transposable elements in the mammalian testes, transposable elements of the female germline give rise to dsRNAs, which can be processed into endo-siRNAs. Thus, endo-siRNAs contribute to maintenance of genomic stability in the female germline [54, 55].

1.2.4 Regulatory Functions of Housekeeping ncRNAs

Our understanding of the complexity of housekeeping and regulatory ncRNAs dramatically changed with recent findings suggesting that some of the housekeeping small ncRNAs with well described functions might be processed to small regulatory ncRNAs. While snoRNAs canonically act as guide for chemical modifications of nucleotides on other small housekeeping RNAs, some of them were shown to be processed to small RNAs with post-transcriptional gene silencing functionality similar to miRNAs [56–59].

Fig. 1.1 MicroRNA biogenesis in mammals. In mammals, miRNAs are transcribed by RNA Polymerase II (Pol II) as primary miRNA transcripts (pri-mRNAs). Processing by DROSHA together with DGCR8 results in the formation of miRNA precursors (pre-miRNAs), which get transported to the cytoplasm by exportin5 (XPO5). The RNAse III enzyme DICER, together with TRBP or PACT process the pre-miRNA into an approximately 21nt long dsRNA which gets incorporated into one of the argonaute proteins. One strand of the miRNA acts as a guide to partially complementary regions of target mRNAs. Binding of one of the TNRC6 proteins to argonaute is prerequisite for processes leading to translational repression and degradation of the mRNA. Posttranscriptional control of the miR-17 ~ 92 cluster partially occurs through formation of a processing intermediate called progenitor-miRNA (pro-miRNA). The pri-miR-17 ~ 92a can adopt a RNA conformation blocking microprocessor in embryonic stem cells. Upon embryonic stem cell differentiation, the splicosome-associated protein ISY1, which is required for pro-miRNA biogenesis, promotes processing of pri-miR-17 ~ 92a into the pro-miRNA intermediate with high affinity for DROSHA/DGCR8

Additionally, several snoRNAs were recently shown to be involved in cancer development. *SNORD50A/B* snoRNAs for example are recurrently lost in cancer. The mature snoRNAs bind and suppress the activity of Ras oncoproteins and SNORD50A/B deficiency was shown to enhance the abundance of active K-Ras resulting in hyperactivation oft he ERK1/2 MAPK pathway [60].

Also surprisingly, precursor and mature tRNAs can be processed to tRNA-derived fragments (tRFs) which were recently shown to play important biological roles independent of the canonical, full-length tRNA function [61, 62]. tRNA-glycine-GCC fragments for example can repress expression of transcripts driven by endogenous retroelements [63]. In a different study, a set of tRFs was able to bind to the oncogenic RNA-binding protein YBX1, thus preventing interaction of YBX1 with pro-ongogenic transcripts, which resulted in reduced transcript stability and less metastasis [64]. Whether fragments derived from other small RNAs, such as rRNAs or snoRNAs might function in similar ways to regulate binding of RNA-binding proteins is conceivable but so far not experimentally proven.

1.3 Characteristics and Functions of Long Non-coding RNAs

Long non-coding RNAs represent a highly divers group of regulatory ncRNAs with respect to characteristics, localization and modes of action [65]. Except for the minimum size limit of 200 nt and a lack of protein-coding potential, there are few structural, functional or mechanistic features common to all mammalian lncRNAs. Additionally, only a fraction of the many thousand predicted mammalian lncRNAs have been thoroughly mechanistically characterized to date, with even fewer being functionally verified in vivo [66]. While these circumstances make attempts of comprehensive classification of lncRNAs exceedingly difficult, several review articles categorize functionally analyzed lncRNAs based on similarities in their modes of action [65, 67–71].

1.3.1 LncRNA Modes of Action in the Nucleus

Subcellular localization of lncRNAs can be a good initial indicator to narrow down potential modes of action. Many nuclear lncRNAs are functionally implicated in gene regulatory processes. These can include promoter-specific repression or activation of transcription or epigenetic gene regulation.

1.3.1.1 Transcriptional Regulation by lncRNAs

A lncRNA transcribed upstream of the dihydrofolate reductase (DHFR) gene locus was shown to directly regulate transcription by interacting with the transcription preinitiation complex at the DHFR promoter (Fig. 1.2a). In quiescent cells, binding of this lncRNA to transcription factor IIB results in dissociation of the pre-initiation complex from the major promoter leading to repression of DHFR transcription [18]. Interestingly, the formation of a triplex structure of the lncRNA with the major promoter might be required for this process.

Another lncRNA, EVF2, appears to act as a transcriptional modulator of the two homeodomain transcription factors Dlx5 and Dlx6 in the ventral forebrain of mice to control GABAergic interneuron circuitry. EVF2 is transcribed across an ultraconserved enhancer between both transcription factors and appears to inhibit enhancer methylation in trans to modulate association of transcriptional repressors and activators [72]. Additionally, Dlx6 transcription might be controlled in cis, most likely mediated by the act of EVF2 transcription through competitive antisense inhibition [19, 20].

EVF2 is not the only lncRNA regulating transcriptional activity at enhancer sites. A number of enhancer-associated RNAs were shown to function as transcriptional regulators, mostly by regulating enhancer activity (Fig. 1.2b). A recent report describes a p53-induced lncRNA named LED, which interacts with and activates p53-regulated enhancers, thus regulating expression of so-called enhancer RNAs (eRNAs). Unlike enhancer-associated lncRNAs, these eRNAs are

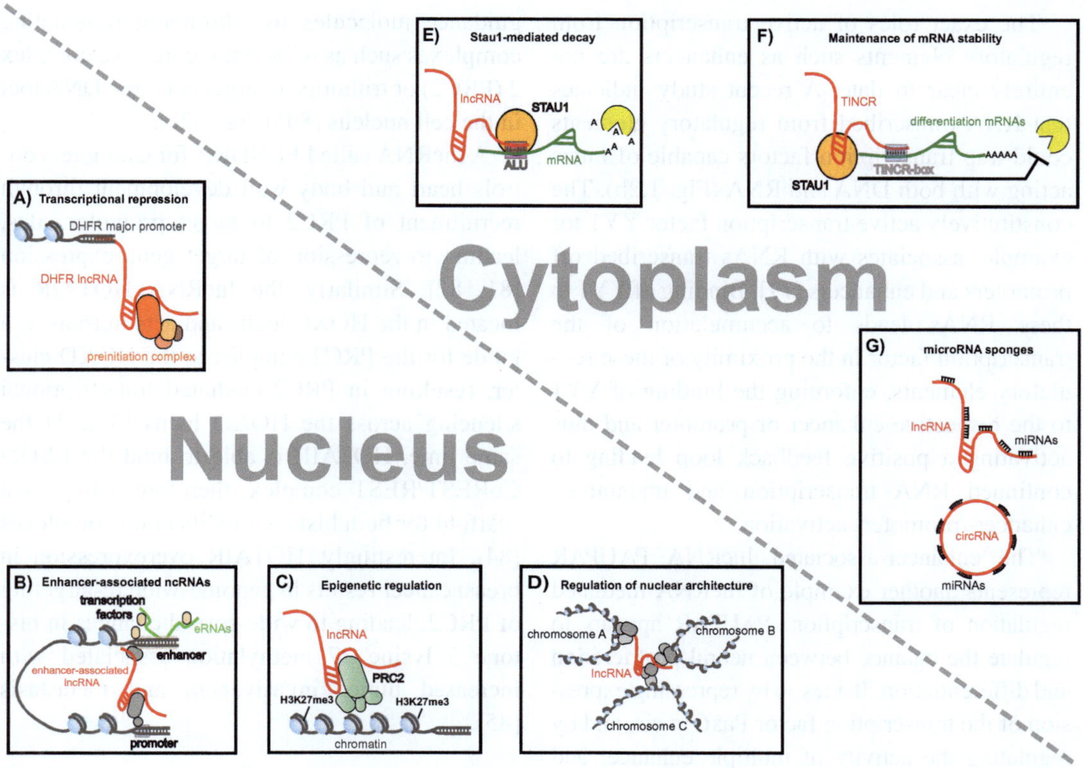

Fig. 1.2 Examples for molecular mechanisms of long non-coding RNA function. A-D: Nuclear functions of lncRNAs. (**a**) A lncRNA transcribed from the DHFR minor promoter can repress DHFR transcription by dissociating the transcription preinitiation complex at the DHFR major promoter. This process likely involves formation of a triplex structure of the lncRNA with the major promoter site. (**b**) Actively transcribed enhancer elements can give rise to enhancer RNAs or enhancer-associated RNAs. eRNAs can trap transcription factors at the enhancer and several enhancer associated lncRNAs were shown to regulate transcription in cis and in trans. (**c**) LncRNAs can regulate epigenetic control of gene expression by acting as guidance molecules for chromatin modifying protein complexes (here shown with polycomb repressive complex 2 (PRC2) as an example). (**d**) LncRNAs are involved in the creation and maintenance of nuclear structures, and the coordination of nuclear architecture across several chromosomes. (**e–g**): Cytoplasmic functions of lncRNAs. (**e**) LncRNAs can target mRNAs for degradation by a process called Staufen 1 (STAU1)-mediated decay. Base-pairing of ALU elements present in the lncRNA and the target mRNA creates a double-stranded STAU1 binding site. (**f**) The lncRNA TINCR regulates epidermal differentiation by binding to STAU1 and differentiation mRNAs containing the TINCR box motif, resulting in increased mRNA stability. (**g**) Several lncRNAs, mRNAs and circular RNAs (circRNAs) can function as competing endogenous RNAs (ceRNAs) by sequestering microRNAs (miRNAs) away from their mRNA targets

relatively short, polymerase II-transcribed, mostly not polyadenylated, bidirectional transcripts, which harbor monomethyl Histone H3 lysine 4 marks [73]. They play important roles in regulating the activity of enhancers to regulate target gene expression, in some cases by supporting the formation of DNA loops to bring the enhancers in close proximity to nearby promoter regions [74, 75]. Activation of LED-mediated enhancer RNA (eRNA) expression is thought to occur through modulation of histone modifications at the enhancer element. LED occupancy leads to acetylation of histon3 at lysine 9 residues, a modification correlated to active transcription [76].

The exact roles of active transcription from regulatory elements such as enhancers are not entirely clear to date. A recent study indicates that RNA transcribed from regulatory elements could trap transcription factors capable of interacting with both DNA and RNA (Fig. 1.2b). The constitutively active transcription factor YY1 for example, associates with RNAs transcribed off promoters and enhancers [77]. Binding of YY1 to these RNAs leads to accumulation of the transcription factor in the proximity of these regulatory elements, enforcing the binding of YY1 to the respective enhancer or promoter and thus activating a positive feedback loop leading to continued RNA transcription and maintained enhancer-/promoter- activation.

The enhancer-associated lncRNA PAUPAR represents another example of lncRNA-mediated regulation of transcription. PAUPAR appears to regulate the balance between neural proliferation and differentiation. It does so by repressing expression of the transcription factor Pax6 in cis, and by regulating the activity of multiple enhancer and repressor elements as well as promoters across multiple chromosomes in trans, in part through direct association with Pax6, thus modulating a complex transcriptional program to control cell cycle and repress neural differentiation [78].

A protein complex called Mediator acts as a transcriptional co-activator by linking transcription factors to basal transcription machinery. This process was recently shown to involve a set of lncRNAs with enhancer-like function [79]. Interaction of these lncRNAs with the Mediator complex regulates its chromatin localization (possibly by facilitating chromatin looping) and kinase activity.

1.3.1.2 Epigenetic Control of Gene Regulation Mediated by lncRNAs

Besides their involvement in promoter specific regulation of transcription, lncRNAs are also implicated in epigenetic gene regulatory processes, activating or repressing multiple gene loci at once through modulating the accessibility of chromatin. In fact, a large number of lncRNAs functionally characterized to date appear to act as guidance molecules for chromatin remodeling complexes such as polycomb repressive complex 2 (PRC2) or trithorax to target genomic DNA loci in the cell nucleus [80] (Fig. 1.2c).

A lncRNA called FENDRR for example, controls heart and body wall development through recruitment of PRC2 to target promoter sites, leading to repression of target gene expression [81, 82]. Similarly, the lncRNA HOTAIR is located in the HOXC locus and acts in trans as a guide for the PRC2 complex to the HOXD cluster, resulting in PRC2-mediated transcriptional silencing across the HOXD locus [83]. At the same time, HOTAIR is able to bind the LSD1/CoREST/REST complex, therefore acting as a scaffold for both histone modification complexes [84]. Interestingly, HOTAIR overexpression in breast cancer results in genome-wide re-targeting of PRC2, leading to wide-spread changes in histone 3 lysine 27 methylation associated with increased tumor invasiveness and metastasis [85].

1.3.1.3 Regulation of Nuclear Architecture by lncRNAs

Several nuclear lncRNAs appear to have roles not directly connected to control of gene expression, but were found to regulate the maintenance of nuclear architecture. Paraspeckles are nuclear structures believed to contribute to the nuclear retention of mRNAs that have undergone adenosine to inosine hyperediting and are known to contain the lncRNA NEAT1 [86]. Loss of NEAT1 results in loss of paraspeckles and induction of cytoplasmic export of mRNAs containing inverted Alu repeats, (more then 90 % of editing events mediated by dsRNA-dependent adenosine deaminases occur in inverted repeated Alu elements) [87]. Correspondingly, overexpression of NEAT1 increases paraspeckle abundance, with new paraspeckles solely originating from NEAT1 transcription sites, indicating a prominent role for NEAT1 in paraspeckle formation [88]. Indeed, live-cell imaging analysis of paraspeckle de novo assembly showed that this process appears to be dependent on NEAT1 transcription, and the lncRNA itself might serve as recruitment platform for paraspeckle proteins [29].

The X-linked lncRNA FIRRE represents another example for a lncRNA functioning as modulator of nuclear architecture. In male cells, FIRRE together with the nuclear matrix protein hnRNP-U appears to coordinate the nuclear architecture across chromosomes, potentially acting as scaffold for trans-chromosomal interactions (Fig. 1.2d) involved in regulation of pluripotency pathways in male embryonic stem cells [30]. More recently, a direct role for FIRRE in relation to X chromosome inactivation has been discovered. In mammals, one of the two female X chromosomes is randomly silenced to balance the dosage of X-linked gene expression between the sexes. The inactivated X chromosome becomes heterochromatic and gets localized near the nucleolus. Interestingly, this process is involves coating of the inactive X chromosome by yet another lncRNA called XIST, which recruits the PRC2 complex, resulting in tri-methylation of histone 3, lysine 27 (H3K27me3) across the whole chromosome– an epigenetic mark required for chromatin compaction [89]. FIRRE is located on the X chromosome and expressed on both X chromosomes before and after X chromosome inactivation. It is involved in positioning of the inactivated X chromosome near the nucleolus and helps maintaining H3K27me3 marks through a mechanism not yet completely understood [31].

1.3.2 LncRNA Modes of Action in the Cytoplasm

LncRNAs localized in the cytoplasm were shown to be involved in post-transcriptional gene regulatory processes, such as modulation of mRNA stability or regulation of miRNA accessibility, translation and signal transduction pathways [65, 69, 90].

1.3.2.1 Control of mRNA Stability by lncRNAs

The dsRNA-binding protein Staufen-1 (STAU1) was recently shown to promote mRNA degradation by a process called Staufen-1 mediated mRNA decay. Prerequisite for STAU1 targeting is the presence of a double stranded binding region within the target mRNA. In some cases, such STAU1 binding sites can be created through involvement of ALU repeat element-containing lncRNAs (Fig. 1.2e). Imperfect base-pairing between ALU elements of the lncRNA and a mRNA target of STAU1 created a double-stranded RNA region functioning as a binding site for STAU1 and resulting in STAU1 mediated degradation of the mRNA [91, 92]. Thus, the lncRNA appears to act as a specificity factor for targeting of mRNAs to STAU1.

At the same time, association of STAU1 with a lncRNA involved in epidermal tissue differentiation can result in stabilization of target mRNAs. The epidermis is a stratified surface epithelium that provides a barrier to the external environment. A precise balance between the progenitor compartment and terminally differentiated layers is needed to ensure formation of a functional epidermis with an intact water barrier [93]. The lncRNA TINCR is mainly located in the cytoplasm of highly differentiated keratinocytes and required for induction of key differentiation genes in epidermal tissue, including genes mutated in human skin diseases characterized by disrupted epidermal barrier formation [94]. TINCR directly interacts with STAU1 protein as well as differentiation-specific mRNAs through a 25-nucleotide motif strongly enriched in interacting RNAs as well as TINCR itself. The lncRNA TINCR together with the STAU1 protein appears to stabilize a set of associated differentiation mRNAs (Fig. 1.2f). Similar to the role of lncRNAs in STAU1-mediated decay, TINCR might act as a guidance molecule and thus provide specificity for mRNAs to be targeted to the STAU1 protein [95]. Interestingly, TINCR also appears to stabilize mRNAs of MAF and MAFB, coding for two transcription factors, as well as CALML5, all acting as key regulators of epidermal differentiation [96, 97]. The mechanism by which stabilization of these mRNAs occurs, and whether there is a direct involvement in the translation process, remains unclear to date.

The lncRNA-p21 on the other hand, was shown to directly impair translation of JUNB as well as CTNNB1 mRNAs [98]. This process likely involves the RNA-binding protein HuR which promotes translation of CTNNB1 and JUNB mRNAs and at the same time renders lincRNA-p21 unstable through recruitment of a Argonaute 2 / let-7 complex. In the absence of HuR, the lncRNA acts as a translational repressor for both mRNAs.

1.3.2.2 LncRNAs Functioning as miRNA Sponges

Several lncRNAs fine-tune regulation of gene expression through association with miRNAs by acting as competing endogenous RNAs (ceRNAs), also termed miRNA sponges [99–102]. These lncRNAs contain multiple binding sites for one or several miRNAs, and regulate target mRNA expression by titrating the miRNA away from its actual mRNA targets, thus modulating miRNA-mediated post-transcriptional silencing (Fig. 1.2g).

The tumor suppressor PTEN is a phosphatase negatively regulating PI3K/AKT signaling, and is frequently mutated in multiple cancer types. Transcription of the PTEN pseudogene (PTENP1) was shown to promote expression of the PTEN mRNA by acting as molecular decoy for miRNAs targeting PTEN [103]. Correspondingly, several protein coding mRNA transcripts sharing common miRNA recognition elements with PTEN mRNA, can also act as ceRNAs by titrating away miRNAs and thus fine-tuning PTEN expression [104–107].

In a similar fashion, linc-MD1, a lncRNA expressed during early muscle differentiation, contains consensus sites for miRNA-135 as well as miRNA-133, two microRNAs important for regulation of muscle differentiation. Linc-MD1 acts as a sponge by sequestering these microRNAs away from their target transcription factor mRNAs, thus promoting the transition to later stages of muscle differentiation [108]. Correspondingly, ectopic expression or siRNA-mediated knock-down of linc-MD1 resulted in an increase or delay of myogenesis. Interestingly, linc-MD1 itself is the host-RNA for miRNA-133. The biogenesis of miRNA-133 itself is controlled by the RNA-binding protein HUR, which is under the repressive control of miRNA-133, thus generating a feed forward regulatory loop involved in linc-MD1-mediated regulation of myogenesis [109]. H19, an imprinted lncRNA was also shown to modulate muscle differentiation by acting as a molecular sponge for let-7 miRNAs [110].

CeRNAs appear to also be involved in regulating pluripotency of embryonic stem cells. Along these lines, linc-RoR regulates expression of pluripotency transcription factors Oct4, NANOG and Sox2 by sharing response elements for several miRNAs with these core transcription factors [101].

Interestingly, a novel class of circular RNAs (circRNAs) has recently gained a lot of attention. CircRNAs result from a non-canonical form of alterative splicing, form a closed, continuous loop and are widely expressed in eukaryotes [111]. So far, their functional roles are mostly unknown, but two circRNAs were found to both act as miRNA sponges [112, 113] (Fig. 1.2g).

1.4 Future Perspectives

Recent discovery of novel classes of short and long regulatory non-coding RNAs revealed a staggering complexity of RNA-mediated regulation involved in nearly all biological processes. Additionally, long known housekeeping ncRNAs such as tRNAs or snoRNAs reveal a multitude of novel gene regulatory functions. Also, the vast majority of the many thousand mammalian long non-coding RNAs identified to date remains completely uncharacterized, suggesting that we are far from grasping the full range of mechanisms these molecules employ to regulate biological processes.

The tremendous recent developments in the field of ncRNA biology clearly indicate that several previous dogmas about the nature of genome composition have to be adapted in order to acknowledge the increased complexity of RNA species and gene loci. The average gene locus is much more complex than previously thought. This transcriptional complexity is partially reflected by recent reports indicating that more than 50 % of protein coding genes have antisense non-coding transcription activity. Alternative splicing, presence of multiple transcription initiation and termination sites, as well as the occurrence of intronic short and long non-coding transcripts or retained introns exceedingly complicate the composition and regulation of gene loci [68, 114–116]. This complexity makes the characterization of protein-coding and especially non-coding RNAs significantly more challenging and requires the careful design of functional studies aimed to analyze the roles of long and short ncRNAs in an isoform-specific manner.

Importantly, not all RNAs with a mode of action dependent on the RNA itself are purely non-coding. Several lncRNAs for example were shown to encode a functional, small peptide [117, 118]. Additionally, a number of protein-coding transcripts have additional roles apart from serving as a template for translation of the protein. Protein-coding transcripts acting as ceRNAs are one example for this phenomenon [107]. Another example is the RNA APELA, which possesses protein-coding ability, but the RNA itself is sufficient to mediate DNA damage-induced apoptosis in embryonic stem cells [119].

In addition to subcellular localization, identifying the interactome of ncRNAs of interest can greatly facilitate discovery of their modes of action. Correspondingly, the mechanisms of many of the functional lncRNAs characterized so far were revealed by analysis of interacting proteins, chromatin or RNA. Significant progress has been made recently in the discovery of methods for the large-scale identification of RNA, chromatin or proteins interacting with endogenous lncRNAs [94, 120–126]. Availability and further development of such biochemical as well as bioinformatics techniques will clearly accelerate our progress in dissecting the highly diverse roles of ncRNAs and will eventually enable us to acquire a more complete understanding of this truly heterogeneous class of lncRNAs.

References

1. Matera AG, Terns RM, Terns MP. Non-coding RNAs: lessons from the small nuclear and small nucleolar RNAs. Nat Rev Mol Cell Biol. 2007;8:209–20.
2. Rocheleau CE, Downs WD, Lin R, Wittmann C, Bei Y, Cha Y-H, et al. Wnt Signaling and an APC-Related Gene Specify Endoderm in Early C. elegans Embryos. Cell. 1997;90:707–16.
3. Fire A, Xu S, Montgomery MK, Kostas SA, Driver SE, Mello CC. Potent and specific genetic interference by double-stranded RNA in Caenorhabditis elegans. Nature. 1998;391:806–11.
4. Czech B, Hannon GJ. Small RNA sorting: matchmaking for Argonautes. Nat Rev Genet. 2011;12:19–31.
5. Meister G, Tuschl T. Mechanisms of gene silencing by double-stranded RNA. Nature. 2004;431:343–9.
6. Farazi TA, Juranek SA, Tuschl T. The growing catalog of small RNAs and their association with distinct Argonaute/Piwi family members. Development. 2008;135:1201–14.
7. Siomi H, Siomi MC. On the road to reading the RNA-interference code. Nature. 2009;457:396–404.
8. Bartel DP. MicroRNAs: target recognition and regulatory functions. Cell. 2009;136:215–33.
9. Carthew RW, Sontheimer EJ. Origins and mechanisms of miRNAs and siRNAs. Cell. 2009;136:642–55.
10. Siomi MC, Sato K, Pezic D, Aravin AA. PIWI-interacting small RNAs: the vanguard of genome defence. Nat Rev Mol Cell Biol. 2011;12:246–58.
11. Mendell JT, Olson EN. MicroRNAs in stress signaling and human disease. Cell. 2012;148:1172–87.
12. Dueck A, Meister G. Assembly and function of small RNA – Argonaute protein complexes. Biol Chem. 2014;395:611–29.
13. Mattick JS. Non-coding RNAs: the architects of eukaryotic complexity. EMBO Rep. 2001;2:986–91.
14. Consortium IHGS. Finishing the euchromatic sequence of the human genome. Nature. 2004;431:931–45.
15. Djebali S, Davis CA, Merkel A, Dobin A, Lassmann T, Mortazavi A, et al. Landscape of transcription in human cells. Nature. 2012;489:101–8.
16. Jia H, Osak M, Bogu GK, Stanton LW, Johnson R, Lipovich L. Genome-wide computational identification and manual annotation of human long noncoding RNA genes. RNA. 2010;16:1478–87.
17. Cabili MN, Trapnell C, Goff L, Koziol M, Tazon-Vega B, Regev A, et al. Integrative annotation of human large intergenic noncoding RNAs reveals global properties and specific subclasses. Genes Dev. 2011;25:1915–27.
18. Martianov I, Ramadass A, Serra Barros A, Chow N, Akoulitchev A. Repression of the human dihydrofo-

late reductase gene by a non-coding interfering transcript. Nature. 2007;445:666–70.
19. Feng J, Bi C, Clark B, Mady R, Shah P, Kohtz J. The Evf-2 noncoding RNA is transcribed from the Dlx-5/6 ultraconserved region and functions as a Dlx-2 transcriptional coactivator. Genes Dev. 2006;20:1470–84.
20. Bond AM, VanGompel MJW, Sametsky EA, Clark MF, Savage JC, Disterhoft JF, et al. Balanced gene regulation by an embryonic brain ncRNA is critical for adult hippocampal GABA circuitry. Nat Neurosci. 2009;12:1020–7.
21. Tian D, Sun S, Lee J. The long noncoding RNA, jpx, is a molecular switch for x chromosome inactivation. Cell. 2010;143:390–403.
22. Lee J. Lessons from X-chromosome inactivation: long ncRNA as guides and tethers to the epigenome. Genes Dev. 2009;23:1831–42.
23. Lee JT, Bartolomei MS. X-inactivation, imprinting, and long noncoding RNAs in health and disease. Cell. 2013;152:1308–23.
24. Bartolomei MS, Zemel S, Tilghman SM. Parental imprinting of the mouse H19 gene. Nature. 1991;351:153–5. Publ. Online 09 May 1991 Doi101038351153a0.
25. Lyle R, Watanabe D, te Vruchte D, Lerchner W, Smrzka OW, Wutz A, et al. The imprinted antisense RNA at the Igf2r locus overlaps but does not imprint Mas1. Nat Genet. 2000;25:19–21.
26. Sleutels F, Zwart R, Barlow D. The non-coding Air RNA is required for silencing autosomal imprinted genes. Nature. 2002;415:810–3.
27. Mancini-Dinardo D, Steele S, Levorse J, Ingram R, Tilghman S. Elongation of the Kcnq1ot1 transcript is required for genomic imprinting of neighboring genes. Genes Dev. 2006;20:1268–82.
28. Williamson CM, Ball ST, Dawson C, Mehta S, Beechey CV, Fray M, et al. Uncoupling antisense-mediated silencing and DNA methylation in the imprinted Gnas cluster. PLoS Genet. 2011;7:e1001347.
29. Mao Y, Sunwoo H, Zhang B, Spector D. Direct visualization of the co-transcriptional assembly of a nuclear body by noncoding RNAs. Nat Cell Biol. 2011;13:95–101.
30. Hacisuleyman E, Goff LA, Trapnell C, Williams A, Henao-Mejia J, Sun L, et al. Topological organization of multichromosomal regions by the long intergenic noncoding RNA Firre. Nat Struct Mol Biol. 2014;21:198–206.
31. Yang F, Deng X, Ma W, Berletch JB, Rabaia N, Wei G, et al. The lncRNA Firre anchors the inactive X chromosome to the nucleolus by binding CTCF and maintains H3K27me3 methylation. Genome Biol. 2015. doi:10.1186/s13059-015-0618-0.
32. Krol J, Loedige I, Filipowicz W. The widespread regulation of microRNA biogenesis, function and decay. Nat Rev Genet. 2010;11:597–610.
33. Esteller M. Non-coding RNAs in human disease. Nat Rev Genet. 2011;12:861–74.
34. Gutschner T, Diederichs S. The hallmarks of cancer. RNA Biol. 2012;9:703–19.
35. Huntzinger E, Izaurralde E. Gene silencing by microRNAs: contributions of translational repression and mRNA decay. Nat Rev Genet. 2011;12:99–110.
36. Sayed D, Abdellatif M. MicroRNAs in development and disease. Physiol Rev. 2011;91:827–87.
37. Friedman RC, Farh KK-H, Burge CB, Bartel DP. Most mammalian mRNAs are conserved targets of microRNAs. Genome Res. 2009;19:92–105.
38. Lin S, Gregory RI. MicroRNA biogenesis pathways in cancer. Nat Rev Cancer. 2015;15:321–33.
39. Liu N, Olson EN. MicroRNA regulatory networks in cardiovascular development. Dev Cell. 2010;18:510–25.
40. Olson EN. MicroRNAs as therapeutic targets and biomarkers of cardiovascular disease. Sci Transl Med. 2014;6:239ps3.
41. Issler O, Chen A. Determining the role of microRNAs in psychiatric disorders. Nat Rev Neurosci. 2015;16:201–12.
42. Treiber T, Treiber N, Meister G. Regulation of microRNA biogenesis and function. Thromb Haemost. 2012;107:605–10.
43. Jonas S, Izaurralde E. Towards a molecular understanding of microRNA-mediated gene silencing. Nat Rev Genet. 2015;16:421–33.
44. Du P, Wang L, Sliz P, Gregory RI. A biogenesis step upstream of microprocessor controls miR-17 ~ 92 expression. Cell. 2015;162:885–99.
45. Luteijn MJ, Ketting RF. PIWI-interacting RNAs: from generation to transgenerational epigenetics. Nat Rev Genet. 2013;14:523–34.
46. Claycomb JM. Ancient endo-siRNA pathways reveal new tricks. Curr Biol. 2014;24:R703–15.
47. Nayak A, Tassetto M, Kunitomi M, Andino R. RNA interference-mediated intrinsic antiviral immunity in invertebrates. In: Cullen BR, editor. Intrinsic immun. Berlin/Heidelberg: Springer; 2013. p. 183–200.
48. Szittya G, Burgyán J. RNA interference-mediated intrinsic antiviral immunity in plants. In: Cullen BR, editor. Intrinsic immun. Berlin /Heidelberg: Springer; 2013. p. 153–81.
49. García-Sastre A. Induction and evasion of type I interferon responses by influenza viruses. Virus Res. 2011;162:12–8.
50. Billy E, Brondani V, Zhang H, Müller U, Filipowicz W. Specific interference with gene expression induced by long, double-stranded RNA in mouse embryonal teratocarcinoma cell lines. Proc Natl Acad Sci. 2001;98:14428–33.
51. Li Y, Lu J, Han Y, Fan X, Ding S-W. RNA interference functions as an antiviral immunity mechanism in mammals. Science. 2013;342:231–4.
52. Maillard PV, Ciaudo C, Marchais A, Li Y, Jay F, Ding SW, et al. Antiviral RNA interference in mammalian cells. Science. 2013;342:235–8.
53. Pare JM, Sullivan CS. Distinct antiviral responses in pluripotent versus differentiated cells. PLoS Pathog. 2014. doi:10.1371/journal.ppat.1003865.

54. Tam OH, Aravin AA, Stein P, Girard A, Murchison EP, Cheloufi S, et al. Pseudogene-derived small interfering RNAs regulate gene expression in mouse oocytes. Nature. 2008;453:534–8.
55. Watanabe T, Totoki Y, Toyoda A, Kaneda M, Kuramochi-Miyagawa S, Obata Y, et al. Endogenous siRNAs from naturally formed dsRNAs regulate transcripts in mouse oocytes. Nature. 2008;453:539–43.
56. Ender C, Krek A, Friedländer MR, Beitzinger M, Weinmann L, Chen W, et al. A human snoRNA with microRNA-like functions. Mol Cell. 2008;32:519–28.
57. Brameier M, Herwig A, Reinhardt R, Walter L, Gruber J. Human box C/D snoRNAs with miRNA like functions: expanding the range of regulatory RNAs. Nucleic Acids Res. 2011;39:675–86.
58. Röther S, Meister G. Small RNAs derived from longer non-coding RNAs. Biochimie. 2011;93:1905–15.
59. Bratkovič T, Rogelj B. The many faces of small nucleolar RNAs. Biochim Biophys Acta BBA - Gene Regul Mech. 1839;2014:438–43.
60. Siprashvili Z, Webster DE, Johnston D, Shenoy RM, Ungewickell AJ, Bhaduri A, et al. The noncoding RNAs SNORD50A and SNORD50B bind K-Ras and are recurrently deleted in human cancer. Nat Genet. 2016;48:53–8.
61. Haussecker D, Huang Y, Lau A, Parameswaran P, Fire AZ, Kay MA. Human tRNA-derived small RNAs in the global regulation of RNA silencing. RNA. 2010;16:673–95.
62. Gebetsberger J, Polacek N. Slicing tRNAs to boost functional ncRNA diversity. RNA Biol. 2013;10:1798–806.
63. Sharma U, Conine CC, Shea JM, Boskovic A, Derr AG, Bing XY, et al. Biogenesis and function of tRNA fragments during sperm maturation and fertilization in mammals. Science. 2016;351:391–6.
64. Goodarzi H, Liu X, Nguyen HCB, Zhang S, Fish L, Tavazoie SF. Endogenous tRNA-derived fragments suppress breast cancer progression via YBX1 displacement. Cell. 2015;161:790–802.
65. Kung JTY, Colognori D, Lee JT. Long noncoding RNAs: past, present, and future. Genetics. 2013;193:651–69.
66. Bassett AR, Akhtar A, Barlow DP, Bird AP, Brockdorff N, Duboule D, et al. Considerations when investigating lncRNA function in vivo. eLife. 2014. doi:10.7554/eLife.03058.
67. Guttman M, Rinn J. Modular regulatory principles of large non-coding RNAs. Nature. 2012;482:339–46.
68. Mercer TR, Mattick JS. Structure and function of long noncoding RNAs in epigenetic regulation. Nat Struct Mol Biol. 2013;20:300–7.
69. Batista PJ, Chang HY. Long noncoding RNAs: cellular address codes in development and disease. Cell. 2013;152:1298–307.
70. Morris KV, Mattick JS. The rise of regulatory RNA. Nat Rev Genet. 2014;15:423–37.
71. Wang K, Chang H. Molecular mechanisms of long noncoding RNAs. Mol Cell. 2014;43:904–14.
72. Berghoff EG, Clark MF, Chen S, Cajigas I, Leib DE, Kohtz JD. Evf2 (Dlx6as) lncRNA regulates ultraconserved enhancer methylation and the differential transcriptional control of adjacent genes. Development. 2013;140:4407–16.
73. Kim T-K, Hemberg M, Gray JM, Costa AM, Bear DM, Wu J, et al. Widespread transcription at neuronal activity-regulated enhancers. Nature. 2010;465:182–7.
74. Melo CA, Drost J, Wijchers PJ, van de Werken H, de Wit E, Vrielink JAFO, et al. eRNAs are required for p53-dependent enhancer activity and gene transcription. Mol Cell. 2013;49:524–35.
75. Arner E, Daub CO, Vitting-Seerup K, Andersson R, Lilje B, Drabløs F, et al. Transcribed enhancers lead waves of coordinated transcription in transitioning mammalian cells. Science. 2015;347:1010–4.
76. Léveillé N, Melo CA, Rooijers K, Díaz-Lagares A, Melo SA, Korkmaz G, et al. Genome-wide profiling of p53-regulated enhancer RNAs uncovers a subset of enhancers controlled by a lncRNA. Nat Commun. 2015;6:6520.
77. Sigova AA, Abraham BJ, Ji X, Molinie B, Hannett NM, Guo YE, et al. Transcription factor trapping by RNA in gene regulatory elements. Science. 2015;350:978–81.
78. Vance KW, Sansom SN, Lee S, Chalei V, Kong L, Cooper SE, et al. The long non-coding RNA Paupar regulates the expression of both local and distal genes. EMBO J. 2014;33:296–311.
79. Lai F, Orom UA, Cesaroni M, Beringer M, Taatjes DJ, Blobel GA, et al. Activating RNAs associate with mediator to enhance chromatin architecture and transcription. Nature. 2013;494:497–501.
80. Khalil A, Guttman M, Huarte M, Garber M, Raj A, Rivea Morales D, et al. Many human large intergenic noncoding RNAs associate with chromatin-modifying complexes and affect gene expression. Proc Natl Acad Sci U S A. 2009;106:11667–72.
81. Grote P, Wittler L, Hendrix D, Koch F, Währisch S, Beisaw A, et al. The tissue-specific lncRNA Fendrr is an essential regulator of heart and body wall development in the mouse. Dev Cell. 2013;24:206–14.
82. Grote P, Herrmann BG. The long non-coding RNA Fendrr links epigenetic control mechanisms to gene regulatory networks in mammalian embryogenesis. RNA Biol. 2013;10.
83. Rinn J, Kertesz M, Wang J, Squazzo S, Xu X, Brugmann S, et al. Functional demarcation of active and silent chromatin domains in human HOX loci by noncoding RNAs. Cell. 2007;129:1311–23.
84. Tsai M, Manor O, Wan Y, Mosammaparast N, Wang J, Lan F, et al. Long noncoding RNA as modular

scaffold of histone modification complexes. Science. 2010;329:689–93.
85. Gupta RA, Shah N, Wang KC, Kim J, Horlings HM, Wong DJ, et al. Long non-coding RNA HOTAIR reprograms chromatin state to promote cancer metastasis. Nature. 2010;464:1071–6.
86. Sunwoo H, Dinger ME, Wilusz JE, Amaral PP, Mattick JS, Spector DL. MEN ε/β nuclear-retained non-coding RNAs are up-regulated upon muscle differentiation and are essential components of paraspeckles. Genome Res. 2009;19:347–59.
87. Chen L-L, Carmichael GG. Altered nuclear retention of mRNAs containing inverted repeats in human embryonic stem cells: functional role of a nuclear noncoding RNA. Mol Cell. 2009;35:467–78.
88. Clemson CM, Hutchinson JN, Sara SA, Ensminger AW, Fox AH, Chess A, et al. An architectural role for a nuclear noncoding RNA: NEAT1 RNA is essential for the structure of paraspeckles. Mol Cell. 2009;33:717–26.
89. Zhao J, Sun B, Erwin J, Song J, Lee J. Polycomb proteins targeted by a short repeat RNA to the mouse X chromosome. Science. 2008;322:750–6.
90. Yoon J-H, Abdelmohsen K, Gorospe M. Posttranscriptional gene regulation by long noncoding RNA. J Mol Biol. 2013;425:3723–30.
91. Gong C, Maquat L. lncRNAs transactivate STAU1-mediated mRNA decay by duplexing with 3′ UTRs via Alu elements. Nature. 2011;470:284–8.
92. Schoenberg D, Maquat L. Regulation of cytoplasmic mRNA decay. Nat Rev Genet. 2012;1–14.
93. Blanpain C, Fuchs E. Epidermal homeostasis: a balancing act of stem cells in the skin. Nat Rev Mol Cell Biol. 2009;10:207–17.
94. Kretz M, Siprashvili Z, Chu C, Webster DE, Zehnder A, Qu K, et al. Control of somatic tissue differentiation by the long non-coding RNA TINCR. Nature. 2013;493:231–5.
95. Hombach S, Kretz M. The non-coding skin: exploring the roles of long non-coding RNAs in epidermal homeostasis and disease. BioEssays. 2013;35:1093–100.
96. Lopez-Pajares V, Qu K, Zhang J, Webster DE, Barajas BC, Siprashvili Z, et al. A LncRNA-MAF:MAFB transcription factor network regulates epidermal differentiation. Dev Cell. 2015;32:693–706.
97. Sun BK, Boxer LD, Ransohoff JD, Siprashvili Z, Qu K, Lopez-Pajares V, et al. CALML5 is a ZNF750- and TINCR-induced protein that binds stratifin to regulate epidermal differentiation. Genes Dev. 2015;29:2225–30.
98. Yoon J-H, Abdelmohsen K, Srikantan S, Yang X, Martindale JL, De S, et al. LincRNA-p21 suppresses target mRNA translation. Mol Cell. 2012;47:648–55.
99. Ebert M, Sharp P. Emerging roles for natural microRNA sponges. Curr Biol CB. 2010;20:R858–61.
100. Salmena L, Poliseno L, Tay Y, Kats L, Pandolfi P. A ceRNA hypothesis: the rosetta stone of a hidden RNA language? Cell. 2011;146:353–8.
101. Wang Y, Xu Z, Jiang J, Xu C, Kang J, Xiao L, et al. Endogenous miRNA sponge lincRNA-RoR regulates Oct4, Nanog, and Sox2 in human embryonic stem cell self-renewal. Dev Cell. 2013;25:69–80.
102. Tay Y, Rinn J, Pandolfi PP. The multilayered complexity of ceRNA crosstalk and competition. Nature. 2014;505:344–52.
103. Poliseno L, Salmena L, Zhang J, Carver B, Haveman WJ, Pandolfi PP. A coding-independent function of gene and pseudogene mRNAs regulates tumour biology. Nature. 2010;465:1033–8.
104. Lee DY, Jeyapalan Z, Fang L, Yang J, Zhang Y, Yee AY, et al. Expression of versican 3′-untranslated region modulates endogenous microRNA functions. PLoS One. 2010;5:e13599.
105. Karreth FA, Tay Y, Perna D, Ala U, Tan SM, Rust AG, et al. In vivo identification of tumor- suppressive PTEN ceRNAs in an oncogenic BRAF-induced mouse model of melanoma. Cell. 2011;147:382–95.
106. Tay Y, Kats L, Salmena L, Weiss D, Tan SM, Ala U, et al. Coding-independent regulation of the tumor suppressor PTEN by competing endogenous mRNAs. Cell. 2011;147:344–57.
107. Poliseno L, Pandolfi PP. PTEN ceRNA networks in human cancer. Methods. 2015;77–78:41–50.
108. Cesana M, Cacchiarelli D, Legnini I, Santini T, Sthandier O, Chinappi M, et al. A long noncoding RNA controls muscle differentiation by functioning as a competing endogenous RNA. Cell. 2011;147:358–69.
109. Legnini I, Morlando M, Mangiavacchi A, Fatica A, Bozzoni I. A feedforward regulatory loop between HuR and the long noncoding RNA linc-MD1 controls early phases of myogenesis. Mol Cell. 2014;53:506–14.
110. Kallen AN, Zhou X-B, Xu J, Qiao C, Ma J, Yan L, et al. The imprinted H19 LncRNA antagonizes Let-7 MicroRNAs. Mol Cell. 2013;52:101–12.
111. Jeck WR, Sharpless NE. Detecting and characterizing circular RNAs. Nat Biotechnol. 2014;32:453–61.
112. Hansen TB, Jensen TI, Clausen BH, Bramsen JB, Finsen B, Damgaard CK, et al. Natural RNA circles function as efficient microRNA sponges. Nature. 2013;495:384–8.
113. Memczak S, Jens M, Elefsinioti A, Torti F, Krueger J, Rybak A, et al. Circular RNAs are a large class of animal RNAs with regulatory potency. Nature. 2013;495:333–8.
114. Carninci P, Kasukawa T, Katayama S, Gough J, Frith M, Maeda N, et al. The transcriptional landscape of the mammalian genome. Science. 2005;309:1559–63.
115. Consortium RGERG and GSG (Genome NPCG and the Katayama S, Tomaru Y, Kasukawa T, Waki K, Nakanishi M, et al. Antisense transcription in the mammalian transcriptome. Science. 2005;309:1564–6.

116. Kawaji H, Severin J, Lizio M, Waterhouse A, Katayama S, Irvine K, et al. The FANTOM web resource: from mammalian transcriptional landscape to its dynamic regulation. Genome Biol. 2009;10:R40.
117. Kondo T, Plaza S, Zanet J, Benrabah E, Valenti P, Hashimoto Y, et al. Small peptides switch the transcriptional activity of Shavenbaby during Drosophila embryogenesis. Science. 2010;329:336–9.
118. Anderson DM, Anderson KM, Chang C-L, Makarewich CA, Nelson BR, McAnally JR, et al. A micropeptide encoded by a putative long noncoding RNA regulates muscle performance. Cell. 2015;160:595–606.
119. Li M, Gou H, Tripathi BK, Huang J, Jiang S, Dubois W, et al. An Apela RNA-containing negative feedback loop regulates p53-mediated apoptosis in embryonic stem cells. Cell Stem Cell. 2015;16:669–83.
120. Chu C, Qu K, Zhong FL, Artandi SE, Chang HY. Genomic maps of long noncoding RNA occupancy reveal principles of RNA-chromatin interactions. Mol Cell. 2011;44:667–78.
121. Simon MD, Wang CI, Kharchenko PV, West JA, Chapman BA, Alekseyenko AA, et al. The genomic binding sites of a noncoding RNA. Proc Natl Acad Sci. 2011;108:20497–502.
122. Engreitz JM, Pandya-Jones A, McDonel P, Shishkin A, Sirokman K, Surka C, et al. The Xist lncRNA exploits three-dimensional genome architecture to spread across the X chromosome. Science. 2013;341:1237973.
123. Chu C, Spitale RC, Chang HY. Technologies to probe functions and mechanisms of long noncoding RNAs. Nat Struct Mol Biol. 2015;22:29–35.
124. Chu C, Zhang QC, da Rocha ST, Flynn RA, Bharadwaj M, Calabrese JM, et al. Systematic discovery of Xist RNA binding proteins. Cell. 2015;161:404–16.
125. McHugh CA, Chen C-K, Chow A, Surka CF, Tran C, McDonel P, et al. The Xist lncRNA interacts directly with SHARP to silence transcription through HDAC3. Nature. 2015;521:232–6.
126. Minajigi A, Froberg JE, Wei C, Sunwoo H, Kesner B, Colognori D, et al. A comprehensive Xist interactome reveals cohesin repulsion and an RNA-directed chromosome conformation. Science. 2015;349:aab2276.

Involvement of Non-coding RNAs in the Signaling Pathways of Colorectal Cancer

2

Yinxue Yang, Yong Du, Xiaoming Liu, and William C. Cho

Abstract

Colorectal cancer (CRC) is one of the most common diagnosed cancers worldwide. The metastasis and development of resistance to anti-cancer treatment are major challenges in the treatment of CRC. Understanding mechanisms underpinning the pathogenesis is therefore critical in developing novel agents for CRC treatments. A large number of evidence has demonstrated that non-coding RNAs (ncRNAs), including microRNAs (miRNAs), long non-coding RNAs (lncRNAs), and circular RNAs have functional roles in both the physiological and pathological processes by regulating the expression of their target genes. These molecules are engaged in the pathobiology of neoplastic diseases and are targets for the diagnosis, prognosis and therapy of a variety of cancers, including CRC. In this regard, ncRNAs have emerged as one of the hallmarks of CRC pathogenesis and they also play key roles in metastasis, drug resistance and the stemness of CRC stem cell by regulating various signaling networks. Therefore, a better understanding the ncRNAs involved in the signaling pathways of CRC may lead to the development of novel strategy for diagnosis, prognosis and treatment of CRC. In this chapter, we summarize the latest findings on ncRNAs, with a focus on miRNAs and lncRNAs involving in signaling networks and in the regulation of pathogenic signaling pathways in CRC.

Keywords

Colorectal cancer (CRC) • Long non-coding RNA (lncRNA) • MicroRNA (miRNA) • Non-coding RNAs (ncRNAs) • Signaling pathway

Y. Yang • Y. Du • X. Liu
The General Hospital, Ningxia Medical University, Yinchuan 750004, Ningxia, China
e-mail: liuxiaoming@nxmu.edu.cn

W.C. Cho (✉)
Department of Clinical Oncology, Queen Elizabeth Hospital, Hong Kong, China
e-mail: chocs@ha.org.hk

O. Slaby, G.A. Calin (eds.), *Non-coding RNAs in Colorectal Cancer*, Advances in Experimental Medicine and Biology 937, DOI 10.1007/978-3-319-42059-2_2

2.1 Introduction

Colorectal cancer (CRC) is one of the most commonly diagnosed cancers with approximately one-third of patients with colon cancer are synchronous or metachronous metastasis. With regard to pathogenesis, CRC is one of the best-characterized cancers and a leading cause of cancer death worldwide [1]. Despite many novel approaches have been implemented for cancer treatment, the majority of CRC patients eventually succumb to metastatic disease after the surgical resection of tumors [2]. The progression of cancer metastasis is a complex process, which ultimately leads to the cancer cells extravasate through circulatory or lymphatic system to distant tissues where they eventually colonize and develop tumors. Currently, the 5-year overall survival rate of patients with primary CRC can be up to 80–90 %, but it will be reduced to 40–60 % in patients with advanced non-metastatic tumors, and can be further decreased to 5–10 % in patients with metastatic tumors [3], owing to an incomplete understanding of the molecular mechanisms underpinning its pathogenesis, the high relapse rate, the development of drug resistant cancer cells and the presence of CRC stem/stem-like cells (CRSCs) (also known as tumor initiating cells), particularly, the metastasis and development of drug resistance remain the major obstacles toward a successful treatment for CRC [4, 5].

Human transcriptome analysis using high-throughput sequencing technologies has revealed that the majority of human genome (~90 %) is dynamically and pervasively transcribed as non-coding RNAs (ncRNAs) [6], and an increasing evidence has confirmed that ncRNAs are overtly involved in the complex molecular signaling needed to regulate the structures and functions of cells and developmental contexts [7]. Therefore, a dysregulation of ncRNAs may result in the development and progression of many pathological conditions, including cancer [6, 7].

ncRNAs are a class of functional RNA molecules that regulate gene expression at the transcriptional and post-transcriptional levels, which can be categorized into two main groups; the short (small) ncRNAs (<30 nts) and the long ncRNAs (lncRNAs) (>200 nts). The microRNAs (miRNAs), short interfering RNAs (siRNAs), and piwi-interacting RNAs (piRNAs) representing classes of well-known short ncRNAs. lncRNAs are a group of long RNA transcripts with no apparent protein-coding role, in which circular RNAs (circRNAs) are a class of lncRNAs that have special 5′- and 3′-end processing [8]. Among these ncRNAs, miRNAs and lncRNAs have gained the most attentions and their pathogenic roles have been extensively investigated in a variety of cancers.

High-throughput sequencing and/or microarray analysis have revealed the alterations of ncRNA profiling as a hallmark of many types of cancers, including CRC, which has remarkably improved our understanding in tumor biology and genetics [8]. In this context, ncRNA profiling is correlated with neoplastic phenotypes and/or disease progression, suggesting that ncRNAs are biomarkers for the diagnosis and prognosis, and are targets for developing novel agents for cancer treatments [6, 7]. In this chapter, we summarize recent understanding in the involvement of ncRNAs in CRC pathobiology, with a focus on miRNAs and lncRNAs in the signaling networks of CRC development, progression and metastasis.

2.2 MicroRNAs as a Hallmark in the Development and Progression of Colorectal Cancer

miRNAs are a class of non-coding, small RNA molecules found in both prokaryotes and eukaryotes, which are able to negatively regulate the gene expressions of target mRNAs at the post-transcriptional level. Recent studies have reported that the alterations of miRNA expression profile in tumors compared with adjacent normal tissues have been observed in a variety of cancers. The altered miRNA expression profile has been proposed to correlate with the stages and survivals in patients with tumors, including CRC, in which miRNAs can exert their regulatory roles by directly targeting genes in the key steps of metastatic processes and acquired drug resistance [9–12]. In this context, miRNAs can play a functionality of either tumor-suppressors or oncogenes (oncomirs) [13, 14].

In addition to their roles in cancer initiation and development, alterations of miRNAs have a great implication in cancer drug resistance through a non-genetically mutational mechanism. The dysregulation of miRNAs has been involved in the regulation of gene function that contributed to metastasis and acquisition of chemoresistant phenotype [15]. Furthermore, miRNAs are also the key players in maintaining the characteristics of cancer stem cells (CSCs) for self-renewal, proliferation, differentiation and chemoresistance, and they can be used for diagnostic, prognostic and therapeutic targets for the metastasis, drug response and treatment of cancers [4, 15–20]. Recently, an increasing number of evidence has indicated that miRNAs are involved in the process of metastasis of CRC, and several miRNAs have been identified as regulators in CRC relapse by targeting metastatic signaling pathways (Table 2.1) [9, 12].

By examining the expression profiling of miRNAs in CRC using a variety of techniques including global miRNA expression profiling with deep sequencing or miRNA microarrays. These studies also tested the selected miRNAs with quantitative reverse transcriptase polymerase chain reaction (qRT-PCR) and confirmed alterations of miRNA profiles in tumor tissues and/or sera of CRC patients compared to that of adjacent normal tissues and/or non-CRC patients. Interestingly, an alteration of majority of miRNA expression was found to be globally elevated in CRC [12]. This view was supported by a recent review on analysis of the profiling of miRNAs in CRC, in which retrospective data indicated that approximately 2/3 of the 164 altered miRNAs were elevated in tumors. This finding suggested that the miRNA processing machinery was not compromised in CRC [12, 126]. Among them, miR-21 is an extensively studied oncogene capable of targeting multiple tumor suppressor genes including phosphatase and tensin homolog (PTEN) and programmed cell death protein 4 (PDCD4), and activating Wnt/β-catenin signaling [127, 128]. Other important miRNAs which have been found to be altered in CRC includes the miR-17-92 cluster, miR-31, miR-29a, miR-135, miR-143, miR-145, miR-181b, miR-183, the miR-200a/b/c family, miR-221 and miR-222 [12]. For instance, miR-143 and miR-145 were down-regulated in CRC and thus suggested that they are tumor suppressors. miR-143 was a well-defined miRNA that associated with CRC metastasis, less abundant miR-143 was found to associate with larger tumor size and longer disease-free interval in colon cancer. While an increased expression of miR-143 could attenuate migration and invasion of CRC cells [129]. Mechanistically, miR-143 was identified to target metastasis-associated in colon cancer-1 (MACC1), a novel prognostic biomarker for metastasis occurrence [129]. Consistently, overexpression of miR-145 could reduce cell capacity of migration and invasion by targeting paxillin in human CRC cells [130]. By comparative analysis of miRNA profiling from colon tissues of 54 CRC patients and 42 normal colon tissue samples, Kara et al. [131] found miR-17, miR-21-5p, miR-27a-3p, miR-29a-3p, miR-29b-3p, miR-34a, miR-34c, miR-96, miR-130a-3p, miR-132-3p, miR-133b, miR-155, miR-193b-3p, miR-203a, miR-205, miR-222-3p, miR-301a-3p and miR-378a-3p were significantly deregulated in CRC.

The metastatic progression of CRC is a complex process, including the angiogenesis of adjacent tumor tissue, migration, and invasion which ultimately leads to the intravasation and fluid transportation of cancer cells through circulatory or lymphatic system and extravasation to distant tissues where they eventually colonize and develop tumors. The angiogenesis is an essential step for the growth of both primary and metastatic tumors with bloodstream. Several lines of evidence have demonstrated that miRNAs could exert either a pro-angiogenic or an anti-angiogenic effect in angiogenesis [132, 133]. For examples, miR-221 and miR-222, whose expressions are related to the TNM stage and local invasion of cancer, and are frequently elevated in colon cancer, they were able to inhibit angiogenic activities in HUVEC (human umbilical vein

Table 2.1 A list of microRNAs involved in the pathogenesis and metastasis of colorectal cancer

Expression	MicroRNA	Target	Signaling pathway	Process in colorectal cancer	Reference
Down-regulated	let-7	HMGA2	EMT	Cell proliferation, migration, invasion, metastasis	[21–24]
	miR-1	MACC1	MET	Cell proliferation, invasion, migration	[25]
	miR-16	PTGS2/COX2	PGE2/COX2	Cell proliferation	[26]
	miR-18a	hnRNP A1, CDC42	Autophagy pathway, PI3K	Cell proliferation, migration	[27, 28]
	miR-23b	FZD7, MEKK1, PAK2, TGFBR2, RRAS2, PLAU,VEGF	Wnt, TGF and VEGF signaling	Cell migration, invasion, angiogenesis	[29]
	miR-27a	SGPP1, SMAD2	TGFβ/EMT	Migration, invasion, metastasis, EMT	[13]
	miR-29c	GNA13, PTP4A	Wnt/β-catenin	Migration, invasion, metastasis	[30]
	miR-30a	PI3KCD	PI3K	Invasion, metastasis	[31]
	miR-34a	Fra-1, E2F, SIRT1-p53, FMNL2, IL6R, ZNF281, MET, SNAIL, CTNNB1, SLUG, ZEB1	Multiple pathways	Migration, invasion, metastasis, EMT	[32–34]
	miR-101b	COX2, EP4, PTGS2	COX2/PGE2 angiogenic pathway	Cell proliferation, motility, invasion	[9, 35–37]
	miR-107	CCND1	Cell cycle pathway	Cell proliferation	[38]
	miR-124a	CDK6, Rb	Notch/Cdk6, Rb/E2F pathways	Cell proliferation, migration, invasion	[39]
	miR-125b	Mcl-1, Bcl-w, IL-6R	Apoptotic pathway	Cell proliferation, cell invasion	[40]
	miR-126	RhoA, VEGF, PI3KR, SPRED1	PI3K/AKT, VEGF, ROCK	Migration, invasion, metastasis, EMT	[41, 42]
	miR-128	IRS1	IGF/IRS1/Akt pathway	Migration, invasion, metastasis	[43]
	miR-129	CDK6	Cell cycle signaling	Cell proliferation, cell invasion	[44]
	miR-132	ZEB2	EMT	Migration, invasion, metastasis, EMT	[45]
	miR-133b	c-Met, K-Ras	ROCK/Kras	Migration, invasion, metastasis, EMT	[39, 46]
	miR-137	Cdc42, LSD-1, TGF2I	TGFβ, Rac/Cdc42 cell survival pathways	Proliferation, invasion, metastasis, EMT	[47–49]
	miR-143/145	KLF5, KRAS, ERK5, BRAF KRAS, MACC1, IGF1R, DNMT3A, MYC, CDK6, E2F1, CCND2	MAPK/p53, EGFR, Wnt, DNA methylation	Cell proliferation, invasion	[50, 51]
	miR-144	mTOR, GSTP1	AKT/mTOR, GSTP1/MMP28	Invasion, metastasis, EMT	[52, 53]
	miR-148a	BCL2	Apoptotic pathway	Cell proliferation	[54]
	miR-149	FOXM1	PI3K/FOXM1	Cell proliferation, invasion	[55]

(continued)

Table 2.1 (continued)

Expression	MicroRNA	Target	Signaling pathway	Process in colorectal cancer	Reference
	miR-185	HIF-2α, PCNA, MMP2	HIF signaling	Invasion, metastasis, EMT	[56]
	miR-192	DHFR	DNA methylation pathway	Cell proliferation	[57, 58]
	miR-200c	ZEB1, ETS1 FLT1, CDH1, VIM	EMT	Cell proliferation, invasion, migration, EMT, metastasis	[59, 60]
	miR-203	AKT2	AKT/p53	Cell proliferation, invasion, migration	[61]
	miR-206	NOTCH3, FMNL2	Notch signaling,	Cell proliferation	[62, 63]
	miR-212	MnSOD	MAPK/PI3K/MnSOD	Cell invasion, migration, metastasis	[64]
	miR-214	TP53, β-catenin, TGFR2, BAX, CDKN2b, EGFR, TFAP2C	Wnt/β-catenin, EGFR, apoptotic pathways	Cell proliferation, invasion, metastasis	[39, 65]
	miR-223	FOXO1, RASA1	Cell cycle pathway	Cell proliferation, invasion, metastasis	[66, 67]
	miR-320a	CTNNB1, RAC1, NRP1	Wnt signaling	Cell proliferation, migration, invasion, metastasis, EMT	[68–70]
	miR-335	ZEB2	EMT pathway	Metastasis, EMT	[71]
	miR-361-5p	SND1	Migration/invasion pathways	Cell migration, invasion	[72]
	miR-409-3p	GAB1	GAB1/PI3K	Cell proliferation, invasion	[73]
	miR-449b	E2F3, CCND1	MAPK/p53 pathway	Cell proliferation, invasion	[74]
	miR-497	IGF1R	IGF1/PI3K/mTOR, IGF1/Kras/MEK	Cell proliferation, migration, invasion	[75–77]
	miR-429	Onecut2	EMT	Cell migration, invasion, metastasis, EMT	[78]
	miR-520d-5p	CTHRC1	CTHRC1/ERK	Cell proliferation, migration, invasion, metastasis	[79]
	miR-520a/525a	PI3KCA, VEGFR1	PI3K/AKT, VEGF	Cell proliferation, migration, invasion, metastasis	[80, 81]
	miR-612	AKT2	ATK/PI3K	Cell proliferation, invasion, metastasis, EMT	[82]
	miR-622	Kras	Rho/ras	Cell migration, invasion	[83]
	miR-638	SOX2, TSPAN1	Sox2/Wnt, FGF signaling	Cell invasion, migration	[84, 85]

(continued)

Table 2.1 (continued)

Expression	MicroRNA	Target	Signaling pathway	Process in colorectal cancer	Reference
Up-regulated	miR-15a/16-1	AP4	p53/EMT	Cell proliferation, invasion, EMT	[86]
	miR-17-92 cluster	E2F1, PTEN, BCL2L11, CDKN1A, TSP-1, CTGF, E2F1, E2F2, E2F3, TGFBR2, CDKN1A, BIM	E2F1/p53, PTEN/PI3K, apoptotic pathways	Angiogenesis, proliferation, metastasis	[87–89]
	miR-19	TG2	EMT	Metastasis, EMT	[90]
	miR-19a	ND	TNFα/miR-19a/EMT	Metastasis, EMT	[91]
	miR-21	TGFβR2, PTEN, PDCD4, CCL20, Cdc25A, RHOB, RASA1	Multiple pathways	Cell proliferation, migration, invasion, metastasis, stemness	[92, 93]
	miR-31	CDKN2B, RASA1	Cell cycle	Cell proliferation, invasion, migration	[94–96]
	miR-32	PTEN	PTEN/PDCD4	Cell proliferation, migration, invasion, metastasis	[97]
	miR-96	FOXO1, FOXO3A, p53	PI3K/FOXO, p53 pathway	Cell proliferation, invasion, migration	[98]
	miR-103	DAPK, KLF4, RB1, TGFBR2	DAPK, Wnt/KLF4 pathways	Invasion, migration, metastasis	[99]
	miR-106b	DLC1,RB1, TGFBR2	DLC1/RhoA	Cell proliferation, migration, invasion, metastasis	[100]
	miR-107	DAPK, KLF4, RB1, TGFBR2	RB1, TGFBR2	Invasion, migration, metastasis, EMT	[99]
	miR-122	CAT1, ADAM17, cyclin-G, Bcl-W	c-MET/STAT3/ERK pathway	Cell proliferation, migration, invasion, metastasis	[101, 102]
	miR-132	ZEB2	EMT	Invasion, migration, metastasis, EMT	
	miR-135b	β-catenin, PTEN, TGFβR2	Wnt/β-catenin, PTEN/PI3K, TGFβ	Invasion, migration, metastasis, EMT	[103–105]
	miR-141	ZEB1, ETS1	EMT	Metastasis, EMT	[106]
	miR-155	E2F2, MSH2, CLDN1, MSH6, MCH1	Cell cycle pathway	Cell proliferation, migration, invasion, chemoresistance	[107–109]
	miR-181a	PTEN, WIF1	Wnt, PTEN/AKT signaling	Cell proliferation, invasion, metastasis, EMT	[110, 111]
	miR-182	ENTPD5, IGFR1	EGFR/Akt pathway	Cell proliferation, migration, invasion, metastasis	[37, 39, 112]
	miR-196a	HoxA7, HoxB8, HoxC8, HoxD8	AKT signaling	Cell proliferation, migration, invasion	[113]
	miR-200	ZEB1	EMT pathway	Metastasis, EMT	[114]
	miR-210	K-Ras	Rho/Kras	Cell proliferation, migration, invasion, metastasis	[77, 115]

(continued)

Table 2.1 (continued)

Expression	MicroRNA	Target	Signaling pathway	Process in colorectal cancer	Reference
	miR-221	c-Kit, Stat5A, ETS1, ENOS	Stat/PI3K/Akt/mTOR, Ras/ERK pathways	Cell proliferation, migration, invasion, metastasis	[77, 115]
	miR-224	CDS2, HSPC159, SMAD4	Cell cycle and EMT pathways	Cell proliferation, EMT	[116–118]
	miR-301a	TGFβR2	TGFβ signaling	Cell proliferation, EMT	[119]
	miR-320b	miR-320a	β-catenin, Neuropilin-1 and Rac-1	Cell proliferation, migration, invasion, metastasis	[120]
	miR-372	TXNIP, LATS2	MAPK/ERK	Cell proliferation	[121]
	miR-451	MIF	MIF/Src pathway	Cell migration, invasion, metastasis	[9, 48]
	miR-495	PTEN, PDCD4	PTEN/PI3K/Akt	Cell proliferation, migration, invasion, metastasis	[56]
	miR-525	PI3K	Akt/PI3K	Cell migration, invasion, metastasis	[56]
	miR-625-3p	SCAI	SCAI/E-cadherin/MMP-9 pathway	Cell proliferation, migration, invasion, metastasis, EMT	[122]
	miR-675	Rb	Rb/E2F pathway	Cell proliferation	[123]
	miR-720	STARD13	MAPK/ERK	Cell proliferation, invasion, migration	[121, 124]
	miR-1269a	TGFβR2, SMAD7, HOXD10	TGFβ/Smad	Cell proliferation, migration, invasion, metastasis, EMT	[125]

EMT: Epithelial to mesenchymal transition; ND: Undefined

endothelial cells) by directly targeting angiogenic genes of c-Kit (v-kit Hardy-Zuckerman 4 feline sarcoma viral oncogene homolog), Stat5A (signal transducer and activator of transcription 5A), ENOS (endothelial nitric oxide synthase) and ETS1 (v-ets erythroblastosis virus E26 oncogene homolog 1) [115]. In contrast, miR-497 is down-regulated in CRC, which is capable of inhibiting cancer cell survival, proliferation and invasion [76], by targeting IGF1R (insulin-like growth factor 1 receptor gene), an angiogenic activator that contributes to angiogenesis in tumors [75]. Other miRNAs, such as miR-194 is also down-regulated in CRC, which can directly targeting an inhibitor of angiogenesis by binding to the 3′ UTR of THBS1 mRNA that encodes thrombospondin-1 (TSP-1) [134]. These studies suggest that miRNAs may play a paradoxical role in tumor angiogenesis through regulating the expression of inhibitors or activators of angiogenesis [9].

Consistently, several miRNAs, including miR-29a, miR-31, miR-103 and miR-107 have been reported to exert effects on the invasion of CRC cells *in vitro* and *in vivo* [135]. For example, the expression of miR-103 and miR-107 were up-regulated in colon cancer cells [99], both of them were able to directly modulate the expression of DAPK1 (death-associated protein kinase 1) and KLF4 (Krüppel-like factor 4), and sequentially led an increased cell motility and suppression of cell-cell adhesion. Such an inhibitory role of miRNA in KLF4 expression of CRC was also found in miR-29a, in which more abundant miR-29a transcript could be detected in colon cancer with liver metastasis as compared to non-metastatic cancer, it thus was suggested as a sensitive and potential marker for colon cancer metastasis [136]. On the other hand, miR-132 and miR-335 have been reported that these miRNAs inhibit colon cancer invasion and metastasis *via* directly targeting ZEB2 [136]. miR-552 and miR-592 were both overexpressed in primary CRC, which could distinguish metastases in the lung between primary lung adenocarcinoma and CRC [136].

With respect to the colonization of cancer cells from primary sites to distant tissues or organs, the process of epithelial to mesenchymal transition (EMT) is the mission-critical step in the metastatic cascade, which is an evolutionarily conserved program of gene expression during which epithelial cells adopt characteristics of mesenchymal cells. A numbers of studies have demonstrated that the EMT is regulated by a variety of signaling pathways, including transforming growth factor-beta (TGF-β), hepatocyte growth factor (HGF), platelet-derived growth factor (PDGF) and epidermal growth factor (EGF). Studies on miRNA expression patterns have been conducted to identify microRNAs with possible roles in TGF-β-induced EMT. It is reported that miR-21 and miR-31 facilitate TGF-β-induced EMT by targeting T-lymphoma invasion and metastasis 1 (TIAM1), repressing its translation rather than inducing mRNA degradation [99].

2.3 Involvement of MicroRNAs in the Signaling Pathways Related to Colorectal Cancer Pathogenesis

An increasing number of studies has suggested that miRNAs can modulate tumor progression process mainly by targeting certain genes in critical signaling in CRC metastasis, such as the Wnt/β-catenin, epithelial growth factor receptor (EGFR), TGF-β, p53 and PTEN/phosphatidylinositol-3-kinase (PI3K) pathways (Table 2.1 and Fig. 2.1) [9, 12, 20, 41, 42, 80, 97, 137–143].

2.3.1 Wnt/β-Catenin Signaling

A hyperactivated canonical Wnt signaling has usually been found to be associated with CRC initiation and progression, implying that the regulatory role of miRNAs in CRC by targeting Wnt signaling [144]. Mutations of adenomatous polyposis coli (APC) occur in more than 60 % of colon cancers, which lead to an activation of canonical Wnt/β-catenin signaling. The canonical Wnt pathway has been recognized to associate with early colon cancer development, suggesting that miRNAs correlated with regulation of Wnt signaling may play a role in colon cancer formation. Indeed, miR-135 was found to be up-regulated in colon tumors and correlated with low level of APC, which could exert an effect on colon cancer *via* regulating Wnt/β-catenin signaling pathway in colon cancer [145]. In addition to miR-135 family, miR-93 and miR-145 were also able to regulate Wnt/β-catenin signaling by targeting catenin gene [146] and Smad7 that can subsequently down-regulated Wnt/β-catenin signaling [147].

2.3.2 EGFR Signaling

EGFR is a member of the human epidermal growth factor receptor or ErbB family of receptor tyrosine kinases. This trans-membrane glycoprotein may be activated through the binding of related ligands, which leads to EGFR forming homodimers or heterodimers with its family members such as ErbB2/neu, ErbB3/HER3 and ErbB4/HER4. This process can promote autophosphorylation of the intracellular domain through tyrosine kinase activity and stimulation of two major downstream signaling pathways, KRAS/RAF/ERK and PI3K/AKT. EGFR signaling is a well-characterized pathway that plays a critical role in the survival, proliferation, migration, angiogenesis, and apoptosis of cancer cells, and a dysregulation of this signaling frequently occurs in several types of epithelial cancers, including the CRC [148]. A series of investigations has revealed that miRNAs were extensively

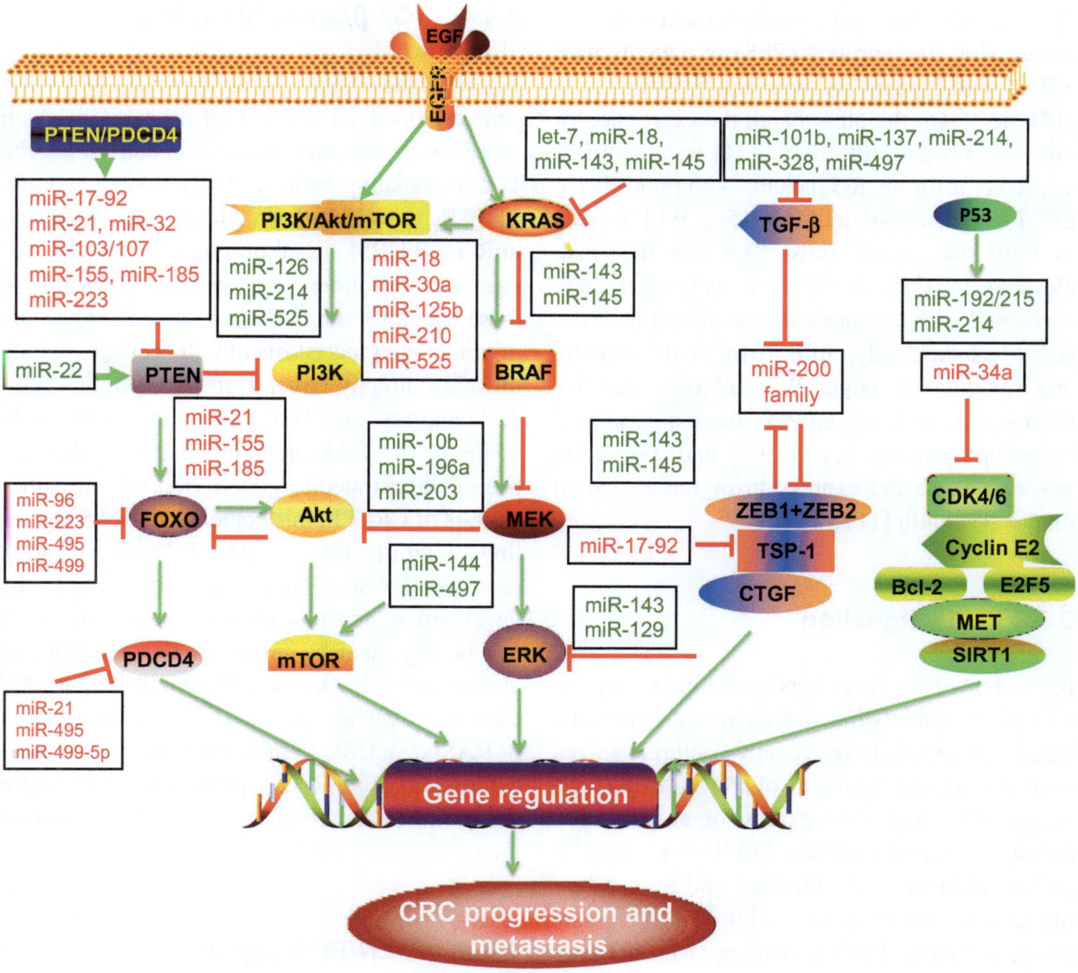

Fig. 2.1 An illustration represents the overview of microRNAs (miRNAs) and their targets involving in the key signaling pathways in colorectal cancer (CRC) metastasis. The depicted miRNAs affect the important factors of colon cancer development and malignancy, such as PTEN/PDCD4, EGFR/KRAS, EGFR/mTOR, TGF-β, p53 and EMT transcription factors. miRNAs that labeled in red fonts are oncomirs upregulated in CRC; whereas miRNAs that labeled in green fonts are tumor suppressor miRNAs downregulated in CRC

involved in the regulation of EGFR signaling, and could serve as promising predictive biomarkers to anti-EGFR therapy [17]. In order to demonstrate the regulatory roles of miRNAs in anti-EGFR treatment of CRC patients, Mosakhani et al. [149] first analyzed the miRNA profiling to predict overall survival (OS) of metastatic CRC (mCRC) patients in anti-EGFR antibody therapy, and found that an up-regulation of let-7 family and miR-140-5p, along with a down-regulation of miR-1224-5p were associated poor OS in anti-EGFR treatment. Interestingly, Ruzzo et al. [150] has previously demonstrated that let-7 could down-regulate KRAS with anti-cancer effects in the presence of activating KRAS mutations, a higher let-7a levels were significantly associated with better survival outcomes in patients who were KRAS-mutated CRC and underwent a treatment of cetuximab plus irinotecan, which implying that let-7 might restore anti-EGFR therapy effects in patients with chemotherapy-refractory metastatic disease. In addition, miR-31 was also recently identified as a predictive marker for progression-free survival (PFS) in patients with KRAS wild-type mCRC with anti-EGFR therapy [151, 152]. On the other hand, the expression of miR-7 [153] and miR-181a [154] was identified as predictor for mCRC patients with poor PFS in EGFR-targeted therapy [153, 154]. With respect to CRC metastasis, miR-181a was the most elevated in CRC with liver metastases, and correlated with advanced stages, distant metastasis. Mechanistically, miR-181a could directly and functionally target Wnt inhibitor factor-1 (WIF-1), as well as suppress the expression of epithelial markers E-cadherin and β-catenin, while increase the expression of mesenchymal marker vimentin [111].

2.3.3 P53 Signaling

p53 is one of the most important tumor suppressors that frequently inactivated in gastrointestinal cancers. miRNAs have recently been recognized as mediators and regulators of p53 signaling, *vice versa*, p53 can alter the expression and/or maturation of several miRNAs [155]. For instance, p53 could induce miR-15a/16-1 and form a double-negative feedback loop with AP4 to regulate the epithelial to EMT and metastasis of CRCs [86]. In this context, the transcription factor AP4 played a key role in EMT, which was down-regulated by DNA damage in a p53-dependent manner in CRCs. On the other hand, the p53-induced miR-15a/16-1 could in turn directly target AP4 and induce mesenchymal to epithelial transition (MET), accordingly inhibited CRC cell migration and invasion [86]. This finding was similar to a function of miR-34a, a tumor suppressor that contributes to apoptosis and acute senescence of cancer cells. miR-34a can be induced by p53, and form a p53/miR-34a axis to regulate a Snail 1-dependent cancer cell EMT [156]. Other miRNAs include let-7a [157], miR-16 [158], miR-133a [159], miR-192/215 [57] and miR-194 [134] were also found to be induced by p53 in CRC. Similarly, miR-96 also has been suggested to able to target p53 pathway and promote CRC cell proliferation and tumor progression [98].

2.3.4 TGF-β/Smad Signaling

TGF-β/Smad signaling is an important molecular pathway involved in EMT of cancers, in which miRNAs are crucial regulators in controlling the TGF-β signaling pathway [160]. Oncomir miR-21 [93], miR-135b [105], miR-301a [119] and miR-1269 [125] are all up-regulated in CRC tissues and cell lines, and an overexpression of these oncomirs can further promote the invasion, migration and metastasis of CRC cells, by which miRNAs directly target TGF-β2 receptor (TGF-β2R) and regulate TGF-β signaling pathway. In addition, the RAS signaling pathway also has been demonstrated to play a vital role in pathogenesis of CRC. In this context, and RAS signalling terminators, RAS-GTPase-activating proteins (RASGAPs) are correlated with the development and progression of cancers, which can be regulated by miRNAs. For example, an aberrant miR-223 transcript was detected in CRC tissues, which was involved in down-regulation of RASA1 in CRC tissues. Furthermore, an over-expression of miR-223 promoted CRC tumor growth and an inhibition of miR-223 repressed the tumor growth [67].

2.3.5 PTEN/PI3K Signaling

The phosphatase and tensin homolog (PTEN) is a phosphatase related to the phosphatidylinositol-3-kinase (PI3K) pathway, which is involved in angiogenesis of tumors [137]. The PTEN/PI3K signaling has been demonstrated to be involved in angiogenesis of many types of cancers. Several miRNAs have been identified for targeting PETN/PI3K pathway in CRC. miR-17-92 cluster, also known as oncomir-1, one of its target is PTEN, which can promote chemotherapeutic drug resistance and metastasis in CRC by targeting PTEN [138]. In addition, Dews et al. [139] discovered that the miR-17-92 cluster could mediate MYC-dependent tumor promoting effects by suppressing the expression of TSP-1 and CTGF (connective tissue growth factor), which are anti-angiogenic factors. miR-32 was another miRNA identified to regulate PTEN expression and promote the growth, migration, and invasion of CRC cells [97]. Other miRNAs include miR-21 [140], miR-22 [141] miR-30a [80], miR-126 [142], miR-153 [143] and miR-520a [80] were also identified to be able to directly target PTEN/PI3K in CRCs. miR-126 is another example, which is down-regulated in primary CRC cancer. The miR-126 can activate vascular endothelial to growth factor (VEGF) pathway by modulating the expression of sprouty-related protein SPRED1 and PIK3R2 (PI3K regulatory subunit 2). In addition, mice knockdown of miR-126 exhibit phenotypes including a loss of vascular integrity and an inhibition of endothelial cell migration and angiogenesis [41]. Moreover, miR-126 is also able to bind to the 3′-UTR of p85beta (phosphatidylinositol-3-kinase regulatory subunit beta, PI3Kβ) mRNA and modulates its expression. PI3Kβ is a regulatory subunit involved in stabilization and propagation of PI3K pathway [142]. Apart from its regulatory role in PI3K pathway, the miR-126 was recently found to exert a role of tumor suppressor by inhibiting RhoA/ROCK signaling pathway through repressing RhoA expression. The activity of ROCK is involved in the invasion and metastasis of tumor cells including the CRC, in which ROCK is the main RhoA downstream effector [42].

In addition to directly target cell signaling molecules, miRNAs also can modulate signaling activity by targeting their homolog miRNAs. For example, miR-320a is a tumor suppressor with single nucleotide different from its homolog miR-320b. The latter was found to be up-regulated tumor from CRC patients with liver metastasis, which showed an opposite function of miR-320a. miR-320b is able to promote CRC cell proliferation and invasion by competing its homolog miR-320a, and its overexpression leads to up-regulation of the target genes of miR-320a including β-catenin, Neuropilin-1 and Rac-1 [120]. Several miRNAs identified in CRC metastasis and their targets and regulated signaling pathways are listed in the Table 2.1.

2.4 MicroRNAs Targeting Signaling Pathways in the Metastasis of Colorectal Cancer

Recently, miRNAs have been suggested to involving in the acquisition of acquire stem-cell-like properties for cancer cells by regulating EMT signaling. For example, Hur et al. [59] found that miR-200c was aberrantly expressed in metastatic colon tumor tissues and colon cancer cells, and this up-regulated miR-200c was correlated with an reduction of the expression of its target genes: zinc finger E-box binding homeobox 1 (ZEB1), ETS1 and fms-related tyrosine kinase 1 (FLT1), which in turn up-regulates E-cadherin and down-regulate the expression of vimentin, sequentially led an activation of EMT signaling pathway (Fig. 2.1). This observation was in line with a study by Korpal et al. [161], in which the authors demonstrated that the effect of down- or up-regulation of miR-200 family members caused a downstream increase/decrease of expression of ZEB1 and ZEB2, and then modulated the EMT pathway. These studies demonstrate that miRNAs may play an important role in mediating EMT and metastatic behavior in the colon cancer.

Another well-defined miRNA that associated with colon cancer metastasis is miR-143. The miR-143 was down-regulated in colon cancer and liver metastasis, and a less abundant miR-143 was found to associate with larger tumor size and longer disease-free interval in colon cancer, and an enhanced expression of miR-143 attenuates migration and invasion in colon cancer [162]. Mechanistically, miR-143 was identified to target metastasis-associated in colon cancer-1 (MACC1), a novel prognostic biomarker for metastasis occurrence, which was over-expressed in colon cancer and other cancer types [129]. Therefore, a down-regulation of miR-143 could enhance colon cancer metastasis through the MACC1-induced HGF-MET signaling pathway [77, 163]. Similarly, a down-regulation of miR-

34a and miR-145 were also found in colon cancer [164, 165]. miR-145 can act as a suppressor of tumor by inhibiting activities of KRAS and BRAF [50], while miR-34a may play a role as a tumor suppressor by regulating the Sirtuin 1 (SIRT1)-p53 pathway. In this context, miR-34a and p53 signaling can form a positive feedback loop, and the miR-34a inhibits the expression of SIRT1 [165]. In addition, miR-622 was found to down-regulated in metastatic CRC tissues and cell lines, which showed a potential to suppress tumor proliferation and migration *in vitro*, by targeting KRAS [83]. Recently, Huang et al. [91] found that miR-19a was up-regulated in CRC tissues, and the elevated miR-19a transcript was strongly associated with lymph node metastasis of CRC. Mechanistically, an overexpression of miR-19a in human CRC cells enhanced the capacity of cell invasion and EMT that was induced by TNF-α. Of note, miR-19a transcription could be up-regulated by TNF-α, and miR-19a was required for TNF-α-induced EMT and metastasis in CRC cells.

Since EMT and MET are well-established biological events that play pivotal roles in the homeostasis and pathogenesis of colon during CRC progression and development of chemoresistance, during which the TGF-β signaling is the key player. Therefore, miRNAs involved in the TGF-β signaling may be targets for reversing drug resistance. Indeed, miR-147 has been shown an ability to reverse anti-EGFR TKI resistance in CRC cells by inducing EMT to MET [166]. Intriguingly, a recent novel therapeutic strategy by reprogramming CRSCs using miRNAs targeting key transcription factors for stemness led an enhanced chemosensitivity in CRC cells [167]. In this study, the authors reprogramed CRC cells (DLD-1, RKO and HCT116) by targeting key transcription factors (Oct3/4, Sox2, c-Myc and Klf4) using a mixture of mature miR-200c, miR-302a-d, miR-369-3p and miR-369-5p. In this regard, Oct3/4, Sox2, c-Myc and Klf4 are known transcription factors essentially for maintenance of the stemness. The CRC cells introduced with the miRNA mixture exhibited an embryonic stem cell-like morphology and expressed the undifferentiated markers of Nanog, Oct3/4, Sox2 and Klf4 but decrease of c-Myc. Notably, the miRNA-transfected DLD-1 cells displayed a reduced proliferative capacity along with an increased expression of the tumor suppressor genes $p16^{ink4a}$ and $p21^{waf1}$, accompanied with an enhanced sensitivity to 5-FU, possibly by down-regulating multidrug resistant protein 8 [167]. All together, these studies highlight an important role of miRNAs in the connection of CSC, EMT and drug resistance, which may offer novel targets to maximize the effects of conventional cancer therapies [144].

2.5 Long Non-coding RNAs as Hallmarks in the Development and Progression of Colorectal Cancer

lncRNAs are a class of non-protein-coding RNAs with length greater than 200 nucleotides, which comprise transcripts resided in long intergenic non-coding RNAs (lincRNAs) and introns of protein coding genes, transcripts of pseudogenes and ultra-conserved regions (tUCRs), as well as transcripts that partially overlapping the UTRs or promoters of protein coding genes [8]. Functionally, increasing lines of evidences have suggested that lncRNAs are involved in a broad spectrum of biological processes, including cell proliferation, differentiation, cell apoptosis, and stem cell self-renewal in a developmental and tissue-specific manner, through a broad range of mechanisms including regulations of epigenetic modification, alternation of RNA splicing, modulating protein localization and activity, in part owing to their abilities to bind DNA, other RNAs and proteins [8, 168, 169]. To date, a compelling body of studies has evidenced the involvement of lncRNAs as hallmarks in neoplastic diseases, including the CRC by targeting various signaling pathways [169]. In this context, lncRNAs can function as oncogenes and/or tumor suppressors by interacting with other regulatory molecules, such as DNA, RNA and proteins related to signaling pathways involved in CRC pathology [168, 169]. By employing high-throughput sequencing technologies and/or microarray analysis, an expanding list of dysregulated lncRNAs has been identified in CRC tissues and/or cell lines, which are hallmarks involved in the pathogenesis of this neoplastic disease, and can be potentially used as diagnostic and prognostic biomarkers for CRC, despite the underlying mechanisms of their functions in CRC biology remain largely unknown (Table 2.2) [168–170].

A genome-wide analysis recently performed by Xue et al. [202] using the high-throughput microarray assay of human CRC tumor tissues and their matched adjacent normal tissues, the investigators found a series of differentially expressed lncRNAs CRC, and in two of the lncRNAs, HOTAIR and lncRNA-422 were further confirmed in 90 paired clinical samples, bioinformatics analysis suggested that that these two lncRNAs might be involved in the pathogenesis by regulating protein coding genes relevant to cancers. Metastasis is a main cause of cancer relapse and development in CRC, several lines of studies thus examined the metastasis associated lncRNAs in metastatic lymph nodes and livers of CRC patients by a microarray analysis [250–252]. By comparing the expression of lncRNAs between metastatic lymph nodes (MLN), normal lymph nodes (NLNs) and CRC tumor tissues, Han et al. [250] found that 1133 of the 33,045 screened lncRNAs were differentially expressed in MLN compared with NLN, of which 260 were up-regulated and 873 down-regulated; 545 lncRNAs were differentially expressed in MLN compared with CRC tumor tissues, of which 460 were up-regulated and 85 down-regulated; in addition, 14 lncRNAs were specifically up-regulated and 5 specifically down-regulated in MLN as compared with the cancer tissues. The expression of lncRNAs AK307796, AK025180 and AK021444 were further confirmed using a qRT-PCR assay in 26 paired clinical samples by the same group [251]. Consistently, Ye et al. [252] determined the expression of ncRNAs in colorectal liver metastasis (CLM) and found that 1332 lncRNAs were differentially expressed in CLM tissues, 40 differentially expressed lncRNAs that potentially related to CLM were further examined in an expanded set of clinical samples, and three lncRNAs, CLMAT1-3 were verified. Clinically, an aberrantly up-regulated CLMAT3 was strongly correlated with CLM and MLN, CRC patients with a high CLMAT3 expression exhibited a shorter median overall survival (OS) duration than those who had a low level of CLMAT3 expression [252].

The chemoresistance causes drug treatment failures that ultimately lead to the cancer recurrence and death. In order to interrogate the mechanisms underpinning resistance development, Sun et al. recently analyzed the expression profile of lncRNAs associated with vincristine (VCR) resistance in HCT-8 colon cancer cells by next-generation sequencing they determined that 23 lncRNAs were up-regulated and 20 lncRNAs were down-regulated with a fold change greater than 10 in VCR-resistant cells in comparison with the VCR-sensitive cells [253]. By using a similar approach, Xiong et al. [188] also profiled the differential expression of lncRNAs between 5-fluorouracil (5-FU)-resistant and non-resistant HCT116 CRC cells using a microarray analysis. A total of 2662 lncRNAs was differentially expressed in 5-FU-resistant HCT116 cells when compared with those in parental HCT116 cells, of these 6 of lncRNAs, including TCONS00026506, ENST00000468960, NR038990, ENST00000575202, ENST00000539009 and ENST00000544591 were further validated by a qRT-PCR assay [188]. Of note, DNA methylation also is an important epigenetic modification for CRC, and 761 lncRNA genes with DNA hypermethylation in CRC have been recently identified using a MethylCap-seq dataset. By integrating the lncRNA profile and methylation datasets, the authors further demonstrated that the expression of lncRNAs was inversely correlated with DNA methylation [254]. Genetically, genetic variants in chromosome 8q24 where the lncRNA prostate cancer-associated ncRNA 1 (PRNCR1) is located, has been recognized to be able to confer the susceptibility to CRC [238]. In addition, genetic variants of LNC00964-3 [255], and HOX transcript antisense RNA (HOTAIR) [201] have also been uncovered to be associated with risks of CRC. These studies

Table 2.2 Long non-coding RNAs implicated in colorectal cancer pathogenesis

Long non-coding RNA	Gene size (kb)	Locus	Expression	Potential mechanism/target	Function	Reference
AK123657	5.3	Chr11q	Down	ND	As a tumor suppressor inhibits cell proliferation and invasion	[171]
BA318C17.1	0.7	Chr20p12	Down	ND	ND	[171]
BACE1AS	2.7	Chr11	Down	ND	Involved in the tumorigenesis and progression of CRC	[172]
BANCR	10.3	Chr9	Up/Down	Regulates MEK and p21 signaling	Induces EMT and cell proliferation	[173–175]
BX648207	5.0	Chr12q	Down	ND	As a tumor suppressor inhibits cell proliferation and invasion	[176]
BX649059	11.4	Chr12	Down	ND	As a tumor suppressor inhibits cell proliferation and invasion	[171]
CAHM	0.88	Chr6	Frequent hypermethylated	ND	Methylated CAHM DNA in CRC plasma is a biomarker for CRC	[177]
CCAL	ND	ND	Up	Activates Wnt/β-catenin signaling by targeting AP2α	Induces multidrug resistance in CRC	[178]
CCAT1	2.6	Chr8q24.21		ND	Increases cell proliferation and invasion	[171, 179]
CCAT1-L	2.6	Chr8q24	Up	Regulates Myc expression	An oncogene	[180]
CCAT2	0.4	Chr8q24	Up	Regulates Wnt and Myc	Increase proliferation and metastasis	[181]
CCAT3	1.6	Chr14	Up	Regulate Myc target genes	Increase proliferation and metastasis	[182]
CCAT4	1.3	Chr12	Up	Regulate Myc target genes	Increase proliferation and metastasis	[182]
CCAT6	3.6	Chr7	Up	Regulate Myc target genes	Increase proliferation and metastasis	[182]
CCAT7	0.65	Chr20	Up	Regulate Myc target genes	Increase proliferation and metastasis	[182]
CCAT8	2.32	Chr9	Up	Regulate Myc target genes	Increase proliferation and metastasis	[182]
CRNDE	~10	Chr16	Up	As scaffolds for regulatory complexes	Increases Warburg effects and risk of CRC	[183–185]
DANCR	7.95	Chr4	Up	ND	Associated with poor prognosis in CRC	[186]
E2F4 antisense	~5.0	Chr16q21	Up	Induced by Wnt/β-catenin signaling, targets E2F4	Inhibits E2F4 expression	[187]
ENST00000468960	15.2	Chr7	Down	ND	Involved in chemoresistance	[188]
ENST00000539009	5.43	Chr12	Down	ND	Involved in chemoresistance	[188]

(continued)

ENST00000544591	4.7	Chr12	Down	ND	Involved in chemoresistance	[188]
ENST00000575202	57.5	Chr17	Down	ND	Involved in chemoresistance	[188]
FIRRE	128	ChrX	Down	ND	CRC patients with higher expression have a favor prognosis	[189]
FTX	330	ChrX	Up	ND	Associated with poor OS in CRC	[190]
GAS5	0.6	Chr1q25.1	Down	As a tumor suppressor by targeting GR	Induces apoptosis	[191, 192]
H19	6.3	Chr11p15.5	Up	Imprinting control	Increases proliferation	[123, 193–198]
HOTAIR	12.65	Chr12q13	Up	Binds to PRC2 and LSD1 to induce gene silencing	Enhances EMT and metastatic potential	[199–204]
HOTTIP	8.7	Chr7	Up	ND	Promotes cell proliferation and inhibits cell apoptosis	[205, 206]
HULC	0.5	Chr6p24.3	Up	RNA-DNA (CREB) binding	ND	[207]
LINC00152	151	Chr2	Down	ND	CRC patients with higher expression have a favor prognosis	[189]
LINC00964-3	ND	ND	Down	ND	ND	[202]
LINC01021	24.1	Chr5	Up	Regulates p53 signaling	ND	[208]
LINC01296	53	Chr14	Down	ND	CRC patients with higher expression have a favor prognosis	[189]
LINC-ROR	17.6	Chr18	Up	Activates TESC promoter by histone modification	Induces tumor growth and metastasis	[209]
KCNQ1OT1/ LIT1	91	Chr11q15	LOI frequently occur	LOI	ND	[210]
lncRNA-422	ND	ND	Up	ND	ND	[202]
lncRNA-ATB	ND	ND	Up	ND	Involved in the progression and metastasis of CRC	[211]
lncRNA-LET/ NPTN-IT1	2.6	Chr15q24	Down	Hypoxia signaling Regulator	Enhances metastatic potential	[212]
lincRNA-p21	~3.0	ND	Up/Down	Binds to hnRNP K to repress p53-dependent pathways, activates Wnt signaling	Increases Warburg effect, invasive capacity and radiation sensitivity	[213–216]
LOC152578	25.5	Chr4	Up	ND	ND	[217]
LOC554202	104	Chr9	Down	Activates specific caspase cleavage cascades	Induces cell apoptosis in CRC	[218]

(continued)

Table 2.2 (continued)

Long non-coding RNA	Gene size (kb)	Locus	Expression	Potential mechanism/target	Function	Reference
LOC285194/TUSC7	2.1	Chr3q13	Down	Acts an endogenous sponge of oncomirs	Enhances metastatic potential	[219, 220]
LSINCT5	2.65	Chr5	Up	ND	Involved in cancer cell proliferation	[221]
MALAT1	~7.0	Chr11q13	Up	RNA splicing, small RNA biogenesis, protein interaction, target AKAP-9, SFPQ/PTBP2	Enhances proliferation, migration and invasion and metastatic capacities	[222–226]
MEG3	1.6–1.8	Chr14q32	Down	Suppresses MDM2 to increase p53	Increases apoptosis and inhibits proliferation	[227, 228]
MNX1-AS1	5.6	Chr7	Up	Regulate Myc target genes	Increase proliferation and metastasis	[182]
MYLKP1	106	Chr3q12	Up	Pseudogene of MYLK	Increases proliferation	[229]
ncNRFR	1.4	Chr1q13	Up	Inhibits the function of the tumor suppressor let-7	An oncogene	[230]
ncRAN	2.3	Chr17q25	Down	ND	Enhances migration and invasive capacities	[231, 232]
ncRuPAR	0.4	Chr5q13	Down	Targets PAR-1	Inhibits tumor progression as a tumor suppressor	[233]
NEAT1	22.8	Chr11	Up	ND	Involved in CRC differentiation, invasion and metastasis	[234]
NR_038990	29	Chr5	Down	ND	Involved in chemoresistance	[188]
OCC-1	1.2–1.3	Chr12q24	Up	ND	ND	[235]
PCAT1	1.9	Chr8q24	Up	Inhibits BRCA2	Associates with an enhanced proliferative potential and poor prognosis	[236]
pou5f1p1	0.4	Chr8q24	Up	Pseudogene of pou5f1	Increasing risk of CRC	[237]
PRNCR1	13	Chr8q24	Up	ND	Enhanced proliferative capacity	[238–240]

PTENP1	3.9	Chr9q13.3	Down	Pseudogene of PTEN	Decreases proliferation	[133, 241]
PVT1	>300	Chr8q24	Up	A p53-inducible target miR-1204	Enhance invasion, decreases apoptosis, and associated with poor prognosis	[242]
RP11-462C24.1	82.3	Chr4q25	Down	ND	Involved in the tumorigenesis and progression of CRC	[243]
snaR-A1	0.12	Chr19	Down	ND	Associated with chemoresistance in CRC	[172]
TCONS_00026506	ND	ND	Up	ND	Involved in chemoresistance	[188]
UC.388	0.2–0.8	ND	Down	ND	Decreases metastatic capacity	[244]
UC.73A	0.2	ND	Up	ND	Increases proliferation, decreases apoptosis	[244]
UCA1/CUDR	2.3	Chr19p13	Up	ND	Induces the proliferation, apoptosis and cell cycle progression of CRC cells	[245–247]
XLOC_000303	181.8	Chr1	Up	ND	A potential biomarker for the tumorigenesis	[217]
XLOC_00684	ND	ND	Up	ND	A potential biomarker for the tumorigenesis	[217]
XIST	19.3	ChrXq13	Up	Increased in MSI sCRC	NA	[248]
ZFAS1	11	Chr20q13	Up	Interacts with CDK1 and is involved in p53-dependent pathway	Regulated cell cycle control and apoptosis in CRC	[249]

ND not determined, *CCAL* colorectal cancer-associated lncRNA, *CCAT* CRC-associated transcript, *CLM* colorectal liver metastasis, *CRC* colorectal cancer, *CREB* cAMP response element-binding protein, *CRNDE* colorectal neoplasia differentially expressed, *hnRNP* heterogeneous ribonucleoprotein particle, *GAS5* growth arrest-specific transcript 5 gene, *HOTAIR* HOX transcript antisense RNA, *HOTTIP* HOXA transcript at the distal tip, *HULC* highly upregulated in liver cancer, *LIT1* long QT intronic transcript 1, *LET* low expression in tumor, *lincRNA* long intergenic non-coding RNA, *LMN* liver metastatic nodules, *lncRNA* long non-coding RNA, *LOI* loss of imprinting, *LSD1* lysine-specific demethylase 1, *MALAT1* metastasis-associated lung adenocarcinoma transcript 1, *MEG3* maternally-expressed gene 3, *miRNA* microRNA, *MSI* microsatellite instable, *MYLKP1* myosin light chain kinase pseudogene 1, *ncNRFR* non-coding Nras functional RNA, *ncRAN* non-coding RNA expressed in aggressive neuroblastoma, *OCC-1* overexpressed in colon carcinoma-1, *PAR-1* protease activated receptor-1, *PCAT1* prostate cancer-associated transcript 1, *PRC2* polycomb repressive complex 2, *PRNCR1* prostate cancer-associated non-coding RNA 1, *PTENP1* phosphatase and tensin homolog 1, *PVT1* plasmacytoma variant translocation 1, *sCRC* sporadic colorectal cancers, *TUSC7* tumor suppressor candidate 7, *UCA1* urothelial cancer-associated 1 clearly suggest that a dysregulated expression of lncRNAs is a hallmark of CRC pathogenesis. Indeed, increasing lines of study have identified several lncRNAs that have importantly regulatory roles in CRC pathology, such as BRAF-activated non-protein coding RNA (BANCR), colorectal neoplasia differentially expressed (CRNDE), H19, HOTAIR, Metastasis-Associated Lung Adenocarcinoma transcript 1 (MALAT1), H19, lincRNA-p21 and PRNCR1 (Table 2.2).

2.6 Long Non-coding RNAs Targeting Signaling Pathways Related to Colorectal Cancer Pathogenesis

HOTAIR is one of the most intensively studied lncRNAs located within the Homeobox C (HOXC) gene cluster on chromosome 12 and co-expressed with the HOXC genes, which is an oncogenic lincRNA and functions as a molecular scaffold to assemble the two histone modification complexes polycomb repressive complex 2 (PRC2) and lysine (K)-specific demethylase 1A (LSD1) on the HOXD gene cluster, induces histone H3 lysine-27 (H3K27) trimethylation and H3K4 demethylation for epigenetic gene silencing to promote cancer metastasis [256, 257]. It is pronouncedly overexpressed in most solid cancers and correlated with tumor invasion, progression, metastasis, and poor prognosis, which was also required for EMT and stemness maintenance in cancer cell lines [200]. Mechanistically, HOTAIR can epigenetically regulate HOXD expression and promotes cancer metastasis in breast cancer by silencing multiple metastasis suppressor genes, such as HOXD10, PGR, and the protocadherin gene family in breast cancer [258]. In CRC, the higher abundance of HOTAIR transcripts was determined in CRC tissues relative to the matched normal tissues, which was correlated with the levels of members of the PRC2 complex H3K27me3, SUZ12 and EZH2, and was involved in maintaining the mesenchymal and undifferentiated status in CRC cells [199]. In addition, CRC patients with abundant HOTAIR transcript in both primary tumor tissues and blood exhibited a relatively unfavorable prognosis [199, 203, 204]. Intriguingly, a recent study in evaluation of the expression of HOTAIR in colon cancer tissues revealed that an increased abundance of HOTAIR transcript was significantly correlated with the extent of tumor invasion, metastasis, histological differentiation and advanced stages, i.e. patients with a higher HOTAIR expression had higher recurrence rates and less metastasis-free and shorter overall survival relative to those who with lower level of HOTAIR expression. More interestingly, the increased expression of HOTAIR had a limited effect on cell proliferation *in vitro*, despite it could significantly promote cancer cell migration and invasion. A depletion of HOTAIR induced the expression of E-cadherin while concomitantly suppressing the expression of vimentin and MMP9, suggesting that HOTAIR may play a crucial role in EMT of colon cancer [204]. These studies indicate that HOTAIR may serve as potential prognostic marker serve as potential surrogate prognostic marker for patients with metastatic CRC. Similarly to HOTAIR, also the lncRNA colorectal neoplasia differentially expressed (CRNDE) that also elevated in early stages of CRC tissues [183], can physically and functionally associate to PRC2 [259]. Knockdown of CRNDE and PRC2 exhibited alterations of the expression of a list of overlapped genes, suggesting an involvement of CRNDE in the epigenetic remodeling of chromatin, in part through the down-regulation of gene expression by targeting histone methylation *via* the PRC2 complex [259]. In addition, knockdown of CRNDE showed a decrease of the expression of several pluripotency markers, such as SOX2, KLF4, NANOG and OCT4), possibly due to some pluripotency-related transcription factors (such as Myc) could bind to the CRNDE promoter [260]. Interesting, CRNDE also exhibited an ability to promote glioma cell growth and invasion through a mammalian target of rapamycin (mTOR) signaling pathway [185].

MALAT1 is another well known cancer related lncRNAs, which is significantly up-regulated in metastases of various cancers, including the CRC [223–226]. MALAT1 could promote

CRC tumor growth and metastasis by regulating tumor suppressor proteins SFPQ (also known as PSF (PTB-associated splicing factor)), and releasing oncogene PTBP2 [also known as PTB (polypyrimidine-tract-binding protein)] to form SFPQ/PTBP2 complex to activate E2F1, a pivotal transcription factor for cell cycle progression [225, 261], and negatively regulating the expression of EMT-associated ZEB1, ZEB2 and SNAI2 genes, but positively regulating E-cadherin gene expression [262]. A most recently study by Yang et al. [226] further confirmed that MALAT1 was up-regulated in human primary CRC tissues with lymph node metastasis, and overexpression of MALAT1 enhanced capacities of CRC cell proliferation, invasion and migration *in vitro*, and promoted tumor growth and metastasis in mice *in vivo*. In contrast, knockdown of MALAT1 exhibited an opposite effect on CRC cells and tumors *in vitro* and *in vivo*. Importantly, the authors further identified that MALAT1 could significantly induce the expression of PRKA kinase anchor protein 9 (AKAP-9). Interestingly, AKAP-9 was also highly expressed in CRC cells with metastatic potential and human primary CRC tissues with lymph node metastasis, but not in normal cells or tissues. Of note, knockdown of AKAP-9 could suppress the MALAT1-mediated cell proliferation, migration and invasion in CRC. These data indicate that MALAT1 may promote CRC tumor development via its target protein AKAP-9. Association analysis of the MALAT1 expression and CRC clinicopathological parameters further suggested that patients with tumors expressing higher level of MALAT1 had a significantly worse prognosis with a hazard ratio (HR) of 2.863 (95 % CI, 1.659–4.943; $P<0.001$) for disease-free survival (DFS) and 3.968 (95 % CI, 1.665–9.456; $P=0.002$) for OS. These studies suggest that MALAT-1 played a pivotal role in CRC metastasis, and may serve as a negative prognostic marker in advanced stage of CRC patients [224]. In addition, gene functional analysis revealed that the 3′ end of MALAT-1 gene was the functional motif that contributed to the biological processes of cell proliferation, migration and invasion in CRC [222]. Equally noteworthy, MALAT1 may serve as a novel target for development of anti-CRC agents. For example, resveratrol showed an ability to down-regulate MALAT1 expression, which in turn reduced the abundance of nuclear localization of beta-catenin thus attenuated Wnt/beta-catenin signaling activity, consequently led to the inhibition of CRC invasion and metastasis [223].

BRAF-activated non-coding RNA (BANCR) is a lncRNA originally identified in melanoma cells, which is pronouncedly expressed melanoma cells and contributes to cell migration [263]. However, the expression of BANCR in CRC cells was in controversy [173–175]. In this respect, Guo et al. [174] reported that BANCR expression was upregulated in CRC tissues, which was significantly correlated with CRC lymph node metastasis and tumor stage. Mechanistically, BANCR could induce EMT by regulating the expression of epithelial and mesenchymal markers through a mitogen-activated protein kinase 7 (MEK)/ERK dependent mechanism [174]. Conversely, more recently studies by Li et al. [175] and Shi et al. [173] reported a down-regulation of BANCR in CRC, which was consistent with a finding in lung cancer in which BANCR was dramatically decreased and severed as a regulator of EMT signaling during cancer metastasis [264]. In line with this finding, Shi et al. also revealed that BANCR expression was significantly down-regulated in CRC tissues as compared with normal tissues, and an enforced expression of BANCR resulted in a decreased CRC cell proliferation *in vitro* and tumor growth *in vivo*, by which BANCR induced cell arrest and apoptosis by interacting with p21 protein [173]. Consistently, Li. et al. [175] also found that BANCR expression was decreased in CRC cells in a study of investigating the anti-cancer effect of fentanyl. They found that fentanyl could inhibit cell migration, invasion and clonogenesis *in vitro*, along with an up-regulation of BANCR and a down-regulation of transcriptional factor Ets-1 in CRC cells. Interestingly, the Ets-1 could in turn negatively regulate BANCR expression and reverse the fentanyl-induced anti-cancer effect by deacetylating histones H3 within BANCR promoter [175].

2.7 Interaction Between Long Non-coding RNAs and microRNAs in Colorectal Cancer Pathogenesis

Of great interest, accumulating studies have suggested the importance of interactions between lncRNAs and miRNAs during the pathogenesis of cancers [265]. For instance, indeed, previous genome-wide association studies (GWAS) have identified that a set of risk loci for different diseases (including CRC) were located within human chromosome 8q24 [266], within which several lncRNAs, including colon cancer associated transcript 1 (CCAT1), CCAT2, PCAT2 and PRNCR1 were also mapped, and the single nucleotide polymorphism (SNP) rs6983267 was resided in this region, which was strongly associated with an increased risk of CRC [267, 268]. The expression of CCAT1 was significantly higher in CRC tumor tissues than that in normal tissues, and its expression level was closely correlated with local infiltration depth, tumor staging, vascular invasion and CA19-9 level in CRC patients [269]. Genetically, DNA enhancer elements of Wnt signaling activator was also located in the rs6983267 spanned genomic region [270]. Intriguingly, the rs6983267 was encompassed in lncRNA CCAT2 genomic region, in which CCAT2 interacted with TCF7L2 and up-regulates Wnt target gene MYC, miR-17-5p and miR-20a, and activated Wnt signaling pathway [181]. Of note, CCAT2 itself is a target gene of Wnt signaling cascade downstream, which may imply the existence of a positive feedback loop in this genomic region [181]. Plasmacytoma Variant Translocation 1 (PVT1) is another lncRNA harbored at 8q24, which is located in the downstream of MYC gene and produces a wide variety of spliced RNAs, such as a cluster of six annotated miRNAs (i.e. miR-1204, miR-1205, miR-1206, miR-1207-3p, miR-1207-5p and miR-1208), in which p53 could bind and activate a canonical response element within the vicinity of miR-1204, and induces the endogenous PVT1 transcripts and consequent up-regulation of miR-1204, which in turn activated p53 signaling and induced cell death in a partially p53-dependent manner [271]. The elevated expression of PVT1 was observed in CRC, owing to a copy number amplification of chromosome 8q24 in CRC patients, and an increased PVT1 expression was essential for high MYC protein levels in 8q24-amplified CRC cells [272]. In addition, a suppression of PVT1 expression could remarkably reduce the capacity of cell proliferation and invasion in CRC cells through activation of TGF-β and apoptotic signaling [242].

Recently, numerous lncRNAs have been shown to post-transcriptionally regulate gene expression by competing binding with miRNAs, by which lncRNAs can interact with miRNAs by complementary sequence to act as a miRNA decoy or sponge [273]. The lncRNA phosphatase and tensin homolog pseudogene 1 (PTENP1) is an example, which can function as a decoy for PTEN-targeting miRNAs (such as miR-19b and miR-20a) in tumor suppression [241]. The H19 is a highly conserved oncogenic lncRNA, which was the precursor of miR-675. Both H19 and miR-675 were up-regulated in human CRC cells primary human CRC tissues. The tumor suppressor retinoblastoma (RB) was a direct target of miR-675, and the H19-derived miR-675 could promote CRC cell proliferation by targeting the RB [123]. Interestingly, the transcript of H19 also harbors canonical and non-canonical binding sites for some members of the let-7 family, which involves in the regulation of let-7 expression, and H19 can down-regulate let-7 expression by exerting its role of a molecular sponge [274, 275]. Such a sponge role for miRNAs was also recently in CRC, in which the lncRNA H19 was highly elevated expression in mesenchymal-like cancer cells and primary CRC tissues, and the expression level of H19 was tightly correlated CRC progression [195]. An overexpression of H19 led a significant progression of EMT and accelerated tumor growth *in vivo*. Importantly, the H19 could functioned as a miRNA sponge for endogenous miR-138 and miR-200a, a reduced these miRNAs led to an increased expression of their target genes, Vimentin, ZEB1 and ZEB2, all of which were core markers of EMT progression [195].

In addition, the lncRNA loc285194 (also known as tumor suppressor candidate 7 (TUSC7)) is a p53-regulated tumor suppressor, which is another lncRNA identified as a sponge or competing endogenous RNA (CeTNA) for miR-211, and was down-regulated in CRC [219, 220]. A low abundance of LOC285194 transcript was correlated with larger tumor size, higher tumor stage, and more distant metastasis, as well as a poor DFS in patients with CRC [220]. Mechanistically, loc285194 is a p53 transcription target, and is able to inhibit tumor cell growth in part by specific suppressing miR-211 and activating p53 signaling, since two miR-211 binding sites were found to be located in an active region of loc285194 gene [219]. These studies strongly suggest that lncRNAs are hallmarks involved in CRC development and progression by targeting key signaling pathways. Additional information regarding lncRNAs related to CRC are summarized in Table 2.2, and some information is also available from URL: http://www.bio-bigdata.com/lnc2cancer/main1.jsp?cancer=colorectal%20cancer. The potential mechanism of lncRNAs in the progression of CRC is also illustrated in Fig. 2.2.

2.8 Circular RNAs as a New Class of Non-coding RNAs in Colorectal Cancer Pathogenesis

Circular RNAs (circRNAs) is a new category of regulatory ncRNAs that have been recently identified, which represent a new layer of posttranscriptional regulation of gene expression [276, 277]. An exponentially increased number of circRNAs has recently been discovered [278, 279]. Functionally, circRNAs exert their regulatory roles in gene expression as miRNA sponges, competing with other RNAs for binding to miRNAs and RNA binding proteins (RBPs) to modulate the local concentration of RBPs and RNAs, as part of the competing endogenous RNA (ceRNA) network [280]. Of note, circRNAs

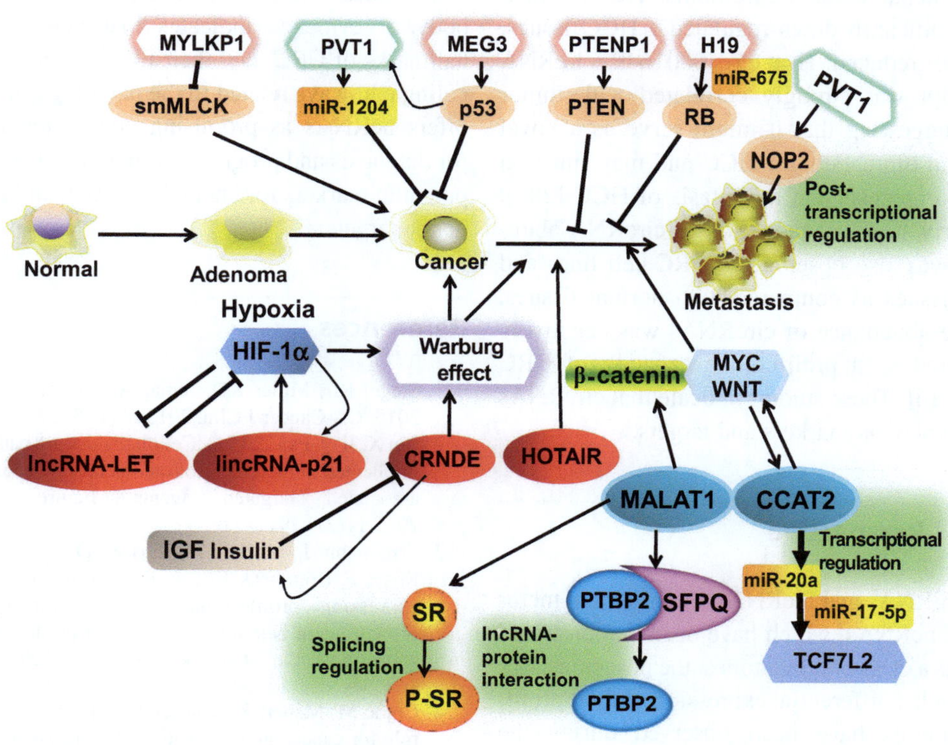

Fig. 2.2 An illustration represents the involvements and potential mechanism of several lncRNAs in the pathogenesis of colorectal cancer

have no accessible termini, this feature differs from classical ceRNAs and makes circRNAs resistant to miRNA-mediated RNA degradation or other exonucleolytic activities, and thus are relatively stable in cytoplasm. Importantly, several lines of evidence have revealed that circRNAs are associated with cancer-related miRNAs and involved in regulations of cancer-related pathways [276, 281–285]. Among which the ciRS-7 [also termed CDR1as (cerebellar degeneration-related protein 1 antisense)] is one of the most studied circRNAs. ciRS-7 contains more than 70 selectively conserved miRNA target sites, and it is highly and widely associated with Argonaute (AGO) proteins in a miR-7-dependent manner [281]. ciRS-7 acts as an inhibitor/sponge for miR-7, accordingly reduces miR-7 activity and increases miR-7 target gene expression [281]. The miR-7 is a tumor suppressor that inhibits the expression of several oncogenes, an impairing miR-7 function may thus enhance the potential of tumorigenicity. Hsa_circ_0001649 is a circRNA recently identified in human hepatocellular carcinoma (HCC), which was significantly down-regulated in HCC tissues, and the reduced Hsa_circ_0001649 circRNA transcript was strongly correlated with tumor size, suggesting that it might serve as a novel potential biomarker for HCC and may function in tumorigenesis and metastasis of HCC [284]. Similarly, a global reduction of circRNA abundance was also observed in CRC cell lines and CRC tissues as compared with normal tissues, and the abundance of circRNAs was negatively correlated with proliferative capacity of CRC cells [286]. These studies indicate that circRNAs may be new biomarkers and targets.

2.9 Perspective

Both miRNAs and lncRNAs are important members of ncRNAs, which have been demonstrated to act as a class of regulators at the post-transcription level, a differential expression of their profiling patterns have been observed during the progression of a variety of cancers, including the CRC. The alteration of ncRNA expression profile can be determined in CRC tissues and in circulating specimens. In the therapeutic standpoint, ncRNAs are mainly involved in anti-cancer therapy by restoring tumor suppressor genes or inhibiting oncogenes. They may enhance the sensitivity of tumor cells to chemotherapy or inhibit cell stemness.

It is encouraging that the identification of ncRNAs and their involvements in signaling pathways related to the progression, metastasis and drug resistance is currently no longer a significant bottleneck, owing to the applications of bioinformatics tools and high-throughput assays. However, there are a number of obstacles that need to be circumvented before ncRNAs can be safely applied as biomarkers for diagnosis, prognosis, and as therapeutic targets in clinical settings. For example, the off-targets of miRNAs and the effective delivery of these molecules with low side effect *in vivo* remain main challenges for miRNA-based therapies. In addition, most of current data are from studies *in vitro*, which need further validations in animal experiments or clinical studies. Nevertheless, there is a growing body of evidence indicates that ncRNAs are hallmarks of CRC, and their involvement in signaling pathways related to the pathology of CRC offers ncRNAs as promising novel biomarkers for diagnosis and prognosis of in CRC, as well as potential markers for chemoprediction and therapeutic targets.

References

1. Siegel RL, Miller KD, Jemal A. Cancer statistics, 2015. CA Cancer J Clin. 2015;65(1):5–29.
2. Liu X, Ji Q, Fan Z, Li Q. Cellular signaling pathways implicated in metastasis of colorectal cancer and the associated targeted agents. Future Oncol. 2015;11(21):2911–22.
3. Fritzmann J, Morkel M, Besser D, Budczies J, Kosel F, Brembeck FH, et al. A colorectal cancer expression profile that includes transforming growth factor beta inhibitor BAMBI predicts metastatic potential. Gastroenterology. 2009;137(1):165–75.
4. Cojoc M, Mabert K, Muders MH, Dubrovska A. A role for cancer stem cells in therapy resistance: cellular and molecular mechanisms. Semin Cancer Biol. 2015;31:16–27.

5. Kartal-Yandim M, Adan-Gokbulut A, Baran Y. Molecular mechanisms of drug resistance and its reversal in cancer. Crit Rev Biotechnol. 2016;36(4):716–26.
6. Venkatesh T, Suresh PS, Tsutsumi R. Non-coding RNAs: functions and applications in endocrine-related cancer. Mol Cell Endocrinol. 2015;416:88–96.
7. Cowie P, Hay EA, MacKenzie A. The noncoding human genome and the future of personalised medicine. Expert Rev Mol Med. 2015;17:e4.
8. Quinn JJ, Chang HY. Unique features of long non-coding RNA biogenesis and function. Nat Rev Genet. 2015;17(1):47–62.
9. Muhammad S, Kaur K, Huang R, Zhang Q, Kaur P, Yazdani HO, et al. MicroRNAs in colorectal cancer: role in metastasis and clinical perspectives. World J Gastroenterol WJG. 2014;20(45):17011–9.
10. Nicoloso MS, Spizzo R, Shimizu M, Rossi S, Calin GA. MicroRNAs – the micro steering wheel of tumour metastases. Nat Rev Cancer. 2009;9(4):293–302.
11. Slaby O, Svoboda M, Michalek J, Vyzula R. MicroRNAs in colorectal cancer: translation of molecular biology into clinical application. Mol Cancer. 2009;8:102.
12. Wang J, Du Y, Liu X, Cho WC, Yang Y. MicroRNAs as regulator of signaling networks in metastatic colon cancer. BioMed Res Int. 2015;2015:823620.
13. Cho WC. Oncomirs: The discovery and progress of microRNAs in cancers. Mol Cancer. 2007;6:60.
14. Cho WC. MicroRNAs: Potential biomarkers for cancer diagnosis, prognosis and targets for therapy. Int J Biochem Cell Biol. 2010;42(8):1273–81.
15. Garofalo M, Croce CM. MicroRNAs as therapeutic targets in chemoresistance. Drug Resist Updat. 2013;16(3–5):47–59.
16. Garg M. Targeting microRNAs in epithelial-to-mesenchymal transition-induced cancer stem cells: therapeutic approaches in cancer. Expert Opin Ther Targets. 2015;19(2):285–97.
17. Mlcochova J, Faltejskova P, Nemecek R, Svoboda M, Slaby O. MicroRNAs targeting EGFR signalling pathway in colorectal cancer. J Cancer Res Clin Oncol. 2013;139(10):1615–24.
18. Stiegelbauer V, Perakis S, Deutsch A, Ling H, Gerger A, Pichler M. MicroRNAs as novel predictive biomarkers and therapeutic targets in colorectal cancer. World J Gastroenterol: WJG. 2014;20(33):11727–35.
19. Caruso S, Bazan V, Rolfo C, Insalaco L, Fanale D, Bronte G, et al. MicroRNAs in colorectal cancer stem cells: new regulators of cancer stemness? Oncogenesis. 2012;1:e32.
20. Garofalo M, Croce CM. Role of microRNAs in maintaining cancer stem cells. Adv Drug Deliv Rev. 2015;81:53–61.
21. Shell S, Park SM, Radjabi AR, Schickel R, Kistner EO, Jewell DA, et al. Let-7 expression defines two differentiation stages of cancer. Proc Natl Acad Sci U S A. 2007;104(27):11400–5.
22. Akao Y, Nakagawa Y, Naoe T. let-7 microRNA functions as a potential growth suppressor in human colon cancer cells. Biol Pharm Bull. 2006;29(5):903–6.
23. Han HB, Gu J, Zuo HJ, Chen ZG, Zhao W, Li M, et al. Let-7c functions as a metastasis suppressor by targeting MMP11 and PBX3 in colorectal cancer. J Pathol. 2012;226(3):544–55.
24. Kjersem JB, Ikdahl T, Guren T, Skovlund E, Sorbye H, Hamfjord J, et al. Let-7 miRNA-binding site polymorphism in the KRAS 3'UTR; colorectal cancer screening population prevalence and influence on clinical outcome in patients with metastatic colorectal cancer treated with 5-fluorouracil and oxaliplatin +/− cetuximab. BMC Cancer. 2012;12:534.
25. Migliore C, Martin V, Leoni VP, Restivo A, Atzori L, Petrelli A, et al. MiR-1 downregulation cooperates with MACC1 in promoting MET overexpression in human colon cancer. Clin Cancer Res . 2012;18(3):737–47.
26. Young LE, Moore AE, Sokol L, Meisner-Kober N, Dixon DA. The mRNA stability factor HuR inhibits microRNA-16 targeting of COX-2. Mol Cancer Res: MCR. 2012;10(1):167–80.
27. Fujiya M, Konishi H, Mohamed Kamel MK, Ueno N, Inaba Y, Moriichi K, et al. microRNA-18a induces apoptosis in colon cancer cells via the autophagolysosomal degradation of oncogenic heterogeneous nuclear ribonucleoprotein A1. Oncogene. 2014;33(40):4847–56.
28. Humphreys KJ, McKinnon RA, Michael MZ. miR-18a inhibits CDC42 and plays a tumour suppressor role in colorectal cancer cells. PLoS One. 2014;9(11):e112288.
29. Zhang H, Hao Y, Yang J, Zhou Y, Li J, Yin S, et al. Genome-wide functional screening of miR-23b as a pleiotropic modulator suppressing cancer metastasis. Nat Commun. 2011;2:554.
30. Zhang JX, Mai SJ, Huang XX, Wang FW, Liao YJ, Lin MC, et al. MiR-29c mediates epithelial-to-mesenchymal transition in human colorectal carcinoma metastasis via PTP4A and GNA13 regulation of beta-catenin signaling. Ann Oncol. 2014;25(11):2196–204.
31. Zhong M, Bian Z, Wu Z. miR-30a suppresses cell migration and invasion through downregulation of PIK3CD in colorectal carcinoma. Cell Physiol Biochem. 2013;31(2–3):209–18.
32. Gao J, Li N, Dong Y, Li S, Xu L, Li X, et al. miR-34a-5p suppresses colorectal cancer metastasis and predicts recurrence in patients with stage II/III colorectal cancer. Oncogene. 2015;34(31):4142–52.
33. Wu J, Wu G, Lv L, Ren YF, Zhang XJ, Xue YF, et al. MicroRNA-34a inhibits migration and invasion of

colon cancer cells via targeting to Fra-1. Carcinogenesis. 2012;33(3):519–28.
34. Lu G, Sun Y, An S, Xin S, Ren X, Zhang D, et al. MicroRNA-34a targets FMNL2 and E2F5 and suppresses the progression of colorectal cancer. Exp Mol Pathol. 2015;99(1):173–9.
35. Schetter AJ, Harris CC. Alterations of microRNAs contribute to colon carcinogenesis. Semin Oncol. 2011;38(6):734–42.
36. Strillacci A, Griffoni C, Sansone P, Paterini P, Piazzi G, Lazzarini G, et al. MiR-101 downregulation is involved in cyclooxygenase-2 overexpression in human colon cancer cells. Exp Cell Res. 2009;315(8):1439–47.
37. Wu WK, Law PT, Lee CW, Cho CH, Fan D, Wu K, et al. MicroRNA in colorectal cancer: from benchtop to bedside. Carcinogenesis. 2011;32(3):247–53.
38. Molina-Pinelo S, Carnero A, Rivera F, Estevez-Garcia P, Bozada JM, Limon ML, et al. MiR-107 and miR-99a-3p predict chemotherapy response in patients with advanced colorectal cancer. BMC Cancer. 2014;14:656.
39. Yang L, Belaguli N, Berger DH. MicroRNA and colorectal cancer. World J Surg. 2009;33(4):638–46.
40. Gong J, Zhang JP, Li B, Zeng C, You K, Chen MX, et al. MicroRNA-125b promotes apoptosis by regulating the expression of Mcl-1, Bcl-w and IL-6R. Oncogene. 2013;32(25):3071–9.
41. Fish JE, Santoro MM, Morton SU, Yu S, Yeh RF, Wythe JD, et al. miR-126 regulates angiogenic signaling and vascular integrity. Dev Cell. 2008;15(2):272–84.
42. Li N, Tang A, Huang S, Li Z, Li X, Shen S, et al. MiR-126 suppresses colon cancer cell proliferation and invasion via inhibiting RhoA/ROCK signaling pathway. Mol Cell Biochem. 2013;380(1–2):107–19.
43. Wu L, Shi B, Huang K, Fan G. MicroRNA-128 suppresses cell growth and metastasis in colorectal carcinoma by targeting IRS1. Oncol Rep. 2015;34(5):2797–805.
44. Wu J, Qian J, Li C, Kwok L, Cheng F, Liu P, et al. miR-129 regulates cell proliferation by downregulating Cdk6 expression. Cell Cycle. 2010;9(9):1809–18.
45. Zheng YB, Luo HP, Shi Q, Hao ZN, Ding Y, Wang QS, et al. miR-132 inhibits colorectal cancer invasion and metastasis via directly targeting ZEB2. World J Gastroenterol. 2014;20(21):6515–22.
46. Ng EK, Chong WW, Jin H, Lam EK, Shin VY, Yu J, et al. Differential expression of microRNAs in plasma of patients with colorectal cancer: a potential marker for colorectal cancer screening. Gut. 2009;58(10):1375–81.
47. Balaguer F, Link A, Lozano JJ, Cuatrecasas M, Nagasaka T, Boland CR, et al. Epigenetic silencing of miR-137 is an early event in colorectal carcinogenesis. Cancer Res. 2010;70(16):6609–18.
48. Bandres E, Bitarte N, Arias F, Agorreta J, Fortes P, Agirre X, et al. microRNA-451 regulates macrophage migration inhibitory factor production and proliferation of gastrointestinal cancer cells. Clin Cancer Res. 2009;15(7):2281–90.
49. Sarver AL, French AJ, Borralho PM, Thayanithy V, Oberg AL, Silverstein KA, et al. Human colon cancer profiles show differential microRNA expression depending on mismatch repair status and are characteristic of undifferentiated proliferative states. BMC Cancer. 2009;9:401.
50. Pagliuca A, Valvo C, Fabrizi E, di Martino S, Biffoni M, Runci D, et al. Analysis of the combined action of miR-143 and miR-145 on oncogenic pathways in colorectal cancer cells reveals a coordinate program of gene repression. Oncogene. 2013;32(40):4806–13.
51. Pekow J, Meckel K, Dougherty U, Butun F, Mustafi R, Lim J, et al. Tumor suppressors miR-143 and miR-145 and predicted target proteins API5, ERK5, K-RAS, and IRS-1 are differentially expressed in proximal and distal colon. Am J Physiol Gastrointest Liver Physiol. 2015;308(3):G179–87.
52. Iwaya T, Yokobori T, Nishida N, Kogo R, Sudo T, Tanaka F, et al. Downregulation of miR-144 is associated with colorectal cancer progression via activation of mTOR signaling pathway. Carcinogenesis. 2012;33(12):2391–7.
53. Xiao R, Li C, Chai B. miRNA-144 suppresses proliferation and migration of colorectal cancer cells through GSPT1. Biomed Pharmacother. 2015;74:138–44.
54. Zhang H, Li Y, Huang Q, Ren X, Hu H, Sheng H, et al. MiR-148a promotes apoptosis by targeting Bcl-2 in colorectal cancer. Cell Death Differ. 2011;18(11):1702–10.
55. Xu K, Liu X, Mao X, Xue L, Wang R, Chen L, et al. MicroRNA-149 suppresses colorectal cancer cell migration and invasion by directly targeting forkhead box transcription factor FOXM1. Cell Physiol Biochem. 2015;35(2):499–515.
56. Lu ZJ, Lu LG, Tao KZ, Chen DF, Xia Q, Weng JJ, et al. MicroRNA-185 suppresses growth and invasion of colon cancer cells through inhibition of the hypoxiainducible factor-2alpha pathway in vitro and in vivo. Mol Med Rep. 2014;10(5):2401–8.
57. Braun CJ, Zhang X, Savelyeva I, Wolff S, Moll UM, Schepeler T, et al. p53-Responsive micrornas 192 and 215 are capable of inducing cell cycle arrest. Cancer Res. 2008;68(24):10094–104.
58. Earle JS, Luthra R, Romans A, Abraham R, Ensor J, Yao H, et al. Association of microRNA expression with microsatellite instability status in colorectal adenocarcinoma. J Mol Diagn. 2010;12(4):433–40.
59. Hur K, Toiyama Y, Takahashi M, Balaguer F, Nagasaka T, Koike J, et al. MicroRNA-200c modulates epithelial-to-mesenchymal transition (EMT) in human colorectal cancer metastasis. Gut. 2013;62(9):1315–26.
60. Chen ML, Liang LS, Wang XK. miR-200c inhibits invasion and migration in human colon cancer cells SW480/620 by targeting ZEB1. Clin Exp Metastasis. 2012;29(5):457–69.
61. Li J, Chen Y, Zhao J, Kong F, Zhang Y. miR-203 reverses chemoresistance in p53-mutated colon can-

cer cells through downregulation of Akt2 expression. Cancer Lett. 2011;304(1):52–9.
62. Wang XW, Xi XQ, Wu J, Wan YY, Hui HX, Cao XF. MicroRNA-206 attenuates tumor proliferation and migration involving the downregulation of NOTCH3 in colorectal cancer. Oncol Rep. 2015;33(3):1402–10.
63. Ren XL, He GY, Li XM, Men H, Yi LZ, Lu GF, et al. MicroRNA-206 functions as a tumor suppressor in colorectal cancer by targeting FMNL2. J Cancer Res Clin Oncol. 2016;142(3)581–92.
64. Meng X, Wu J, Pan C, Wang H, Ying X, Zhou Y, et al. Genetic and epigenetic down-regulation of microRNA-212 promotes colorectal tumor metastasis via dysregulation of MnSOD. Gastroenterology. 2013;145(2):426–36 e1-6.
65. Penna E, Orso F, Cimino D, Tenaglia E, Lembo A, Quaglino E, et al. microRNA-214 contributes to melanoma tumour progression through suppression of TFAP2C. EMBO J. 2011;30(10):1990–2007.
66. Wu L, Li H, Jia CY, Cheng W, Yu M, Peng M, et al. MicroRNA-223 regulates FOXO1 expression and cell proliferation. FEBS Lett. 2012;586(7): 1038–43.
67. Sun D, Wang C, Long S, Ma Y, Guo Y, Huang Z, et al. C/EBP-beta-activated microRNA-223 promotes tumour growth through targeting RASA1 in human colorectal cancer. Br J Cancer. 2015;112(9):1491–500.
68. Sun JY, Huang Y, Li JP, Zhang X, Wang L, Meng YL, et al. MicroRNA-320a suppresses human colon cancer cell proliferation by directly targeting beta-catenin. Biochem Biophys Res Commun. 2012;420(4):787–92.
69. Zhao H, Dong T, Zhou H, Wang L, Huang A, Feng B, et al. miR-320a suppresses colorectal cancer progression by targeting Rac1. Carcinogenesis. 2014;35(4):886–95.
70. Zhang Y, He X, Liu Y, Ye Y, Zhang H, He P, et al. microRNA-320a inhibits tumor invasion by targeting neuropilin 1 and is associated with liver metastasis in colorectal cancer. Oncol Rep. 2012;27(3):685–94.
71. Sun Z, Zhang Z, Liu Z, Qiu B, Liu K, Dong G. MicroRNA-335 inhibits invasion and metastasis of colorectal cancer by targeting ZEB2. Med Oncol. 2014;31(6):982.
72. Ma F, Song H, Guo B, Zhang Y, Zheng Y, Lin C, et al. MiR-361-5p inhibits colorectal and gastric cancer growth and metastasis by targeting staphylococcal nuclease domain containing-1. Oncotarget. 2015;6(19):17404–16.
73. Bai R, Weng C, Dong H, Li S, Chen G, Xu Z. MicroRNA-409-3p suppresses colorectal cancer invasion and metastasis partly by targeting GAB1 expression. Int J Cancer J Int du Cancer. 2015;137(10):2310–22.
74. Fang Y, Gu X, Li Z, Xiang J, Chen Z. miR-449b inhibits the proliferation of SW1116 colon cancer stem cells through downregulation of CCND1 and E2F3 expression. Oncol Rep. 2013;30(1):399–406.
75. Dobrucki LW, Tsutsumi Y, Kalinowski L, Dean J, Gavin M, Sen S, et al. Analysis of angiogenesis induced by local IGF-1 expression after myocardial infarction using microSPECT-CT imaging. J Mol Cell Cardiol. 2010;48(6):1071–9.
76. Guo ST, Jiang CC, Wang GP, Li YP, Wang CY, Guo XY, et al. MicroRNA-497 targets insulin-like growth factor 1 receptor and has a tumour suppressive role in human colorectal cancer. Oncogene. 2013;32(15):1910–20.
77. Tokarz P, Blasiak J. The role of microRNA in metastatic colorectal cancer and its significance in cancer prognosis and treatment. Acta Biochim Pol. 2012;59(4):467–74.
78. Sun Y, Shen S, Liu X, Tang H, Wang Z, Yu Z, et al. MiR-429 inhibits cells growth and invasion and regulates EMT-related marker genes by targeting Onecut2 in colorectal carcinoma. Mol Cell Biochem. 2014;390(1–2):19–30.
79. Yan L, Yu J, Tan F, Ye GT, Shen ZY, Liu H, et al. SP1-mediated microRNA-520d-5p suppresses tumor growth and metastasis in colorectal cancer by targeting CTHRC1. Am J Cancer Res. 2015;5(4):1447–59.
80. Arcaroli JJ, Quackenbush KS, Powell RW, Pitts TM, Spreafico A, Varella-Garcia M, et al. Common PIK3CA mutants and a novel 3′ UTR mutation are associated with increased sensitivity to saracatinib. Clin Cancer Res. 2012;18(9):2704–14.
81. Ye X, Fan F, Bhattacharya R, Bellister S, Boulbes DR, Wang R, et al. VEGFR-1 pseudogene expression and regulatory function in human colorectal cancer cells. Mol Cancer Res MCR. 2015;13(9):1274–82.
82. Sheng L, He P, Yang X, Zhou M, Feng Q. miR-612 negatively regulates colorectal cancer growth and metastasis by targeting AKT2. Cell Death Dis. 2015;6:e1808.
83. Fang Y, Sun B, Li Z, Chen Z, Xiang J. MiR-622 inhibited colorectal cancer occurrence and metastasis by suppressing K-Ras. Mol Carcinog. 2015.
84. Ma K, Pan X, Fan P, He Y, Gu J, Wang W, et al. Loss of miR-638 in vitro promotes cell invasion and a mesenchymal-like transition by influencing SOX2 expression in colorectal carcinoma cells. Mol Cancer. 2014;13:118.
85. Zhang J, Fei B, Wang Q, Song M, Yin Y, Zhang B, et al. MicroRNA-638 inhibits cell proliferation, invasion and regulates cell cycle by targeting tetraspanin 1 in human colorectal carcinoma. Oncotarget. 2014;5(23):12083–96.
86. Shi L, Jackstadt R, Siemens H, Li H, Kirchner T, Hermeking H. p53-induced miR-15a/16-1 and AP4 form a double-negative feedback loop to regulate epithelial-mesenchymal transition and metastasis in colorectal cancer. Cancer Res. 2014;74(2):532–42.
87. Meng WJ, Yang L, Ma Q, Zhang H, Adell G, Arbman G, et al. MicroRNA expression profile reveals miR-17-92 and miR-143-145 cluster in synchronous colorectal cancer. Medicine. 2015;94(32):e1297.

88. Monzo M, Navarro A, Bandres E, Artells R, Moreno I, Gel B, et al. Overlapping expression of microRNAs in human embryonic colon and colorectal cancer. Cell Res. 2008;18(8):823–33.
89. Humphreys KJ, Cobiac L, Le Leu RK, Van der Hoek MB, Michael MZ. Histone deacetylase inhibition in colorectal cancer cells reveals competing roles for members of the oncogenic miR-17-92 cluster. Mol Carcinog. 2013;52(6):459–74.
90. Cellura D, Pickard K, Quaratino S, Parker H, Strefford JC, Thomas GJ, et al. miR-19-mediated inhibition of transglutaminase-2 leads to enhanced invasion and metastasis in colorectal cancer. Mol Cancer Res MCR. 2015;13(7):1095–105.
91. Huang L, Wang X, Wen C, Yang X, Song M, Chen J, et al. Hsa-miR-19a is associated with lymph metastasis and mediates the TNF-alpha induced epithelial-to-mesenchymal transition in colorectal cancer. Sci Rep. 2015;5:13350.
92. Zhang Y, Geng L, Talmon G, Wang J. MicroRNA-520g confers drug resistance by regulating p21 expression in colorectal cancer. J Biol Chem. 2015;290(10):6215–25.
93. Yu Y, Kanwar SS, Patel BB, Oh PS, Nautiyal J, Sarkar FH, et al. MicroRNA-21 induces stemness by downregulating transforming growth factor beta receptor 2 (TGFbetaR2) in colon cancer cells. Carcinogenesis. 2012;33(1):68–76.
94. Wang CJ, Stratmann J, Zhou ZG, Sun XF. Suppression of microRNA-31 increases sensitivity to 5-FU at an early stage, and affects cell migration and invasion in HCT-116 colon cancer cells. BMC Cancer. 2010;10:616.
95. Lei SL, Zhao H, Yao HL, Chen Y, Lei ZD, Liu KJ, et al. Regulatory roles of microRNA-708 and microRNA-31 in proliferation, apoptosis and invasion of colorectal cancer cells. Oncol Lett. 2014;8(4):1768–74.
96. Sun D, Yu F, Ma Y, Zhao R, Chen X, Zhu J, et al. MicroRNA-31 activates the RAS pathway and functions as an oncogenic MicroRNA in human colorectal cancer by repressing RAS p21 GTPase activating protein 1 (RASA1). J Biol Chem. 2013;288(13):9508–18.
97. Wu W, Yang J, Feng X, Wang H, Ye S, Yang P, et al. MicroRNA-32 (miR-32) regulates phosphatase and tensin homologue (PTEN) expression and promotes growth, migration, and invasion in colorectal carcinoma cells. Mol Cancer. 2013;12:30.
98. Gao F, Wang W. MicroRNA-96 promotes the proliferation of colorectal cancer cells and targets tumor protein p53 inducible nuclear protein 1, forkhead box protein O1 (FOXO1) and FOXO3a. Mol Med Rep. 2015;11(2):1200–6.
99. Chen HY, Lin YM, Chung HC, Lang YD, Lin CJ, Huang J, et al. miR-103/107 promote metastasis of colorectal cancer by targeting the metastasis suppressors DAPK and KLF4. Cancer Res. 2012;72(14):3631–41.
100. Zhang GJ, Li JS, Zhou H, Xiao HX, Li Y, Zhou T. MicroRNA-106b promotes colorectal cancer cell migration and invasion by directly targeting DLC1. J Exp Clin Cancer Res: CR. 2015;34(1):73.
101. Iino I, Kikuchi H, Miyazaki S, Hiramatsu Y, Ohta M, Kamiya K, et al. Effect of miR-122 and its target gene cationic amino acid transporter 1 on colorectal liver metastasis. Cancer Sci. 2013;104(5):624–30.
102. Liu L, Chen L, Xu Y, Li R, Du X. microRNA-195 promotes apoptosis and suppresses tumorigenicity of human colorectal cancer cells. Biochem Biophys Res Commun. 2010;400(2):236–40.
103. Khatri R, Subramanian S. MicroRNA-135b and its circuitry networks as potential therapeutic targets in colon cancer. Front Oncol. 2013;3:268.
104. Valeri N, Braconi C, Gasparini P, Murgia C, Lampis A, Paulus-Hock V, et al. MicroRNA-135b promotes cancer progression by acting as a downstream effector of oncogenic pathways in colon cancer. Cancer Cell. 2014;25(4):469–83.
105. Li J, Liang H, Bai M, Ning T, Wang C, Fan Q, et al. miR-135b promotes cancer progression by targeting transforming growth factor beta receptor II (TGFBR2) in colorectal cancer. PLoS One. 2015;10(6):e0130194.
106. Tanaka S, Hosokawa M, Yonezawa T, Hayashi W, Ueda K, Iwakawa S. Induction of epithelial-mesenchymal transition and down-regulation of miR-200c and miR-141 in oxaliplatin-resistant colorectal cancer cells. Biol Pharm Bull. 2015;38(3):435–40.
107. Rossi S, Di Narzo AF, Mestdagh P, Jacobs B, Bosman FT, Gustavsson B, et al. microRNAs in colon cancer: a roadmap for discovery. FEBS Lett. 2012;586(19):3000–7.
108. Valeri N, Gasparini P, Fabbri M, Braconi C, Veronese A, Lovat F, et al. Modulation of mismatch repair and genomic stability by miR-155. Proc Natl Acad Sci U S A. 2010;107(15):6982–7.
109. Li T, Yang J, Lv X, Liu K, Gao C, Xing Y, et al. miR-155 regulates the proliferation and cell cycle of colorectal carcinoma cells by targeting E2F2. Biotechnol Lett. 2014;36(9):1743–52.
110. Wei Z, Cui L, Mei Z, Liu M, Zhang D. miR-181a mediates metabolic shift in colon cancer cells via the PTEN/AKT pathway. FEBS Lett. 2014;588(9):1773–9.
111. Ji D, Chen Z, Li M, Zhan T, Yao Y, Zhang Z, et al. MicroRNA-181a promotes tumor growth and liver metastasis in colorectal cancer by targeting the tumor suppressor WIF-1. Mol Cancer. 2014;13:86.
112. Pizzini S, Bisognin A, Mandruzzato S, Biasiolo M, Facciolli A, Perilli L, et al. Impact of microRNAs on regulatory networks and pathways in human colorectal carcinogenesis and development of metastasis. BMC Genomics. 2013;14:589.
113. Schimanski CC. High miR-196a levels promote the oncogenic phenotype of colorectal cancer cells. World J Gastroenterol. 2009;15(17):2089–96.
114. Sun J, Ding W, Zhi J, Chen W. MiR-200 suppresses metastases of colorectal cancer through

ZEB1. Tumour Biol. 2015. doi:10.1007/s13277-015-3822-3.
115. Ota T, Doi K, Fujimoto T, Tanaka Y, Ogawa M, Matsuzaki H, et al. KRAS up-regulates the expression of miR-181a, miR-200c and miR-210 in a three-dimensional-specific manner in DLD-1 colorectal cancer cells. Anticancer Res. 2012;32(6):2271–5.
116. Amankwatia EB, Chakravarty P, Carey FA, Weidlich S, Steele RJ, Munro AJ, et al. MicroRNA-224 is associated with colorectal cancer progression and response to 5-fluorouracil-based chemotherapy by KRAS-dependent and -independent mechanisms. Br J Cancer. 2015;112(9):1480–90.
117. Mencia N, Selga E, Noe V, Ciudad CJ. Underexpression of miR-224 in methotrexate resistant human colon cancer cells. Biochem Pharmacol. 2011;82(11):1572–82.
118. Ling H, Pickard K, Ivan C, Isella C, Ikuo M, Mitter R, et al. The clinical and biological significance of MIR-224 expression in colorectal cancer metastasis. Gut. 2015.
119. Zhang W, Zhang T, Jin R, Zhao H, Hu J, Feng B, et al. MicroRNA-301a promotes migration and invasion by targeting TGFBR2 in human colorectal cancer. J Exp Clin Cancer Res: CR. 2014;33(1):113.
120. Zhou J, Zhang M, Huang Y, Feng L, Chen H, Hu Y, et al. MicroRNA-320b promotes colorectal cancer proliferation and invasion by competing with its homologous microRNA-320a. Cancer Lett. 2015;356(2 Pt B):669–75.
121. Ragusa M, Statello L, Maugeri M, Majorana A, Barbagallo D, Salito L, et al. Specific alterations of the microRNA transcriptome and global network structure in colorectal cancer after treatment with MAPK/ERK inhibitors. J Mol Med. 2012;90(12):1421–38.
122. Zheng H, Ma R, Wang Q, Zhang P, Li D, Wang Q, et al. MiR-625-3p promotes cell migration and invasion via inhibition of SCAI in colorectal carcinoma cells. Oncotarget. 2015;6(29):27805–15.
123. Tsang WP, Ng EK, Ng SS, Jin H, Yu J, Sung JJ, et al. Oncofetal H19-derived miR-675 regulates tumor suppressor RB in human colorectal cancer. Carcinogenesis. 2010;31(3):350–8.
124. Wang X, Kuang Y, Shen X, Zhou H, Chen Y, Han Y, et al. Evaluation of miR-720 prognostic significance in patients with colorectal cancer. Tumour Biol. 2015;36(2):719–27.
125. Bu P, Wang L, Chen KY, Rakhilin N, Sun J, Closa A, et al. miR-1269 promotes metastasis and forms a positive feedback loop with TGF-beta. Nat Commun. 2015;6:6879.
126. Rokkas T, Kothonas F, Rokka A, Koukoulis G, Symvoulakis E. The role of circulating microRNAs as novel biomarkers in diagnosing colorectal cancer: a meta-analysis. Eur J Gastroenterol Hepatol. 2015;27(7):819–25.
127. Carames C, Cristobal I, Moreno V, Del Puerto L, Moreno I, Rodriguez M, et al. MicroRNA-21 predicts response to preoperative chemoradiotherapy in locally advanced rectal cancer. Int J Color Dis. 2015;30(7):899–906.
128. Li T, Leong MH, Harms B, Kennedy G, Chen L. MicroRNA-21 as a potential colon and rectal cancer biomarker. World J Gastroenterol WJG. 2013;19(34):5615–21.
129. Hu H, Tian D, Chen T, Han R, Sun Y, Wu C. Metastasis-associated in colon cancer 1 is a novel survival-related biomarker for human patients with renal pelvis carcinoma. PLoS One. 2014;9(6):e100161.
130. Qin J, Wang F, Jiang H, Xu J, Jiang Y, Wang Z. MicroRNA-145 suppresses cell migration and invasion by targeting paxillin in human colorectal cancer cells. Int J Clin Exp Pathol. 2015;8(2):1328–40.
131. Kara M, Yumrutas O, Ozcan O, Celik OI, Bozgeyik E, Bozgeyik I, et al. Differential expressions of cancer-associated genes and their regulatory miRNAs in colorectal carcinoma. Gene. 2015;567(1):81–6.
132. Bonauer A, Carmona G, Iwasaki M, Mione M, Koyanagi M, Fischer A, et al. MicroRNA-92a controls angiogenesis and functional recovery of ischemic tissues in mice. Science. 2009;324(5935):1710–3.
133. Poliseno L, Tuccoli A, Mariani L, Evangelista M, Citti L, Woods K, et al. MicroRNAs modulate the angiogenic properties of HUVECs. Blood. 2006;108(9):3068–71.
134. Sundaram P, Hultine S, Smith LM, Dews M, Fox JL, Biyashev D, et al. p53-responsive miR-194 inhibits thrombospondin-1 and promotes angiogenesis in colon cancers. Cancer Res. 2011;71(24):7490–501.
135. Chang KH, Miller N, Kheirelseid EA, Lemetre C, Ball GR, Smith MJ, et al. MicroRNA signature analysis in colorectal cancer: identification of expression profiles in stage II tumors associated with aggressive disease. Int J Color Dis. 2011;26(11):1415–22.
136. Wang X, Wang J, Ma H, Zhang J, Zhou X. Downregulation of miR-195 correlates with lymph node metastasis and poor prognosis in colorectal cancer. Med Oncol. 2012;29(2):919–27.
137. Rodriguez S, Huynh-Do U. The role of PTEN in tumor angiogenesis. J Oncol. 2012;2012:141236.
138. Fang L, Li H, Wang L, Hu J, Jin T, Wang J, et al. MicroRNA-17-5p promotes chemotherapeutic drug resistance and tumour metastasis of colorectal cancer by repressing PTEN expression. Oncotarget. 2014;5(10):2974–87.
139. Dews M, Homayouni A, Yu D, Murphy D, Sevignani C, Wentzel E, et al. Augmentation of tumor angiogenesis by a Myc-activated microRNA cluster. Nat Genet. 2006;38(9):1060–5.
140. Roy S, Yu Y, Padhye SB, Sarkar FH, Majumdar AP. Difluorinated-curcumin (CDF) restores PTEN expression in colon cancer cells by down-regulating miR-21. PLoS One. 2013;8(7):e68543.
141. Li J, Zhang Y, Zhao J, Kong F, Chen Y. Overexpression of miR-22 reverses paclitaxel-induced chemoresistance through activation of PTEN signaling in p53-

mutated colon cancer cells. Mol Cell Biochem. 2011;357(1–2):31–8.

142. Guo C, Sah JF, Beard L, Willson JK, Markowitz SD, Guda K. The noncoding RNA, miR-126, suppresses the growth of neoplastic cells by targeting phosphatidylinositol 3-kinase signaling and is frequently lost in colon cancers. Genes Chromosom Cancer. 2008;47(11):939–46.
143. Wu Z, He B, He J, Mao X. Upregulation of miR-153 promotes cell proliferation via downregulation of the PTEN tumor suppressor gene in human prostate cancer. Prostate. 2013;73(6):596–604.
144. Song L, Li Y, He B, Gong Y. Development of small molecules targeting the Wnt signaling pathway in cancer stem cells for the treatment of colorectal cancer. Clin Colorectal Cancer. 2015;14(3):133–45.
145. Faber C, Kirchner T, Hlubek F. The impact of microRNAs on colorectal cancer. Virchows Arch. 2009;454(4):359–67.
146. Yamada N, Noguchi S, Mori T, Naoe T, Maruo K, Akao Y. Tumor-suppressive microRNA-145 targets catenin delta-1 to regulate Wnt/beta-catenin signaling in human colon cancer cells. Cancer Lett. 2013;335(2):332–42.
147. Tang Q, Zou Z, Zou C, Zhang Q, Huang R, Guan X, et al. MicroRNA-93 suppress colorectal cancer development via Wnt/beta-catenin pathway downregulating. Tumour Biol. 2015;36(3):1701–10.
148. Gomez GG, Wykosky J, Zanca C, Furnari FB, Cavenee WK. Therapeutic resistance in cancer: microRNA regulation of EGFR signaling networks. Cancer Biol Med. 2013;10(4):192–205.
149. Mosakhani N, Lahti L, Borze I, Karjalainen-Lindsberg ML, Sundstrom J, Ristamaki R, et al. MicroRNA profiling predicts survival in anti-EGFR treated chemorefractory metastatic colorectal cancer patients with wild-type KRAS and BRAF. Cancer Genet. 2012;205(11):545–51.
150. Ruzzo A, Graziano F, Vincenzi B, Canestrari E, Perrone G, Galluccio N, et al. High let-7a microRNA levels in KRAS-mutated colorectal carcinomas may rescue anti-EGFR therapy effects in patients with chemotherapy-refractory metastatic disease. Oncologist. 2012;17(6):823–9.
151. Manceau G, Imbeaud S, Thiebaut R, Liebaert F, Fontaine K, Rousseau F, et al. Hsa-miR-31-3p expression is linked to progression-free survival in patients with KRAS wild-type metastatic colorectal cancer treated with anti-EGFR therapy. Clin Cancer Res. 2014;20(12):3338–47.
152. Igarashi H, Kurihara H, Mitsuhashi K, Ito M, Okuda H, Kanno S, et al. Association of microRNA-31-5p with clinical efficacy of anti-EGFR therapy in patients with metastatic colorectal cancer. Ann Surg Oncol. 2015;22(8):2640–8.
153. Suto T, Yokobori T, Yajima R, Morita H, Fujii T, Yamaguchi S, et al. MicroRNA-7 expression in colorectal cancer is associated with poor prognosis and regulates cetuximab sensitivity via EGFR regulation. Carcinogenesis. 2015;36(3):338–45.
154. Pichler M, Winter E, Ress AL, Bauernhofer T, Gerger A, Kiesslich T, et al. miR-181a is associated with poor clinical outcome in patients with colorectal cancer treated with EGFR inhibitor. J Clin Pathol. 2014;67(3):198–203.
155. Rokavec M, Li H, Jiang L, Hermeking H. The p53/microRNA connection in gastrointestinal cancer. Clin Exp Gastroenterol. 2014;7:395–413.
156. Kim NH, Kim HS, Li XY, Lee I, Choi HS, Kang SE, et al. A p53/miRNA-34 axis regulates Snail1-dependent cancer cell epithelial-mesenchymal transition. J Cell Biol. 2011;195(3):417–33.
157. Luu C, Heinrich EL, Duldulao M, Arrington AK, Fakih M, Garcia-Aguilar J, et al. TP53 and let-7a micro-RNA regulate K-Ras activity in HCT116 colorectal cancer cells. PLoS One. 2013;8(8):e70604.
158. Cui X, Witalison EE, Chumanevich AP, Chumanevich AA, Poudyal D, Subramanian V, et al. The induction of microRNA-16 in colon cancer cells by protein arginine deiminase inhibition causes a p53-dependent cell cycle arrest. PLoS One. 2013;8(1):e53791.
159. Dong Y, Zhao J, Wu CW, Zhang L, Liu X, Kang W, et al. Tumor suppressor functions of miR-133a in colorectal cancer. Mol Cancer Res: MCR. 2013;11(9):1051–60.
160. Butz H, Racz K, Hunyady L, Patocs A. Crosstalk between TGF-beta signaling and the microRNA machinery. Trends Pharmacol Sci. 2012;33(7):382–93.
161. Korpal M, Lee ES, Hu G, Kang Y. The miR-200 family inhibits epithelial-mesenchymal transition and cancer cell migration by direct targeting of E-cadherin transcriptional repressors ZEB1 and ZEB2. J Biol Chem. 2008;283(22):14910–4.
162. Wu X, Somlo G, Yu Y, Palomares MR, Li AX, Zhou W, et al. De novo sequencing of circulating miRNAs identifies novel markers predicting clinical outcome of locally advanced breast cancer. J Transl Med. 2012;10:42.
163. Borralho PM, Kren BT, Castro RE, da Silva IB, Steer CJ, Rodrigues CM. MicroRNA-143 reduces viability and increases sensitivity to 5-fluorouracil in HCT116 human colorectal cancer cells. FEBS J. 2009;276(22):6689–700.
164. Arndt GM, Dossey L, Cullen LM, Lai A, Druker R, Eisbacher M, et al. Characterization of global microRNA expression reveals oncogenic potential of miR-145 in metastatic colorectal cancer. BMC Cancer. 2009;9:374.
165. Hudson JD, Shoaibi MA, Maestro R, Carnero A, Hannon GJ, Beach DH. A proinflammatory cytokine inhibits p53 tumor suppressor activity. J Exp Med. 1999;190(10):1375–82.
166. Lee CG, McCarthy S, Gruidl M, Timme C, Yeatman TJ. MicroRNA-147 induces a mesenchymal-to-epi-

thelial transition (MET) and reverses EGFR inhibitor resistance. PLoS One. 2014;9(1):e84597.
167. Miyazaki S, Yamamoto H, Miyoshi N, Wu X, Ogawa H, Uemura M, et al. A cancer reprogramming method using MicroRNAs as a novel therapeutic approach against colon cancer: research for reprogramming of cancer cells by microRNAs. Ann Surg Oncol. 2015;22(Suppl 3):S1394–401.
168. Smolle M, Uranitsch S, Gerger A, Pichler M, Haybaeck J. Current status of long non-coding RNAs in human cancer with specific focus on colorectal cancer. Int J Mol Sci. 2014;15(8):13993–4013.
169. Xie X, Tang B, Xiao YF, Xie R, Li BS, Dong H, et al. Long non-coding RNAs in colorectal cancer. Oncotarget. 2016;7(5):5226–39.
170. Chen H, Xu J, Hong J, Tang R, Zhang X, Fang JY. Long noncoding RNA profiles identify five distinct molecular subtypes of colorectal cancer with clinical relevance. Mol Oncol. 2014;8(8):1393–403.
171. Hu Y, Chen HY, Yu CY, Xu J, Wang JL, Qian J, et al. A long non-coding RNA signature to improve prognosis prediction of colorectal cancer. Oncotarget. 2014;5(8):2230–42.
172. Lee H, Kim C, Ku JL, Kim W, Yoon SK, Kuh HJ, et al. A long non-coding RNA snaR contributes to 5-fluorouracil resistance in human colon cancer cells. Mol Cells. 2014;37(7):540–6.
173. Shi Y, Liu Y, Wang J, Jie D, Yun T, Li W, et al. Downregulated long noncoding RNA BANCR promotes the proliferation of colorectal cancer cells via downregualtion of p21 expression. PLoS One. 2015;10(4):e0122679.
174. Guo Q, Zhao Y, Chen J, Hu J, Wang S, Zhang D, et al. BRAF-activated long non-coding RNA contributes to colorectal cancer migration by inducing epithelial-mesenchymal transition. Oncol Lett. 2014;8(2):869–75.
175. Li AX, Xin WQ, Ma CG. Fentanyl inhibits the invasion and migration of colorectal cancer cells via inhibiting the negative regulation of Ets-1 on BANCR. Biochem Biophys Res Commun. 2015;465(3):594–600.
176. Davison EJ, Tarpey PS, Fiegler H, Tomlinson IP, Carter NP. Deletion at chromosome band 20p12.1 in colorectal cancer revealed by high resolution array comparative genomic hybridization. Genes Chromosom Cancer. 2005;44(4):384–91.
177. Pedersen SK, Mitchell SM, Graham LD, McEvoy A, Thomas ML, Baker RT, et al. CAHM, a long noncoding RNA gene hypermethylated in colorectal neoplasia. Epigenetics. 2014;9(8):1071–82.
178. Ma Y, Yang Y, Wang F, Moyer MP, Wei Q, Zhang P, et al. Long non-coding RNA CCAL regulates colorectal cancer progression by activating Wnt/beta-catenin signalling pathway via suppression of activator protein 2alpha. Gut. 2015;pii:gutjnl–2014–308392.
179. Kam Y, Rubinstein A, Naik S, Djavsarov I, Halle D, Ariel I, et al. Detection of a long non-coding RNA (CCAT1) in living cells and human adenocarcinoma of colon tissues using FIT-PNA molecular beacons. Cancer Lett. 2014;352(1):90–6.
180. Xiang JF, Yin QF, Chen T, Zhang Y, Zhang XO, Wu Z, et al. Human colorectal cancer-specific CCAT1-L lncRNA regulates long-range chromatin interactions at the MYC locus. Cell Res. 2014;24(5):513–31.
181. Ling H, Spizzo R, Atlasi Y, Nicoloso M, Shimizu M, Redis RS, et al. CCAT2, a novel noncoding RNA mapping to 8q24, underlies metastatic progression and chromosomal instability in colon cancer. Genome Res. 2013;23(9):1446–61.
182. Kim T, Jeon YJ, Cui R, Lee JH, Peng Y, Kim SH, et al. Role of MYC-regulated long noncoding RNAs in cell cycle regulation and tumorigenesis. J Nat Cancer Inst. 2015;107(4):dju505.
183. Graham LD, Pedersen SK, Brown GS, Ho T, Kassir Z, Moynihan AT, et al. Colorectal neoplasia differentially expressed (CRNDE), a novel gene with elevated expression in colorectal adenomas and adenocarcinomas. Genes Cancer. 2011;2(8):829–40.
184. Ellis BC, Graham LD, Molloy PL. CRNDE, a long non-coding RNA responsive to insulin/IGF signaling, regulates genes involved in central metabolism. Biochim Biophys Acta. 2014;1843(2):372–86.
185. Wang Y, Wang Y, Li J, Zhang Y, Yin H, Han B. CRNDE, a long-noncoding RNA, promotes glioma cell growth and invasion through mTOR signaling. Cancer Lett. 2015;367(2):122–8.
186. Liu Y, Zhang M, Liang L, Li J, Chen YX. Overexpression of lncRNA DANCR is associated with advanced tumor progression and poor prognosis in patients with colorectal cancer. Int J Clin Exp Pathol. 2015;8(9):11480–4.
187. Yochum GS, Cleland R, McWeeney S, Goodman RH. An antisense transcript induced by Wnt/beta-catenin signaling decreases E2F4. J Biol Chem. 2007;282(2):871–8.
188. Xiong W, Jiang YX, Ai YQ, Liu S, Wu XR, Cui JG, et al. Microarray analysis of long Non-coding RNA expression profile associated with 5-fluorouracil-based chemoradiation resistance in colorectal cancer cells. Asian Pac J Cancer Prev APJCP. 2015;16(8):3395–402.
189. Qiu JJ, Yan JB. Long non-coding RNA LINC01296 is a potential prognostic biomarker in patients with colorectal cancer. Tumour Biol. 2015;36(9):7175–83.
190. Guo XB, Hua Z, Li C, Peng LP, Wang JS, Wang B, et al. Biological significance of long non-coding RNA FTX expression in human colorectal cancer. Int J Clin Exp Med. 2015;8(9):15591–600.
191. Yin D, He X, Zhang E, Kong R, De W, Zhang Z. Long noncoding RNA GAS5 affects cell proliferation and predicts a poor prognosis in patients with colorectal cancer. Med Oncol. 2014;31(11):253.
192. Krell J, Frampton AE, Mirnezami R, Harding V, De Giorgio A, Roca Alonso L, et al. Growth arrest-specific transcript 5 associated snoRNA levels are related to p53 expression and DNA damage in colorectal cancer. PLoS One. 2014;9(6):e98561.

193. Ohana P, Schachter P, Ayesh B, Mizrahi A, Birman T, Schneider T, et al. Regulatory sequences of H19 and IGF2 genes in DNA-based therapy of colorectal rat liver metastases. J Gene Med. 2005;7(3):366–74.
194. Fellig Y, Ariel I, Ohana P, Schachter P, Sinelnikov I, Birman T, et al. H19 expression in hepatic metastases from a range of human carcinomas. J Clin Pathol. 2005;58(10):1064–8.
195. Liang WC, Fu WM, Wong CW, Wang Y, Wang WM, Hu GX, et al. The lncRNA H19 promotes epithelial to mesenchymal transition by functioning as miRNA sponges in colorectal cancer. Oncotarget. 2015;6(26):22513–25.
196. Tian F, Tang Z, Song G, Pan Y, He B, Bao Q, et al. Loss of imprinting of IGF2 correlates with hypomethylation of the H19 differentially methylated region in the tumor tissue of colorectal cancer patients. Mol Med Rep. 2012;5(6):1536–40.
197. Nakagawa H, Chadwick RB, Peltomaki P, Plass C, Nakamura Y, de La Chapelle A. Loss of imprinting of the insulin-like growth factor II gene occurs by biallelic methylation in a core region of H19-associated CTCF-binding sites in colorectal cancer. Proc Natl Acad Sci U S A. 2001;98(2):591–6.
198. Deng Q, He B, Gao T, Pan Y, Sun H, Xu Y, et al. Up-regulation of 91H promotes tumor metastasis and predicts poor prognosis for patients with colorectal cancer. PLoS One. 2014;9(7):e103022.
199. Kogo R, Shimamura T, Mimori K, Kawahara K, Imoto S, Sudo T, et al. Long noncoding RNA HOTAIR regulates polycomb-dependent chromatin modification and is associated with poor prognosis in colorectal cancers. Cancer Res. 2011;71(20):6320–6.
200. Padua Alves C, Fonseca AS, Muys BR, de Barros ELBR, Burger MC, de Souza JE, et al. Brief report: the lincRNA hotair is required for epithelial-to-mesenchymal transition and stemness maintenance of cancer cell lines. Stem Cells. 2013;31(12):2827–32.
201. Xue Y, Gu D, Ma G, Zhu L, Hua Q, Chu H, et al. Genetic variants in lncRNA HOTAIR are associated with risk of colorectal cancer. Mutagenesis. 2015;30(2):303–10.
202. Xue Y, Ma G, Gu D, Zhu L, Hua Q, Du M, et al. Genome-wide analysis of long noncoding RNA signature in human colorectal cancer. Gene. 2015;556(2):227–34.
203. Svoboda M, Slyskova J, Schneiderova M, Makovicky P, Bielik L, Levy M, et al. HOTAIR long non-coding RNA is a negative prognostic factor not only in primary tumors, but also in the blood of colorectal cancer patients. Carcinogenesis. 2014;35(7):1510–5.
204. Wu ZH, Wang XL, Tang HM, Jiang T, Chen J, Lu S, et al. Long non-coding RNA HOTAIR is a powerful predictor of metastasis and poor prognosis and is associated with epithelial-mesenchymal transition in colon cancer. Oncol Rep. 2014;32(1):395–402.
205. Lian Y, Ding J, Zhang Z, Shi Y, Zhu Y, Li J, et al. The long noncoding RNA HOXA transcript at the distal tip promotes colorectal cancer growth partially via silencing of p21 expression. Tumour Biol. 2015;37(6):7431.
206. Ren YK, Xiao Y, Wan XB, Zhao YZ, Li J, Li Y, et al. Association of long non-coding RNA HOTTIP with progression and prognosis in colorectal cancer. Int J Clin Exp Pathol. 2015;8(9):11458–63.
207. Matouk IJ, Abbasi I, Hochberg A, Galun E, Dweik H, Akkawi M. Highly upregulated in liver cancer noncoding RNA is overexpressed in hepatic colorectal metastasis. Eur J Gastroenterol Hepatol. 2009;21(6):688–92.
208. Hunten S, Kaller M, Drepper F, Oeljeklaus S, Bonfert T, Erhard F, et al. p53-regulated networks of protein, mRNA, miRNA, and lncRNA expression revealed by integrated pulsed stable isotope labeling with amino acids in cell culture (pSILAC) and next generation sequencing (NGS) analyses. Mol Cell Proteomic MCP. 2015;14(10):2609–29.
209. Fan J, Xing Y, Wen X, Jia R, Ni H, He J, et al. Long non-coding RNA ROR decoys gene-specific histone methylation to promote tumorigenesis. Genome Biol. 2015;16:139.
210. Nakano S, Murakami K, Meguro M, Soejima H, Higashimoto K, Urano T, et al. Expression profile of LIT1/KCNQ1OT1 and epigenetic status at the KvDMR1 in colorectal cancers. Cancer Sci. 2006;97(11):1147–54.
211. Iguchi T, Uchi R, Nambara S, Saito T, Komatsu H, Hirata H, et al. A long noncoding RNA, lncRNA-ATB, is involved in the progression and prognosis of colorectal cancer. Anticancer Res. 2015;35(3):1385–8.
212. Yang F, Huo XS, Yuan SX, Zhang L, Zhou WP, Wang F, et al. Repression of the long noncoding RNA-LET by histone deacetylase 3 contributes to hypoxia-mediated metastasis. Mol Cell. 2013;49(6):1083–96.
213. Yang F, Zhang H, Mei Y, Wu M. Reciprocal regulation of HIF-1alpha and lincRNA-p21 modulates the Warburg effect. Mol Cell. 2014;53(1):88–100.
214. Zhai H, Fesler A, Schee K, Fodstad O, Flatmark K, Ju J. Clinical significance of long intergenic noncoding RNA-p21 in colorectal cancer. Clin Colorectal Cancer. 2013;12(4):261–6.
215. Wang G, Li Z, Zhao Q, Zhu Y, Zhao C, Li X, et al. LincRNA-p21 enhances the sensitivity of radiotherapy for human colorectal cancer by targeting the Wnt/beta-catenin signaling pathway. Oncol Rep. 2014;31(4):1839–45.
216. Wang J, Lei ZJ, Guo Y, Wang T, Qin ZY, Xiao HL, et al. miRNA-regulated delivery of lincRNA-p21 suppresses beta-catenin signaling and tumorigenicity of colorectal cancer stem cells. Oncotarget. 2015;6(35):37852–70.
217. Shi J, Li X, Zhang F, Zhang C, Guan Q, Cao X, et al. Circulating lncRNAs associated with occurrence of

colorectal cancer progression. Am J Cancer Res. 2015;5(7):2258–65.
218. Ding J, Lu B, Wang J, Wang J, Shi Y, Lian Y, et al. Long non-coding RNA Loc554202 induces apoptosis in colorectal cancer cells via the caspase cleavage cascades. J Exp Clini Cancer Res: CR. 2015;34:100.
219. Liu Q, Huang J, Zhou N, Zhang Z, Zhang A, Lu Z, et al. LncRNA loc285194 is a p53-regulated tumor suppressor. Nucleic Acids Res. 2013;41(9):4976–87.
220. Qi P, Xu MD, Ni SJ, Huang D, Wei P, Tan C, et al. Low expression of LOC285194 is associated with poor prognosis in colorectal cancer. J Transl Med. 2013;11:122.
221. Xu MD, Qi P, Weng WW, Shen XH, Ni SJ, Dong L, et al. Long non-coding RNA LSINCT5 predicts negative prognosis and exhibits oncogenic activity in gastric cancer. Medicine. 2014;93(28):e303.
222. Xu C, Yang M, Tian J, Wang X, Li Z. MALAT-1: a long non-coding RNA and its important 3′end functional motif in colorectal cancer metastasis. Int J Oncol. 2011;39(1):169–75.
223. Ji Q, Liu X, Fu X, Zhang L, Sui H, Zhou L, et al. Resveratrol inhibits invasion and metastasis of colorectal cancer cells via MALAT1 mediated Wnt/beta-catenin signal pathway. PLoS One. 2013;8(11):e78700.
224. Zheng HT, Shi DB, Wang YW, Li XX, Xu Y, Tripathi P, et al. High expression of lncRNA MALAT1 suggests a biomarker of poor prognosis in colorectal cancer. Int J Clin Exp Pathol. 2014;7(6):3174–81.
225. Ji Q, Zhang L, Liu X, Zhou L, Wang W, Han Z, et al. Long non-coding RNA MALAT1 promotes tumour growth and metastasis in colorectal cancer through binding to SFPQ and releasing oncogene PTBP2 from SFPQ/PTBP2 complex. Br J Cancer. 2014;111(4):736–48.
226. Yang MH, Hu ZY, Xu C, Xie LY, Wang XY, Chen SY, et al. MALAT1 promotes colorectal cancer cell proliferation/migration/invasion via PRKA kinase anchor protein 9. Biochim Biophys Acta. 2015;1852(1):166–74.
227. Zhang X, Zhou Y, Mehta KR, Danila DC, Scolavino S, Johnson SR, et al. A pituitary-derived MEG3 isoform functions as a growth suppressor in tumor cells. J Clin Endocrinol Metab. 2003;88(11):5119–26.
228. Yin DD, Liu ZJ, Zhang E, Kong R, Zhang ZH, Guo RH. Decreased expression of long noncoding RNA MEG3 affects cell proliferation and predicts a poor prognosis in patients with colorectal cancer. Tumour Biol. 2015;36(6):4851–9.
229. Han YJ, Ma SF, Yourek G, Park YD, Garcia JG. A transcribed pseudogene of MYLK promotes cell proliferation. FASEB J. 2011;25(7):2305–12.
230. Franklin JL, Rankin CR, Levy S, Snoddy JR, Zhang B, Washington MK, et al. Malignant transformation of colonic epithelial cells by a colon-derived long noncoding RNA. Biochem Biophys Res Commun. 2013;440(1):99–104.
231. Qi P, Xu MD, Ni SJ, Shen XH, Wei P, Huang D, et al. Down-regulation of ncRAN, a long non-coding RNA, contributes to colorectal cancer cell migration and invasion and predicts poor overall survival for colorectal cancer patients. Mol Carcinog. 2015;54(9):742–50.
232. Yu M, Ohira M, Li Y, Niizuma H, Oo ML, Zhu Y, et al. High expression of ncRAN, a novel non-coding RNA mapped to chromosome 17q25.1, is associated with poor prognosis in neuroblastoma. Int J Oncol. 2009;34(4):931–8.
233. Yan B, Gu W, Yang Z, Gu Z, Yue X, Gu Q, et al. Downregulation of a long noncoding RNA-ncRuPAR contributes to tumor inhibition in colorectal cancer. Tumour Biol. 2014;35(11):11329–35.
234. Li Y, Li Y, Chen W, He F, Tan Z, Zheng J, et al. NEAT expression is associated with tumor recurrence and unfavorable prognosis in colorectal cancer. Oncotarget. 2015;6(29):27641–50.
235. Pibouin L, Villaudy J, Ferbus D, Muleris M, Prosperi MT, Remvikos Y, et al. Cloning of the mRNA of overexpression in colon carcinoma-1: a sequence overexpressed in a subset of colon carcinomas. Cancer Genet Cytogenet. 2002;133(1):55–60.
236. Ge X, Chen Y, Liao X, Liu D, Li F, Ruan H, et al. Overexpression of long noncoding RNA PCAT-1 is a novel biomarker of poor prognosis in patients with colorectal cancer. Med Oncol. 2013;30(2):588.
237. Panagopoulos I, Moller E, Collin A, Mertens F. The POU5F1P1 pseudogene encodes a putative protein similar to POU5F1 isoform 1. Oncol Rep. 2008;20(5):1029–33.
238. Li L, Sun R, Liang Y, Pan X, Li Z, Bai P, et al. Association between polymorphisms in long non-coding RNA PRNCR1 in 8q24 and risk of colorectal cancer. J Exp Clin Cancer Res: CR. 2013;32:104.
239. Chung S, Nakagawa H, Uemura M, Piao L, Ashikawa K, Hosono N, et al. Association of a novel long non-coding RNA in 8q24 with prostate cancer susceptibility. Cancer Sci. 2011;102(1):245–52.
240. Yang L, Qiu M, Xu Y, Wang J, Zheng Y, Li M, et al. Upregulation of long non-coding RNA PRNCR1 in colorectal cancer promotes cell proliferation and cell cycle progression. Oncol Rep. 2016;35(1):318–24.
241. Poliseno L, Salmena L, Zhang J, Carver B, Haveman WJ, Pandolfi PP. A coding-independent function of gene and pseudogene mRNAs regulates tumour biology. Nature. 2010;465(7301):1033–8.
242. Takahashi Y, Sawada G, Kurashige J, Uchi R, Matsumura T, Ueo H, et al. Amplification of PVT-1 is involved in poor prognosis via apoptosis inhibition in colorectal cancers. Br J Cancer. 2014;110(1):164–71.
243. Shi D, Zheng H, Zhuo C, Peng J, Li D, Xu Y, et al. Low expression of novel lncRNA RP11-462C24.1 suggests a biomarker of poor prognosis in colorectal cancer. Med Oncol. 2014;31(7):31.
244. Sana J, Hankeova S, Svoboda M, Kiss I, Vyzula R, Slaby O. Expression levels of transcribed ultraconserved regions uc.73 and uc.388 are altered in colorectal cancer. Oncology. 2012;82(2):114–8.
245. Wang F, Li X, Xie X, Zhao L, Chen W. UCA1, a non-protein-coding RNA up-regulated in bladder

carcinoma and embryo, influencing cell growth and promoting invasion. FEBS Lett. 2008;582(13):1919–27.
246. Han Y, Yang YN, Yuan HH, Zhang TT, Sui H, Wei XL, et al. UCA1, a long non-coding RNA up-regulated in colorectal cancer influences cell proliferation, apoptosis and cell cycle distribution. Pathology. 2014;46(5):396–401.
247. Ni B, Yu X, Guo X, Fan X, Yang Z, Wu P, et al. Increased urothelial cancer associated 1 is associated with tumor proliferation and metastasis and predicts poor prognosis in colorectal cancer. Int J Oncol. 2015;47(4):1329–38.
248. Lassmann S, Weis R, Makowiec F, Roth J, Danciu M, Hopt U, et al. Array CGH identifies distinct DNA copy number profiles of oncogenes and tumor suppressor genes in chromosomal- and microsatellite-unstable sporadic colorectal carcinomas. J Mol Med. 2007;85(3):293–304.
249. Thorenoor N, Faltejskova-Vychytilova P, Hombach S, Mlcochova J, Kretz M, Svoboda M, et al. Long non-coding RNA ZFAS1 interacts with CDK1 and is involved in p53-dependent cell cycle control and apoptosis in colorectal cancer. Oncotarget. 2016;7(1):622–37.
250. Han J, Rong LF, Shi CB, Dong XG, Wang J, Wang BL, et al. Screening of lymph nodes metastasis associated lncRNAs in colorectal cancer patients. World J Gastroenterol WJG. 2014;20(25):8139–50.
251. Rui Q, Xu Z, Yang P, He Z. Long noncoding RNA expression patterns in lymph node metastasis in colorectal cancer by microarray. Biomed Pharmacother. 2015;75:12–8.
252. Ye LC, Ren L, Qiu JJ, Zhu DX, Chen T, Chang WJ, et al. Aberrant expression of long noncoding RNAs in colorectal cancer with liver metastasis. Tumour Biol. 2015;36(11):8747–54.
253. Sun QL, Zhao CP, Wang TY, Hao XB, Wang XY, Zhang X, et al. Expression profile analysis of long non-coding RNA associated with vincristine resistance in colon cancer cells by next-generation sequencing. Gene. 2015;572(1):79–86.
254. Liao Q, He W, Liu J, Cen Y, Luo L, Yu C, et al. Identification and functional annotation of lncRNA genes with hypermethylation in colorectal cancer. Gene. 2015;572(2):259–65.
255. Chu H, Xia L, Qiu X, Gu D, Zhu L, Jin J, et al. Genetic variants in noncoding PIWI-interacting RNA and colorectal cancer risk. Cancer. 2015;121(12):2044–52.
256. Wu Y, Zhang L, Wang Y, Li H, Ren X, Wei F, et al. Long noncoding RNA HOTAIR involvement in cancer. Tumour Biol. 2014;35(10):9531–8.
257. Hajjari M, Salavaty A. HOTAIR: an oncogenic long non-coding RNA in different cancers. Cancer Biol Med. 2015;12(1):1–9.
258. Gupta RA, Shah N, Wang KC, Kim J, Horlings HM, Wong DJ, et al. Long non-coding RNA HOTAIR reprograms chromatin state to promote cancer metastasis. Nature. 2010;464(7291):1071–6.
259. Khalil AM, Guttman M, Huarte M, Garber M, Raj A, Rivea Morales D, et al. Many human large intergenic noncoding RNAs associate with chromatin-modifying complexes and affect gene expression. Proc Natl Acad Sci U S A. 2009;106(28):11667–72.
260. Guttman M, Donaghey J, Carey BW, Garber M, Grenier JK, Munson G, et al. lincRNAs act in the circuitry controlling pluripotency and differentiation. Nature. 2011;477(7364):295–300.
261. Yang L, Lin C, Liu W, Zhang J, Ohgi KA, Grinstein JD, et al. ncRNA- and Pc2 methylation-dependent gene relocation between nuclear structures mediates gene activation programs. Cell. 2011;147(4):773–88.
262. Ying L, Chen Q, Wang Y, Zhou Z, Huang Y, Qiu F. Upregulated MALAT-1 contributes to bladder cancer cell migration by inducing epithelial-to-mesenchymal transition. Mol BioSyst. 2012;8(9):2289–94.
263. Flockhart RJ, Webster DE, Qu K, Mascarenhas N, Kovalski J, Kretz M, et al. BRAFV600E remodels the melanocyte transcriptome and induces BANCR to regulate melanoma cell migration. Genome Res. 2012;22(6):1006–14.
264. Sun M, Liu XH, Wang KM, Nie FQ, Kong R, Yang JS, et al. Downregulation of BRAF activated non-coding RNA is associated with poor prognosis for non-small cell lung cancer and promotes metastasis by affecting epithelial-mesenchymal transition. Mol Cancer. 2014;13:68.
265. Zhou X, Ye F, Yin C, Zhuang Y, Yue G, Zhang G. The interaction between MiR-141 and lncRNA-H19 in regulating cell proliferation and migration in gastric cancer. Cell Physiol Biochem. 2015;36(4):1440–52.
266. Grisanzio C, Freedman ML. Chromosome 8q24-associated cancers and MYC. Genes Cancer. 2010;1(6):555–9.
267. Huppi K, Pitt JJ, Wahlberg BM, Caplen NJ. The 8q24 gene desert: an oasis of non-coding transcriptional activity. Front Genet. 2012;3:69.
268. Haiman CA, Le Marchand L, Yamamato J, Stram DO, Sheng X, Kolonel LN, et al. A common genetic risk factor for colorectal and prostate cancer. Nat Genet. 2007;39(8):954–6.
269. Ye Z, Zhou M, Tian B, Wu B, Li J. Expression of lncRNA-CCAT1, E-cadherin and N-cadherin in colorectal cancer and its clinical significance. Int J Clin Exp Med. 2015;8(3):3707–15.
270. Pomerantz MM, Ahmadiyeh N, Jia L, Herman P, Verzi MP, Doddapaneni H, et al. The 8q24 cancer risk variant rs6983267 shows long-range interaction with MYC in colorectal cancer. Nat Genet. 2009;41(8):882–4.
271. Barsotti AM, Beckerman R, Laptenko O, Huppi K, Caplen NJ, Prives C. p53-dependent induction of PVT1 and miR-1204. J Biol Chem. 2012;287(4):2509–19.
272. Tseng YY, Moriarity BS, Gong W, Akiyama R, Tiwari A, Kawakami H, et al. PVT1 dependence in cancer with MYC copy-number increase. Nature. 2014;512(7512):82–6.

273. Gong J, Liu W, Zhang J, Miao X, Guo AY. lncRNASNP: a database of SNPs in lncRNAs and their potential functions in human and mouse. Nucleic Acids Res. 2015;43(Database issue):D181–6.
274. Ma C, Nong K, Zhu H, Wang W, Huang X, Yuan Z, et al. H19 promotes pancreatic cancer metastasis by derepressing let-7's suppression on its target HMGA2-mediated EMT. Tumour Biol. 2014;35(9):9163–9.
275. Kallen AN, Zhou XB, Xu J, Qiao C, Ma J, Yan L, et al. The imprinted H19 lncRNA antagonizes let-7 microRNAs. Mol Cell. 2013;52(1):101–12.
276. Li J, Yang J, Zhou P, Le Y, Zhou C, Wang S, et al. Circular RNAs in cancer: novel insights into origins, properties, functions and implications. Am J Cancer Res. 2015;5(2):472–80.
277. Qu S, Yang X, Li X, Wang J, Gao Y, Shang R, et al. Circular RNA: a new star of noncoding RNAs. Cancer Lett. 2015;365(2):141–8.
278. Cheng J, Metge F, Dieterich C. Specific identification and quantification of circular RNAs from sequencing data. Bioinformatics. 2016;32(7):1094–6.
279. Ebbesen KK, Kjems J, Hansen TB. Circular RNAs: identification, biogenesis and function. Biochim et biophysica acta. 2016;1859(1):163–8.
280. Hansen TB, Jensen TI, Clausen BH, Bramsen JB, Finsen B, Damgaard CK, et al. Natural RNA circles function as efficient microRNA sponges. Nature. 2013;495(7441):384–8.
281. Hansen TB, Kjems J, Damgaard CK. Circular RNA and miR-7 in cancer. Cancer Res. 2013;73(18):5609–12.
282. Li P, Chen S, Chen H, Mo X, Li T, Shao Y, et al. Using circular RNA as a novel type of biomarker in the screening of gastric cancer. Clin Chim Acta. 2015;444:132–6.
283. Peng L, Yuan XQ, Li GC. The emerging landscape of circular RNA ciRS-7 in cancer. Oncol Rep. 2015;33(6):2669–74.
284. Qin M, Liu G, Huo X, Tao X, Sun X, Ge Z, et al. Hsa_circ_0001649: a circular RNA and potential novel biomarker for hepatocellular carcinoma. Cancer Biomarkers. 2015;16(1):161–9.
285. Zhao ZJ, Shen J. Circular RNA participates in the carcinogenesis and the malignant behavior of cancer. RNA Biol. 2015. doi:10.1080/15476286.2015.1122162.
286. Bachmayr-Heyda A, Reiner AT, Auer K, Sukhbaatar N, Aust S, Bachleitner-Hofmann T, et al. Correlation of circular RNA abundance with proliferation – exemplified with colorectal and ovarian cancer, idiopathic lung fibrosis, and normal human tissues. Sci Rep. 2015;5:8057.

3 MicroRNAs and Inflammation in Colorectal Cancer

Claire Josse and Vincent Bours

Abstract

Colorectal cancers (CRC) are known to be related to inflammatory conditions, and inflammatory bowel diseases increase the relative risk for developing CRC. The use of anti-inflammatory drugs prevents the development of colorectal cancer.

Several molecular mediators are connecting the pathways that are involved in inflammatory conditions and in carcinogenesis. By the way these pathways are tightly interwoven, with the consequence that a deregulation at the level of any of these molecular mediators can affect the others.

MiRNAs are demonstrated to be deregulated in inflammatory bowel diseases and in colorectal cancer. Moreover, they target several molecular mediators that connect inflammation to cancer, and they are thus implicated in the route from inflammation to colorectal cancer.

This chapter will focus on the miRNAs that are jointly deregulated in inflammatory bowel disease and in colorectal cancer. Their role on the regulation of the molecular mediators and pathways that link inflammation to cancer will be described.

Keywords

microRNAs • Colorectal cancer • Inflammation • Inflammatory bowel disease • CAC • IBD

3.1 Introduction

MicroRNAs (miRNAs) are endogenous ~22 nt RNAs that can play important regulatory roles in animals and plants by targeting mRNAs for cleavage or destabilisation, or translational repression. MiRNAs are binding their mRNA targets through nucleotides 2–7 of their 5′ region,

C. Josse (✉)
University of Liège, GIGA-Research, Laboratory of Human Genetics, Liège, Belgium
e-mail: c.josse@ulg.ac.be

V. Bours
Human Genetics Department, University Hospital Centre of Liège, CHU, Liège, Belgium

O. Slaby, G.A. Calin (eds.), *Non-coding RNAs in Colorectal Cancer*, Advances in Experimental Medicine and Biology 937, DOI 10.1007/978-3-319-42059-2_3

also called the "miRNA seed". As the shortness of this sequence does not allow a high sequence binding specificity, one miRNAs has hundreds of mRNA targets. Most of the time, miRNAs are binding the 3′UTR part of their mRNA targets, and these 3′UTR are also containing binding sites for numerous miRNAs which may interact with each other by synergism or competition. Currently, around 2000 miRNAs have been described in humans, targeting ~60 % of the coding genes [1]. In the same manner as coding genes, miRNAs can display tissue expression specificity, or their expression can be driven by external stimuli. Their levels of expression are tightly regulated in cellular processes and normal tissue development, but they are also reported as disturbed in many diseases.

Among their regulatory roles, miRNAs are implicated in immunological processes. As reviewed in the article of Sonkoli et al., miRNAs are dynamically regulated during the development of T cells, B cells, and granulocytes. They represent a previously hidden layer of regulators involved in the development of the immune system and in the regulation of immune responses [2]. Many of them are implicated in inflammatory and autoimmune diseases, such as miR-155, miR-146a, or miR-192.

Besides, cancer has been a major focus of miRNA research over the past decade, and many studies have demonstrated the importance of miRNAs in cancer biology to facilitate tumour growth, invasion, angiogenesis, and immune evasion (reviewed in [3]). The dysregulation of miRNAs expression in cancer can occur through multiple mechanisms. At the genomic level this can be due to amplification or deletion of the sequence; mutations can also alter the target site or the processing outcome of the miRNA. MiRNAs implication in cancer development is currently completely acknowledged, and the first miRNA mimic entered the clinic for cancer therapy in 2013 [3].

As the inflammation in the colon observed during inflammatory bowel diseases increases the risk of colon cancer development (§3.2), one could hypothesized that miRNAs that display deregulated levels during inflammatory states could also be implicated in the molecular route from inflammation to cancer development. This hypothesis will be the common thread of the next paragraphs.

3.2 Inflammation and Colorectal Cancer

As early as in the nineteenth century, Rudolf Virchow observed that inflammatory cells infiltrate tumours and that tumours often arise at chronic inflammation sites. He hypothesized that in general, cancer is often linked to inflammatory conditions and indeed, this was confirmed the last two decades. Currently, inflammation is acknowledged as playing a critical role in tumorigenesis. In some types of cancer, inflammatory conditions are present before a malignant change occurs. Conversely, in other types of cancer, an oncogenic change induces an inflammatory microenvironment that promotes the development of tumours. In both cases, inflammation in the tumour microenvironment has many tumour-promoting effects. It facilitates in the proliferation and survival of malignant cells, promotes angiogenesis and metastasis, subverts adaptive immune responses, and alters responses to hormones and chemotherapeutic agents [4].

Some of the most convincing data demonstrating the connection between inflammation and cancer are that certain anti-inflammatory drugs reduce the risk of various cancer [5], and colorectal cancer in particular [6].

Among chronic inflammatory diseases well known to be associated with cancer are the inflammatory bowel diseases (IBD). They are autoimmune disorders that increase the risk of colitis associated colorectal cancer (CAC) development. The two mains sub-groups of IDB are the Crohn's disease (CD) and the ulcerative colitis (UC), presenting overlapping symptoms. Germline genetic predispositions play a major role in IBD. High throughput association studies that are associating the genotype and the phenotype (GWAS) of IBD patients showed that a large proportion of the IBD risk loci are shared with other immune-mediated diseases, primary immune-deficiencies and mycobacterial diseases [7]. However, environmental factors also play a crucial role, and IBD are considered as complex diseases implying links between genes, immunity, epithelial-barrier dysfunction, and the gut flora, this later being, in its turn, regulated by the diet.

Although recent evidence from population-based studies reports a decline in risk, CRC accounts for 10–15 % of all deaths in IBD [8]. The relative risks for developing CRC is about 5.6 and 30 in patients with CD and UC, respectively, in comparison with the general population [9].

The next paragraphs will describe the major molecular mediators that are recognized as a link between inflammation and colorectal cancer. One should note that all these molecules are tightly inter-related, highlighting the fact that a deregulation observed at the level of any of these molecular mediators can affect the others.

3.3 Molecular Mediators Linking Inflammation and Colorectal Cancer

Inflammatory response aims to eliminate harmful pathogens and subsequently restore homeostasis. This process is orchestrated by many different molecular mediators, and need to be finely regulated to avoid tissue injury, necrosis and malignant transformation. A healthy and regulated adaptive immune response is regarded as anti-tumorigenic, whereas an unrestrained innate or inappropriate adaptive response may lead to chronic inflammation and a pro-tumorigenic environment [10].

3.3.1 Cytokines and Growth Factors

At the inflammation site, the different cells such as innate and adaptive immune cells, myofibroblasts, fibroblasts, epithelial and stem cells are interacting in an autocrine and paracrine manner. The main mediators of inflammation regulation are the cytokines and chemokines that these cells secrete. They affect leukocyte recruitment, immune cell activation, angiogenesis, as well as turn-over and differentiation of stem cells. When the process is deregulated, they participate in oncogenesis by sustaining cell proliferation, survival, invasion, and metastasis. The role of inflammatory cytokines in inflammation, IBD and CAC is extensively described in the review by Francescone et al. [11].

3.3.1.1 Cytokines

Tumour necrosis factor-α (TNF-α), interleukin-1 (IL-1), IL-6, the IL-23/IL-17 axis and IL-22 are the main cytokines that link inflammation to colorectal cancer. Those cytokines and their related signaling provide an opportunity for targeted therapies to subdue IBD-linked inflammation. TNF-α has emerged as the most successful cytokine-based clinical target in the treatment of IBD. Other anti-cytokine therapies are also under consideration, and there are a number of ongoing clinical trials targeting proteins such as IL-6, IL17A, IL-12/IL-23, and IL-13 [11].

3.3.1.2 IGF and Insulin

During IBD there is increased insulin resistance as well as increased insulin levels [12]. Moreover, IGF-1 protein accumulates in the intestine with the same distribution as the inflammatory cells infiltrates [13]. Insulin resistance and an increased fat mass create an oxidative stress environment in tissues and increase the expression of various pro-inflammatory cytokines, including TNF-α and IL-6, which stimulate tumour growth and progression [14]. Binding of insulin, IGF-I or IGF-II to the extracellular portion of the IR, IGF-IR or hybrid receptor leads to the activation of the phosphatidylinositol 3-kinase (PI3K) and mitogen activated protein kinase (MAPK) signaling pathways [15].

3.3.2 Toll-Like Receptors

Pathogen associated molecular patterns trigger innate immune response mainly by activating Toll-like receptors (TLR) and intracytoplasmic nucleotide-binding oligomerisation domain-containing protein (NOD)-like receptors. NOD2 is the strongest single genetic susceptibility locus in CD [16]. TLR and NOD-like receptors can trigger downstream signaling pathways such as nuclear factor-kB (NF-κB), and lead to the secretion of inflammatory cytokines.

TLRs are also necessary for proper intestinal homeostasis as they regulate wound healing programs and intestinal barrier integrity. A proper regulation by the TLRs generally protects against injury and colitis but they can exacerbate CAC when deregulated.

3.3.3 PI3K/MAPK Signaling

Inflammatory stimuli described above are transmitted from the extracellular medium to the nucleus and lead to gene transcription by transcription factors that will be described later. At the interface stand the PI3K and MAPK pathways that are connecting the cytokines/chemokines, growth factors and Toll-like receptor pathways, to the NF-κB, p53, cell cycle and apoptosis pathways. The activity of PI3K is regulated by the tumour suppressor gene PTEN. PTEN dephosphorylates PI3K, and inhibits MAPK signaling, cell growth and cell cycle progression, through its interaction with cyclin D [17].

As a consequence, in the IBD and CAC context, the PI3K pathway inhibition is a potential therapeutic target [18].

3.3.4 Transcription Factors

3.3.4.1 NF-κB/STAT3

The transcription factor NF-κB is first associated with innate immunity: inflammatory cytokines such as TNF-α and IL-1β, lead to its activation in immune cells. However, the role played by this transcription factor is not restricted to innate immunity as its downstream regulated genes are not only pro- or anti-inflammation related molecules (IL-8, IL-1, cyclooxygenase 2 (COX2), inhibitor-kB (IκB), A20), but also genes implicated in cell proliferation (c-Myc; cyclin D1), in angiogenesis, in anti-apoptotic response (BCL-XL, Bcl-2), and in the regulation of adhesion molecules (ICAM); and these last functions are clearly deregulated during oncogenesis. The function of NF-κB in linking inflammation to cancer is extensively described in the review by Ben-Neriah and Karin (2011) [19].

In CAC, the proliferative function of NF-κB is indirect and is mediated through IL-6 and related cytokines produced by myeloid cells that lead to the activation of STAT3 in intestinal epithelial cells. NF-κB and STAT3 can both interfere with synthesis of the tumour suppressor p53 and attenuate p53-mediated genomic surveillance [20]. STAT3 activation, downstream of NF-κB, controls also the expression of the proliferation genes c-Myc and cyclin D1.

Moreover, NF-κB displays an oncogenic activity by promoting the polarization of tumour-associated macrophages and by affecting epithelial-to-mesenchymal transition (EMT) through induction of the transcription factors Twist and Snail.

NF-κB is also tightly implicated in oxidative stress linked to chronic inflammation as it induces the expression of key enzymes in the prostaglandin synthase pathway (COX-2), and the nitric oxide (NO) synthase.

3.3.4.2 P53

P53 is a tumour suppressor protein that responds to DNA damage and other cellular stresses to regulate the expression of target genes inducing cell cycle arrest, apoptosis, senescence, DNA repair, or changes in metabolism. Kukitsu and colleagues have disclosed an aberrant crypt foci-dysplasia-cancer sequence in colitis-associated carcinogenesis similar to the aberrant crypt foci -adenoma-carcinoma sequence in sporadic colon carcinogenesis. In this model, there was a significant stepwise increase in p53 mutations in the different progression stages [21]. Concomitantly with the accumulation of mutant p53 gene in cancerous glands, sustained DNA damage and NF-κB activation are observed. Mutant p53 prolongs NF-κB activation by inflammatory stimulus and promotes chronic inflammation and inflammation-associated colorectal cancer [22]. This process is illustrated in Fig. 3.1.

The tumour suppressor p53 negatively regulates a number of genes, including the proto-oncogene c-Myc [23].

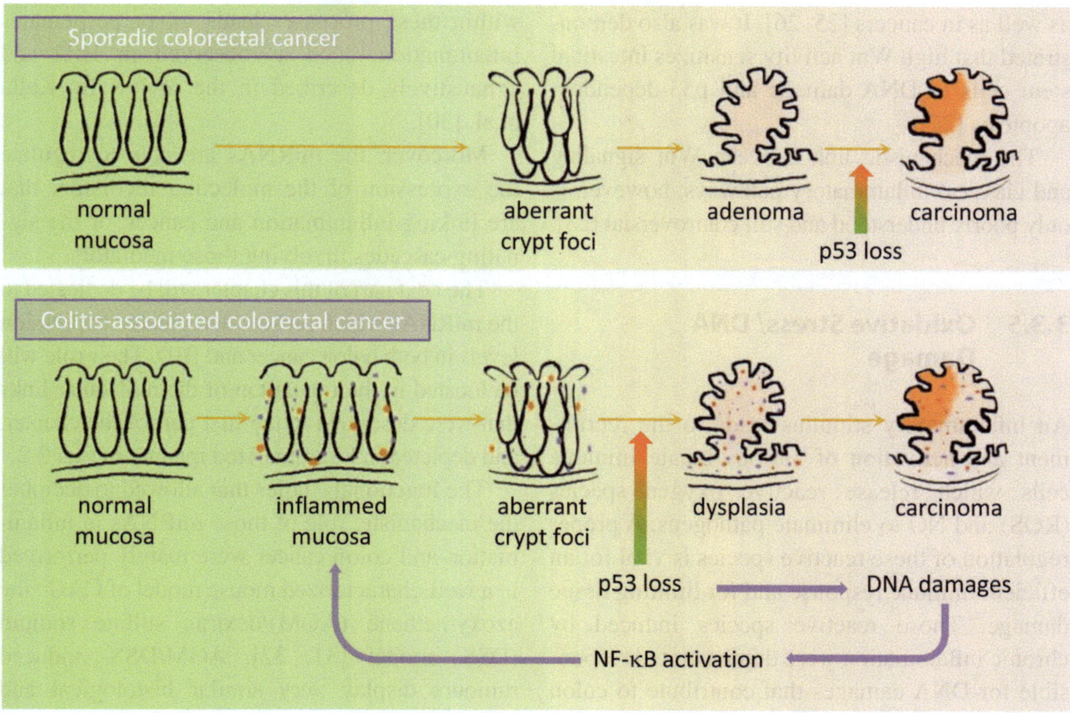

Fig. 3.1 Similarities are observed between colitis-associated carcinogenesis and sporadic colorectal cancer development, such as the sequences from aberrant crypt foci-dysplasia-cancer in colitis-associated cancer that resemble to the aberrant crypt foci-adenoma-carcinoma sequence in sporadic colon carcinogenesis. In the colitis-associated cancer development model, a significant stepwise increase in p53 mutations is observed in the different progression stages [21]. Concomitantly with the accumulation of mutant p53 gene in cancerous glands, sustained DNA damage and NF-κB activation are observed. Mutant p53 prolongs NF-κB activation by inflammatory stimulus and promotes chronic inflammation and inflammation-associated colorectal cancer [22]

3.3.4.3 c-Myc

The protein encoded by this gene is a multifunctional, nuclear phosphoprotein that plays a role in cell cycle progression, apoptosis, epithelial to mesenchymal transition and cellular transformation. Mutations, overexpression, rearrangement and translocation of this gene have been associated with a variety of tumours. During inflammation and CAC, IL-6 induces STAT3 activation in colonic epithelial cells and upregulates c-Myc. On its turn, c-Myc can downregulate p53 transcription factor by enhancing ribosome biogenesis that is responsible for a MDM2-mediated p53 degradation [24].

3.3.4.4 Wnt/β-Catenin

The Drosophila melanogaster wingless gene (Wnt) and nuclear transcription factor β-catenin signaling pathways play a central role in the biology of gastro-intestinal stem cells as it constitutes the major driving force behind homeostatic self-renewal of the crypt. Approximately 80 % of colorectal cancers harbour APC gene mutations and half of the remainder have mutations in CTNNB1 gene encoding β-catenin that, both of which activate Wnt/β-catenin signaling. Wnt activates the accumulation of the β-catenin, and the transcription of downstream target genes such as c-Myc, Cyclin D1 and VEGF, thereby regulating proliferation, apoptosis and angiogenesis. A crucial role of Wnt signaling pathway components is also evident in IBD and the transition to the malignant stage. Several studies found a stage-specific increased or decreased expression of a number of Wnt pathway-related genes in colonic biopsies of subjects with ulcerative colitis or in IBD associated colorectal neoplasia as well as in cancers [25, 26]. It was also demonstrated that high Wnt activity sensitizes intestinal stem cells to DNA damage and p53-dependent apoptosis [27].

The mechanistic link between Wnt signaling and classical inflammatory pathways, however, is only poorly understood and still controversial [28].

3.3.5 Oxidative Stress/ DNA Damage

An inflammatory stimulus leads to the recruitment and activation of various innate immune cells which release reactive oxygen species (ROS) and NO to eliminate pathogens. A proper regulation of these reactive species is vital for an efficient immune response and for limiting tissue damage. Those reactive species induced by chronic inflammation were demonstrated responsible for DNA damages that contribute to colon carcinogenesis in a mouse model [29].

During IBD, NO is produced by the inducible nitric oxide synthase (NOS2) enzyme as a response to inflammatory stimuli. If p53 is active, there is a negative feedback loop between NO and p53 where NO causes the stabilization and accumulation of p53, and activated p53 will then repress NOS2. Therefore, NO leads to increased p53 activity, which in turn promotes apoptosis, cell cycle arrest or senescence in damaged cells. When mutation in p53 occurs, during the aberrant crypt foci-dysplasia-cancer sequence, p53 protein is lost and cells are not as sensitive to NO-induced apoptosis or cell cycle arrest and instead NO can lead to genotoxic stress and cell proliferation [10].

3.4 MicroRNAs as Regulators of the Molecular Mediators Linking Inflammation and Colorectal Cancer

MiRNAs are acknowledged to play regulatory roles in inflammatory conditions and IBD. As they are tightly implicated in the regulation and signaling of the innate and adaptive immune systems, the deregulation of miRNAs expression within these processes leads to, or perpetuates inflammation, such as observed in IBD and exhaustively described in the review by Kalla et al. [30].

Moreover, the miRNAs are able to regulate the expression of the molecular mediators that are linking inflammation and cancer, or the signaling cascades involving those mediators.

The next part of this chapter will be dedicated to the miRNAs which present deregulated expression levels in both colon cancer and IBD. Their role will be located in the regulation of the molecular links that were described in the first part of the chapter, and depicted in a summarized manner in Fig. 3.2.

The functional studies that allowed to decipher the mechanistic role of those miRNAs in inflammation and colon cancer were mainly performed in a well-characterized mouse model of CAC : the azoxymethane (AOM)/dextran sulfate sodium (DSS) model [31, 32]. AOM/DSS- induced tumours display very similar histological and molecular features to human CRC. The tumour-induction method comprises a single intraperitoneal injection of a carcinogen, AOM, followed by one or three cycles of DSS administration via the drinking water. DSS treatment leads to intestinal epithelial barrier disruption and the establishment of chronic inflammation. After 10–20 weeks, and only when AOM is administered, colorectal tumours grow in a manner that recapitulates the aberrant crypt foci-adenoma-carcinoma sequence seen in human CRC [33].

The Table 3.1 is listing the main miRNAs that are deregulated in IBD, in the AOM/DSS mouse model, as well as in colon cancer.

The role of those miRNAs as regulators of the molecular mediators linking inflammation and cancer is summarized in the Table 3.2.

3.4.1 MiRNAs Regulation of Cytokines, Chemokines and Growth factors

3.4.1.1 Cytokines

The cluster 17–92, composed of six miRNAs (miR-17/18/19/20/92/106), that is encoded on human chr13 was demonstrated up-regulated by

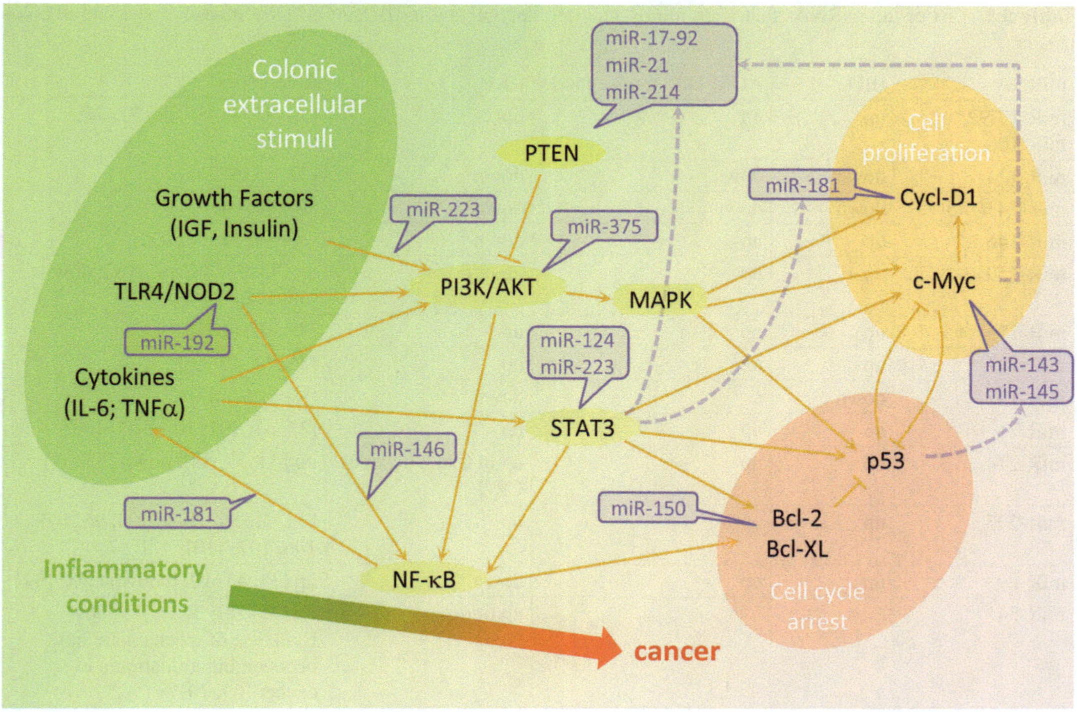

Fig. 3.2 The signaling cascades linking inflammation and cancer are tightly related, and miRNAs regulate the expression of the molecular mediators that are linking these two processes. Moreover, miRNAs are recognized as working in integrated transcriptional regulatory circuits and are frequently reported to be part of regulatory feedback loops of the signaling cascades involving those mediators. The above scheme is dedicated to the miRNAs which present deregulated expression levels in both colon cancer and IBD. Their role is located in the regulation of the molecular links between inflammation and cancer

many authors in colon cancer and in AOM/DSS mouse model, but also in IBD (Table 3.1). The miR-17-92 cluster promotes TH17 differentiation and TH17-related inflammation, and indirectly affects the expression of IL-17A, IL-17 F and IFN-γ expression [34].

MiR-146 is upregulated from non-neoplastic tissue to dysplasia, but downregulated from dysplasia to cancer [35], and accordingly, it was observed as upregulated in IBD and down regulated in CAC (Table 3.1). miR-146b, that modulates the TLR4 signaling pathway by direct targeting several of its downstream elements, is an IL-10–responsive miRNA with an anti-inflammatory activity [36].

Regarding the diet role in IBD and colon cancer, an AOM carcinogenesis model performed on mice fed with a high calorie diet shows that miR-150 expression is down-regulated and associated with an inflammatory pattern of cytokines expression, a high proliferation and a low apoptosis rates [37]. MiR-150 is upregulated in active IBD and in AOM/DSS mouse model, however its status is less clear in colon cancer at it was reported upregulated or down-regulated (Table 3.1). It has been demonstrated that miR-150 is targeting the MYB transcription factor that regulates the anti-apoptotic Bcl2 protein [38]. miR-150 is mainly expressed by immune cells, and is secreted by pro-monocytes in microvesicles and can regulate its target gene, c-Myb, into distant recipient endothelial cells to modulate their migration properties [39].

MiR-155 is up-regulated in various inflammatory disease states, including IBD, and is a positive regulator of T-cell responses. Acute colitis mouse model in miR-155(−/−) animals has allowed to demonstrate that the miR-155 deficiency protects mice from experimental colitis by reducing Th1/Th17 response [40].

Table 3.1 List of the miRNAs that were jointly reported deregulated in IBD, AOM/DSS mouse model and in colon cancer

miRNA	IBD	AOM/DSS mouse model	CRC	References
miR-17-92/ miR-106	up	up	up	[33–35, 41, 55, 56, 77, 78, 92–96]
miR-124	down	down	down	[33, 65, 67, 97]
miR-143/145	down	down	down	[33, 35, 84, 98, 99]
miR-146	up	up	down	[33, 35, 55, 56, 100]
miR-150	up	up	up or down	up [33, 38, 78, 96, 101]; down [56, 102]
miR-155	up	up	up	[75, 103–105]
miR-181	up	/	up	[35, 41, 42, 56]
miR-192	down	/	down	[99, 106, 107]
miR-21	up	up	up	[33, 41, 101, 103, 107–109]
miR-214	up	up	up in CAC; down in CRC	up [33, 57, 58]; down [55–57]
miR-223	up	up	up	[33, 41, 56, 78, 93, 94, 96, 101, 106, 107, 110]
miR-29	up	up	down	up [33, 45, 101]; down CRC [94]
miR-34	/	/	up or down	up [75–78]; down [79, 80]; Evidence of mechanistic link between inflammation and cancer [69, 70]
miR-375	up	up	down	up [33, 101, 108]; down [59, 94, 99]

MiR-21 is upregulated in both IBD and CAC and DSS mouse model, and its levels are related to survival in colon cancer patients (Table 3.1) [41]. MiR-181b is also upregulated in IBD and CAC (Table 3.1). Their transcription is triggered by IL-6 via the STAT3 activation. Those miRNAs are able to downregulate the expression of PTEN and CyclinD1 respectively, and are participating in an epigenetic switch that links inflammation to cancer (see PI3K and NF-κB/STAT3 paragraphs) [42].

MiR-21-knockout mice showed reduced expression of proinflammatory and procarcinogenic cytokines IL-6, IL-23, IL-17A and IL-21 and a decrease in the size and number of tumours in an AOM/DSS mouse model [43].

MiR-223 is up-regulated in IBD and in IL-10–/– mouse model, as well as in AOM/DSS and human colon cancer (Table 3.1). The miR-223 was demonstrated to be implicated in the pathway by witch IL-10 modulates IL-17–mediated inflammation. Indeed, miR-223 is a negative regulator of the Roquin ubiquitin ligase that curtails IL-17A synthesis [44].

MiR-29 is a fibroblast enriched miRNA family first known to control the extracellular matrix deposition [45]. MiR-29 level is upregulated in IBD and in AOM/DSS mouse model, downregulated in CAC tumorous tissue, but upregulated in the plasma of patient suffering of CAC (Table 3.1). Crohn's disease (CD) patient DCs expressing NOD2 polymorphisms fail to induce miR-29 in dendritic cells. MiR-29 is able to downregulate IL-23 by targeting IL-12p40 directly and IL-23p19 indirectly, likely via reduction of ATF2. DSS-induced colitis is worse in miR-29-deficient mice and is associated with elevated IL-23 and T helper 17 signature cytokines in the intestinal mucosa [46].

3.4.1.2 IGF and Insulin

MiR-223 is targeting several proteins of the IGF/insulin pathway. In colonic epithelial cells, surexpression of miR-223 was reported to reduce cell proliferation by directly targeting FOXO1 protein, and subsequently regulating the cyclin D1/p21/p27 [47].

Table 3.2 Pathways and targets of the miRNAs that are regulating the molecular mediators linking inflammation and cancer

miRNA	Target gene	Target molecular mediator or pathway	Reference
miR-17-92/ miR-106	PTEN, IKZF4	PI3K; Cytokine: IL-17 and IFNγ; c-Myc	[34, 81, 82]
miR-124	STAT3, CDK6	NF-κB	[65–67]
miR-143/145	K-Ras, MYC, CCND2, CDK6, E2F3	c-Myc; PI3K; p53	[53, 72, 84]
miR-146	IRAK1/2, TRAF6	Cytokines: IL-10 responsive miRNA; Toll like-receptor; NF-κB	[36, 52]
miR-150	c-MYB; ZEB1	Cytokines : pro-inflammatory pattern; PI3K	[37–39, 54]
miR-155	FADD, TP53INP1	Cytokines : Th1/Th17 response, NF-κB	[40, 63]
miR-181	CYCLD1	Cytokines : IL6; NF-κB /STAT3	[42]
miR-192	NOD2	Toll like-receptor, NF-κB /STAT3	[46, 51]
miR-21	PTEN	Cytokines : IL6, IL-23, IL-17A, IL-21; NF-κB /STAT3, PI3K	[42, 43]
miR-214	PTEN	Cytokine : IL-6, PI3K, NF-κB/STAT3	[57, 58]
miR-223	Roquin, FOXO1, IGF1R, STAT3, STMN1, ATM, MEF2C	Cytokines : IL-17a; IL-6 and IL-1β; IGF/insulin pathway, NF-κB /STAT3, p53, DNA damage	[44, 47–50, 68, 73, 87, 90]
miR-29	IL-12p40	Cytokines : IL-23	[46]
miR-34	IL6R	Cytokine : IL-6, P53, NF-κB /STAT3; DNA damage	[69–71]
miR-375	PIK3CA, KLF4, YAP1	PI3K	[59–61, 85, 86]

MiR-223 was also described to directly target IGF1R and indirectly affect the Akt/mTOR/p70S6K proteins and ERK pathways [33, 48–50].

3.4.2 Toll-Like Receptor

Many microRNA can actually regulate this pathways : NOD2 is targeted by miR-192 and miR-122 while it activates miR-29 expression to limit IL-23 release [46].

MiR-192 is down-regulated in IBD and CAC (Table 3.1), and has been demonstrated to directly target NOD2, thereby down-regulating the activation of NF-κB and the expression of IL-8 and CXCL3 chemokines. A single-nucleotide polymorphism (rs3135500) located in the NOD2 3′-untranslated region significantly reduced miR-192 effects on NOD2 gene expression [51].

MiR-146 controls Toll-like receptor and cytokine signaling through the direct targeting of the MyD88 adaptor proteins IRAK1/2 and TRAF6 [52].

3.4.3 PI3K/MAPK Signaling

The cluster 17–92 that was demonstrated to regulate TH17-related inflammation in IBD is actually directly targeting the PI3K inhibitor, the phosphatase and tensin homolog PTEN [34].

MiR-143/miR-145 downregulation observed in IBD and CAC results in the activation of PI3K/Akt signaling and MAPK signaling [53].

MiR-150 has been reported as directly targeting the major EMT inducer ZEB1, after activation of the AKT kinase by the SPROUTY intracellular modulators of tyrosine kinase receptor signaling [54].

MiR-21, upregulated by IL-6 mediated STAT3 activation, is targeting PTEN and participates in an inflammatory positive feedback loop that mediates the epigenetic switch between non transformed and transformed cells, and that is also implying the regulation of the secretion of let-7 and IL-6 by the NF-κB activation. The miR-21-mediated PTEN/Akt pathway down-regulation leads to an increase of NF-κB activity and IL-6 production [42].

PTEN is also a targeted by miR-214, which levels are up-regulated in IBD and in AOM/DSS mouse model (Table 3.1). Interestingly, in colon cancer, miR-214 is only upregulated in colitis-associated cancer but not in colorectal cancer [55–57]. The inverse correlation between PTEN and miR-214 is observed in human CAC tumorous tissues and in mouse model tumours. miR-214 regulation is also associated with the IL-6-STAT3-NF-κB pathway and is described later in the text [58].

MiR-375 was reported upregulated both in active UC and CD, and in the AOM/DSS mouse model, but down regulated in colon cancer (Table 3.1). MicroRNA-375 has been demonstrated to inhibit colorectal cancer growth by targeting PIK3CA [59].

3.4.4 Transcription Factors

3.4.4.1 NF-κB/STAT3

Polytarchou et al. demonstrated that miR-214 activates an inflammatory response and is amplified through a feedback loop circuit mediated by PTEN. In healthy non-inflamed colonic epithelial cells, PTEN expression suppresses the activation of Akt and NF-κB. During the development of UC, miR-214 targets PTEN to activate Akt and NF-κB. In turn, NF-κB regulates IL6 expression and thus STAT3 activity. STAT3-mediated transcriptional activation of miR-214 creates a positive feedback loop circuit that is attenuated when disease is in an inactive state. In longstanding UC, overexpression of miR-214 and hyperactivation of this inflammatory circuit promotes the development of colorectal cancer.

MiR-375 is directly targeting the KLF4 protein, that impacts the proliferation of colorectal carcinoma [60]. On the other side, Ghaleb et al. showed that genetic deletion of Klf4 in the mouse intestinal epithelium ameliorates DSS-induced colitis by modulating the NF-κB pathway inflammatory response suggesting that it could be involved in the pathogenesis and/or propagation of IBD [61]. Accordingly, KLF4 protein level is down-regulated in colonic tumour, and its level in normal tissue can predict poor survival in colorectal cancer patients [62].

During inflammation and its resolution, miR-155 and miR-146-a/b play important roles as regulators of Toll-like receptor and NF-κB signaling. Their expression is induced by NF-κB and form negative feedback circuits to fine-tune the inflammatory response upon bacterial infection. Oncogenic properties of miR-155 are attributed to its stimulation of cellular proliferation and its inhibition of apoptosis through the down-repression of caspase-3 activity, the targeting of FADD (Fas associated death domain), or the targeting of pro-apoptotic molecules such as TP53INP1 [63].

Upon LPS engagement, monocytes up-regulate miR-146b via an IL-10-mediated STAT3-dependent loop [36], leading to a negative regulation of NF-κB activation. Different studies have shown the pathological relevance of NF-κB/miR-146 in human cancers, however, it is unknown whether miR-146a dysregulation is causal to cancer. Chae et al. have recently reported a miR-146a polymorphism (rs2910164) that predicts risk of and survival from colorectal cancer [64].

NF-κB is the central component of a positive inflammatory loop and an epigenetic switch that link inflammation to cancer. Its activity is increased by the CyclinD1 and PTEN/Akt inhibition mediated by miR-21 and miR-181. In consequence, it activates the secretion of IL-6, and the down-regulation of let-7 by LIN28. All these molecular events lead to the activation of STAT3 and the formation of a positive regulation loop maintaining the cell transformation [42].

CAC model in miR-21-knockout mice shows that the absence of miR-21 increased PDCD4 expression, reduced nuclear factor NF-κB, STAT3 and Bcl-2 expression in tumour and stromal cells, and cause increased apoptosis of tumour cells [43].

MiR-124a, measured as downregulated in UC, CAC and AOM/DSS mouse model (Table 3.1), was recently reported to regulate the expression of STAT3 [65]. miR-124a is downregulated by methylation, and is known to have a tumour-suppressive activity by down-regulating oncogenic cyclin-dependent kinase 6 (CDK6) [66, 67].

In inflammatory conditions, miR-223 is down-regulated in macrophages, and lead to the activation of STAT3, which is directly targeted by miR-223, thus promoting the production of pro-inflammatory cytokines IL-6 and IL-1β, but not TNF-α. Interestingly, IL-6 was found to be a main factor in inducing the decrease in miR-223 expression after LPS stimulation, which formed a positive feedback loop to regulate IL-6 and IL-1β [68].

The link of the well-known onco-suppressor miR-34 with inflammation and cancer is less evident, as there is no report of miR-34 deregulation in IBD. Its status in colon cancer, is also debated (Table 3.1 and § p53). However, there are two reports of miR-34 a as mechanistical link between inflammation and cancer, first at the NF-κB regulation [69] and second at the DNA damage level [70, 71]. In colorectal cancer cells, IL-6 activates STAT3, which directly represses the MIR34A gene via a site in the first intron. Repression of MIR34A is required for IL-6–induced EMT and invasion. Furthermore, IL-6 receptor is also a direct miR-34a target. The resulting IL-6R/STAT3/miR-34a feedback loop was present in primary colorectal tumours; moreover in a mir-34a (−/−) AOM/DSS mouse model, tumours displayed upregulation of p-STAT3, IL-6R, and SNAIL and progressed to invasive carcinomas, which was not observed in WT animals. An active p53 protein interferes with this process [69].

3.4.4.2 P53

One mechanism of the p53-mediated c-Myc repression (see 3.3.4) may involve transcriptional regulation: p53 transcriptionally induces the expression of miR-145 and c-Myc is a direct target for miR-145. This specific silencing of c-Myc by miR-145 after p53 activation accounts at least in part for the miR-145- mediated inhibition of tumour cell growth both in vitro and in vivo [72].

In colon cancer cell lines, mutant p53 was demonstrated to activate the miR-223 promoter. miR-223 is subsequently targeting the stathmin-1 (STMN-1), an oncoprotein known to confer resistance to chemotherapeutic drugs and associated with poor clinical prognosis [73].

MiR-34 is a well-known onco-suppressor, that is under the p53 control [74]. Its deregulation in colon cancer is unclear : upregulated [75–78] or down-regulated [79, 80]. This discrepancy could be related to the stage of the cancer where miR-34a is measured, or more likely to the transactivation potential of p53 [76]. The miR-34 was demonstrated to be mainly expressed by stromal cells in human tumours [76].

3.4.4.3 c-Myc

c-Myc directly activates the expression of the cluster 17–92 [81] and targets a second important factor, the E2F1 protein. Complex positive and negative feedback loops exist among E2F, c-MYC, and the cluster miR-17–92, which ultimately determine the levels of E2F and whether the cells would progress from G1 into S phase, arrest at G1, or undergo apoptosis [82, 83].

The effect on the diet on colonic tumour promotion was investigated in mouse models. Western diet was demonstrated to activate EGFR signaling and down-regulate the expression of the miR-143/145. Indeed, those miRNAs were observed down-regulated in IBD, CAC and DSS mouse-model (Table 3.1). They regulate cell proliferation via the targeting of G1 regulators, K-Ras, MYC, CCND2, CDK6, and E2F3 [84].

3.4.4.4 Wnt/β-Catenin

MiR-375 was recently demonstrated as directly targeting the nuclear effector YAP1 that is playing a key role in intestinal stem regeneration and cancer [85, 86]. Gregorieff et al. have shown that Yap inactivation abolishes adenomas in the Apc(Min) mouse model of colon cancer upon tissue injury and that Yap reprograms Lgr5(+) intestinal stem cells by inhibiting the Wnt homeostatic program, while inducing a regenerative program that includes activation of EGFR signaling [86].

Mouse model of CAC performed in genetically modified miR-21(−/−) mice have shown that the absence of miR-21 resulted in attenuated proliferation of tumour cells with a simultaneous increase in E-cadherin and decreased β-catenin and stem cell markers in tumour tissues [43].

3.4.5 Oxidative Stress/ DNA Damage

It was demonstrated that NO induced apoptosis and stimulated expression of miR-34. In agreement with the link between p53 and NO, the loss of p53 inhibited both consequences. In human colorectal cancer samples, the expression of miR-34 significantly correlated with the level of inducible nitric oxide synthase (iNOS). So, the increased NO production may select cells with low levels of p53-dependent miRNAs which contributes to human colonic carcinogenesis and tumour progression [70, 71].

MiR-223 is able to target ATM expression and sensitizes culture cells to radiation-induced DNA damage [87].

3.5 Concluding Thoughts

As described above, the signaling cascades linking inflammation and cancer are tightly related.

MiRNAs themselves are recognized as working in integrated transcriptional regulatory circuits and are frequently reported to be part of regulatory feed-back loops.

A study that compares the pathways that are targeted by miRNAs in DSS-induced chronic inflammation and in AOM/DSS-induced carcinogenesis shows that both conditions are involving the same pathways and functions. Several miRNAs deregulated by inflammation or carcinogenesis are identical, but not all of them, however both conditions involved pathways in cancer, apoptosis, and proliferation in the same manner [33]. Thus, during IBD, many miRNAs are deregulated by inflammation and, as they are also acting together on pathways involved in carcinogenesis, they participate in the progressive molecular shift observed in the route from inflammation to cancer.

Colon cancer and IBD are rarely studied at the whole organism level. Nevertheless, miRNAs are secreted and can affect their targets genes in distant cells. This was demonstrated for miR-150 and miR-223, that are both mainly expressed in myeloid cells [39, 50, 88, 89]. In an animal model of Th1-mediated inflammatory bowel disease, the upregulation of some miRNAs (including miR-223) in peripheral blood leukocytes precedes their expression in the colon [44]. Myeloid-derived suppressor cells (MDSCs) play a critical role in the suppression of T-cell responses and the induction of T-cell tolerance in cancer. MDSCs are accumulating in bone marrow, spleen and blood of patient carrying cancer, and miR-223 was demonstrated to prevent their differentiation from bone marrow cells, and accumulation in peripheral tissues [90]. MDSCs are recruited in tumours mainly by the CXCL5 chemoattractant secreted during inflammation, and are able to induce EMT and early cancer dissemination [91]. As miR-223 is tightly implicated in the CAC pathology (see Table 3.1 and 3.2), the clarification of its role in the recruitment of MDSCs and in the tumorous immune-suppression at a whole organism level would be of great interest.

Indeed, these observations could lead to preventive and therapeutic measures. Chronic intake of anti-inflammatory drugs reduces the risk of colorectal cancer. A more precise identification of key regulators of the involved pathways might pave the way towards a more specific preventive approach.

References

1. Friedman RC, Farh KK-H, Burge CB, Bartel DP. Most mammalian mRNAs are conserved targets of microRNAs. Genome Res. 2009;19:92–105. doi:10.1101/gr.082701.108.
2. Sonkoly E, Pivarcsi A. microRNAs in inflammation. Int Rev Immunol. 2009;28:535–61. doi:10.3109/08830180903208303.
3. Hayes J, Peruzzi PP, Lawler S. MicroRNAs in cancer: biomarkers, functions and therapy. Trends Mol Med. 2014;1–10. doi:10.1016/j.molmed.2014.06.005.
4. Mantovani A, Allavena P, Sica A, Balkwill F. Cancer-related inflammation. Nature. 2008;454:436–44. doi:10.1038/nature07205.
5. Cuzick J, Otto F, Baron JA, Brown PH, Burn J, Greenwald P, et al. Aspirin and non-steroidal anti-inflammatory drugs for cancer prevention: an international consensus statement. Lancet Oncol. 2009;10:501–7. doi:10.1016/S1470-2045(09)70035-X.
6. Rigas B, Tsioulias GJ. The evolving role of nonsteroidal anti-inflammatory drugs in colon cancer pre-

vention: a cause for optimism. J Pharmacol Exp Ther. 2015;353:2–8. doi:10.1124/jpet.114.220806.
7. Ek WE, D'Amato M, Halfvarson J. The history of genetics in inflammatory bowel disease. Ann Gastroenterol Q Publ Hell Soc Gastroenterol. 2014;27:294–303.
8. Herszényi L, Barabás L, Miheller P, Tulassay Z. Colorectal cancer in patients with inflammatory bowel disease: the true impact of the risk. Dig Dis. 2015;33:52–7. doi:10.1159/000368447.
9. Goel A. MicroRNAs as therapeutic targets in colitis and colitis-associated cancer: tiny players with a giant impact. Gastroenterology. 2015;149:859–61. doi:10.1053/j.gastro.2015.08.041.
10. Schetter AJ, Heegaard NHH, Harris CC. Inflammation and cancer: interweaving microRNA, free radical, cytokine and p53 pathways. Carcinogenesis. 2010;31:37–49. doi:10.1093/carcin/bgp272.
11. Francescone R, Hou V, Grivennikov SI. Cytokines, IBD, and colitis-associated cancer. Inflamm Bowel Dis. 2015;21:409–18. doi:10.1097/MIB.0000000000000236.
12. Bregenzer N, Hartmann A, Strauch U, Schölmerich J, Andus T, Bollheimer LC. Increased insulin resistance and beta cell activity in patients with Crohn's disease. Inflamm Bowel Dis. 2006;12:53–6.
13. Lawrance IC, Maxwell L, Doe W. Inflammation location, but not type, determines the increase in TGF-beta1 and IGF-1 expression and collagen deposition in IBD intestine. Inflamm Bowel Dis. 2001;7:16–26.
14. Shirakami Y, Shimizu M, Kubota M, Araki H, Tanaka T, Moriwaki H, et al. Chemoprevention of colorectal cancer by targeting obesity-related metabolic abnormalities. World J Gastroenterol. 2014;20:8939–46. doi:10.3748/wjg.v20.i27.8939.
15. Gallagher EJ, LeRoith D. The proliferating role of insulin and insulin-like growth factors in cancer. Trends Endocrinol Metab. 2010;21:610–8. doi:10.1016/j.tem.2010.06.007.
16. Cuthbert AP, Fisher SA, Mirza MM, King K, Hampe J, Croucher PJP, et al. The contribution of NOD2 gene mutations to the risk and site of disease in inflammatory bowel disease. Gastroenterology. 2002;122:867–74.
17. Danielsen SA, Eide PW, Nesbakken A, Guren T, Leithe E, Lothe RA. Portrait of the PI3K/AKT pathway in colorectal cancer. Biochim Biophys Acta. 1855;2015:104–21. doi:10.1016/j.bbcan.2014.09.008.
18. Zhang J, Roberts TM, Shivdasani RA. Targeting PI3K signaling as a therapeutic approach for colorectal cancer. Gastroenterology. 2011;141:50–61. doi:10.1053/j.gastro.2011.05.010.
19. Ben-Neriah Y, Karin M. Inflammation meets cancer, with NF-κB as the matchmaker. Nat Immunol. 2011;12:715–23. doi:10.1038/ni.2060.
20. Colotta F, Allavena P, Sica A, Garlanda C, Mantovani A. Cancer-related inflammation, the seventh hallmark of cancer: links to genetic instability. Carcinogenesis. 2009;30:1073–81. doi:10.1093/carcin/bgp127.
21. Kukitsu T, Takayama T, Miyanishi K, Nobuoka A, Katsuki S, Sato Y, et al. Aberrant crypt foci as precursors of the dysplasia-carcinoma sequence in patients with ulcerative colitis. Clin Cancer Res. 2008;14:48–54. doi:10.1158/1078-0432.CCR-07-1835.
22. Cooks T, Pateras IS, Tarcic O, Solomon H, Schetter AJ, Wilder S, et al. Mutant p53 prolongs NF-κB activation and promotes chronic inflammation and inflammation-associated colorectal cancer. Cancer Cell. 2013;23:634–46. doi:10.1016/j.ccr.2013.03.022.
23. Levy N, Yonish-Rouach E, Oren M, Kimchi A. Complementation by wild-type p53 of interleukin-6 effects on M1 cells: induction of cell cycle exit and cooperativity with c-myc suppression. Mol Cell Biol. 1993;13:7942–52.
24. Brighenti E, Calabrese C, Liguori G, Giannone FA, Trerè D, Montanaro L, et al. Interleukin 6 downregulates p53 expression and activity by stimulating ribosome biogenesis: a new pathway connecting inflammation to cancer. Oncogene. 2014;33:4396–406. doi:10.1038/onc.2014.1.
25. Claessen MMH, Schipper MEI, Oldenburg B, Siersema PD, Offerhaus GJA, Vleggaar FP. WNT-pathway activation in IBD-associated colorectal carcinogenesis: potential biomarkers for colonic surveillance. Cell Oncol. 2010;32:303–10. doi:10.3233/CLO-2009-0503.
26. Roos J, Grösch S, Werz O, Schröder P, Ziegler S, Fulda S, et al. Regulation of tumorigenic Wnt signaling by cyclooxygenase-2, 5-lipoxygenase and their pharmacological inhibitors: a basis for novel drugs targeting cancer cells? Pharmacol Ther. 2015. doi:10.1016/j.pharmthera.2015.11.001.
27. Tao S, Tang D, Morita Y, Sperka T, Omrani O, Lechel A, et al. Wnt activity and basal niche position sensitize intestinal stem and progenitor cells to DNA damage. EMBO J. 2015;34:624–40. doi:10.15252/embj.201490700.
28. Yu C-H, Nguyen TTK, Irvine KM, Sweet MJ, Frazer IH, Blumenthal A. Recombinant Wnt3a and Wnt5a elicit macrophage cytokine production and tolerization to microbial stimulation via Toll-like receptor 4. Eur J Immunol. 2014;44:1480–90. doi:10.1002/eji.201343959.
29. Meira LB, Bugni JM, Green SL, Lee C, Pang B, Borenshtein D, et al. DNA damage induced by chronic inflammation contributes to colon carcinogenesis in mice. J Clin Invest. 2008;118:2516–25. doi:10.1172/JCI35073.
30. Kalla R, Ventham NT, Kennedy NA, Quintana JF, Nimmo ER, Buck AH, et al. MicroRNAs: new play-

ers in IBD. Gut. 2015;64:504–13. doi:10.1136/gutjnl-2014-307891.
31. De Robertis M, Massi E, Poeta ML, Carotti S, Morini S, Cecchetelli L, et al. The AOM/DSS murine model for the study of colon carcinogenesis: from pathways to diagnosis and therapy studies [Internet]. J Carcinog 2011. p. 9. doi:10.4103/1477-3163.78279.
32. Greten FR, Eckmann L, Greten TF, Park JM, Li Z-W, Egan LJ, et al. IKKbeta links inflammation and tumorigenesis in a mouse model of colitis-associated cancer. Cell. 2004;118:285–96. doi:10.1016/j.cell.2004.07.013.
33. Josse C, Bouznad N, Geurts P, Irrthum A, Huynh-Thu VA, Servais L, et al. Identification of a microRNA landscape targeting the PI3K/Akt signaling pathway in inflammation-induced colorectal carcinogenesis. Am J Physiol Gastrointest Liver Physiol. 2014;306:G229–43. doi:10.1152/ajpgi.00484.2012.
34. Liu S-Q, Jiang S, Li C, Zhang B, Li Q-J. miR-17-92 cluster targets phosphatase and tensin homology and Ikaros Family Zinc Finger 4 to promote TH17-mediated inflammation. J Biol Chem. 2014;289:12446–56. doi:10.1074/jbc.M114.550723.
35. Kanaan Z, Rai SN, Eichenberger MR, Barnes C, Dworkin AM, Weller C, et al. Differential microRNA expression tracks neoplastic progression in inflammatory bowel disease-associated colorectal cancer. Hum Mutat. 2012;33:551–60. doi:10.1002/humu.22021.
36. Curtale G, Mirolo M, Renzi TA, Rossato M, Bazzoni F, Locati M. Negative regulation of Toll-like receptor 4 signaling by IL-10-dependent microRNA-146b. Proc Natl Acad Sci U S A. 2013;110:11499–504. doi:10.1073/pnas.1219852110.
37. Olivo-Marston SE, Hursting SD, Perkins SN, Schetter A, Khan M, Croce C, et al. Effects of calorie restriction and diet-induced obesity on murine colon carcinogenesis, growth and inflammatory factors, and microRNA expression. PLoS One. 2014;9:e94765. doi:10.1371/journal.pone.0094765.
38. Bian Z, Li L, Cui J, Zhang H, Liu Y, Zhang C-Y, et al. Role of miR-150-targeting c-Myb in colonic epithelial disruption during dextran sulphate sodium-induced murine experimental colitis and human ulcerative colitis. J Pathol. 2011;225:544–53. doi:10.1002/path.2907.
39. Zhang Y, Liu D, Chen X, Li J, Li L, Bian Z, et al. Secreted monocytic miR-150 enhances targeted endothelial cell migration. Mol Cell. 2010;39:133–44. doi:10.1016/j.molcel.2010.06.010.
40. Singh UP, Murphy AE, Enos RT, Shamran HA, Singh NP, Guan H, et al. miR-155 deficiency protects mice from experimental colitis by reducing T helper type 1/type 17 responses. Immunology. 2014;143:478–89. doi:10.1111/imm.12328.
41. Schetter AJ, Leung SY, Sohn JJ, Zanetti KA, Bowman ED, Yanaihara N, et al. MicroRNA expression profiles associated with prognosis and therapeutic outcome in colon adenocarcinoma. JAMA. 2008;299:425–36. doi:10.1001/jama.299.4.425.
42. Iliopoulos D. Jaeger S a, Hirsch H a, Bulyk ML, Struhl K. STAT3 activation of miR-21 and miR-181b-1 via PTEN and CYLD are part of the epigenetic switch linking inflammation to cancer. Mol Cell. 2010;39:493–506. doi:10.1016/j.molcel.2010.07.023.
43. Shi C, Yang Y, Xia Y, Okugawa Y, Yang J, Liang Y, et al. Novel evidence for an oncogenic role of microRNA-21 in colitis-associated colorectal cancer. Gut. 2015;1–12. doi:10.1136/gutjnl-2014-308455
44. Schaefer JS, Montufar-Solis D, Vigneswaran N, Klein JR. Selective upregulation of microRNA expression in peripheral blood leukocytes in IL-10–/– mice precedes expression in the colon. J Immunol. 2011;187:5834–41. doi:10.4049/jimmunol.1100922.
45. Hofsli E, Sjursen W, Prestvik WS, Johansen J, Rye M, Tranø G, et al. Identification of serum microRNA profiles in colon cancer. Br J Cancer. 2013;108:1712–9. doi:10.1038/bjc.2013.121.
46. Brain O, Owens BMJ, Pichulik T, Allan P, Khatamzas E, Leslie A, et al. The intracellular sensor NOD2 induces microRNA-29 expression in human dendritic cells to limit IL-23 release. Immunity. 2013;39:521–36. doi:10.1016/j.immuni.2013.08.035.
47. Wu L, Li H, Jia CY, Cheng W, Yu M, Peng M, et al. MicroRNA-223 regulates FOXO1 expression and cell proliferation. FEBS Lett. 2012;586:1038–43. doi:10.1016/j.febslet.2012.02.050.
48. Jia CY, Li HH, Zhu XC, Dong YW, Fu D, Zhao QL, et al. MiR-223 suppresses cell proliferation by targeting IGF-1R. PLoS One. 2011;6:e27008. doi:10.1371/journal.pone.0027008.
49. Nian W, Ao X, Wu Y, Huang Y, Shao J, Wang Y, et al. miR-223 functions as a potent tumor suppressor of the Lewis lung carcinoma cell line by targeting insulin-like growth factor-1 receptor and cyclin-dependent kinase 2. Oncol Lett. 2013;6:359–66. doi:10.3892/ol.2013.1375.
50. Pan Y, Liang H, Liu H, Li D, Chen X, Li L, et al. Platelet-secreted microRNA-223 promotes endothelial cell apoptosis induced by advanced glycation end products via targeting the insulin-like growth factor 1 receptor. J Immunol. 2014;192:437–46. doi:10.4049/jimmunol.1301790.
51. Chuang AY, Chuang JC, Zhai Z, Wu F, Kwon JH. NOD2 expression is regulated by microRNAs in colonic epithelial HCT116 cells. Inflamm Bowel Dis. 2014;20:126–35. doi:10.1097/01.MIB.0000436954.70596.9b.
52. Taganov KD, Boldin MP, Chang K-J, Baltimore D. NF-kappaB-dependent induction of microRNA miR-146, an inhibitor targeted to signaling proteins of innate immune responses. Proc Natl Acad Sci U S

A. 2006;103:12481–6. doi:10.1073/pnas.0605298103.

53. Noguchi S, Yasui Y, Iwasaki J, Kumazaki M, Yamada N, Naito S, et al. Replacement treatment with microRNA-143 and −145 induces synergistic inhibition of the growth of human bladder cancer cells by regulating PI3K/Akt and MAPK signaling pathways. Cancer Lett. 2013;328:353–61. doi:10.1016/j.canlet.2012.10.017.
54. Barbáchano A, Fernández-Barral A, Pereira F, Segura MF, Ordóñez-Morán P, Carrillo-de Santa Pau E, et al. SPROUTY-2 represses the epithelial phenotype of colon carcinoma cells via upregulation of ZEB1 mediated by ETS1 and miR-200/miR-150. Oncogene. 2015. doi:10.1038/onc.2015.366.
55. Bandrés E, Cubedo E, Agirre X, Malumbres R, Zárate R, Ramirez N, et al. Identification by Real-time PCR of 13 mature microRNAs differentially expressed in colorectal cancer and non-tumoral tissues. Mol Cancer. 2006;5:29. doi:10.1186/1476-4598-5-29.
56. Chen X, Guo X, Zhang H, Xiang Y, Chen J, Yin Y, et al. Role of miR-143 targeting KRAS in colorectal tumorigenesis. Oncogene. 2009;28:1385–92. doi:10.1038/onc.2008.474.
57. Polytarchou C, Hommes DW, Palumbo T, Hatziapostolou M, Koutsioumpa M, Koukos G, et al. MicroRNA214 is associated with progression of ulcerative colitis, and inhibition reduces development of colitis and colitis-associated cancer in mice. Gastroenterology. 2015;49:981–92.e11. doi:10.1053/j.gastro.2015.05.057.
58. Polytarchou C, Iliopoulos D, Struhl K. An integrated transcriptional regulatory circuit that reinforces the breast cancer stem cell state. Proc Natl Acad Sci U S A. 2012;109:14470–5. doi:10.1073/pnas.1212811109.
59. Wang Y, Tang Q, Li M, Jiang S, Wang X. MicroRNA-375 inhibits colorectal cancer growth by targeting PIK3CA. Biochem Biophys Res Commun. 2014;444:199–204. doi:10.1016/j.bbrc.2014.01.028.
60. Mao Q, Quan T, Luo B, Guo X, Liu L, Zheng Q. MiR-375 targets KLF4 and impacts the proliferation of colorectal carcinoma. Tumor Biol. 2015. doi:10.1007/s13277-015-3809-0.
61. Ghaleb AM, Laroui H, Merlin D, Yang VW. Genetic deletion of Klf4 in the mouse intestinal epithelium ameliorates dextran sodium sulfate-induced colitis by modulating the NF-κB pathway inflammatory response. Inflamm Bowel Dis. 2014;20:811–20. doi:10.1097/MIB.0000000000000022.
62. Lee H-Y, Ahn JB, Rha SY, Chung HC, Park KH, Kim TS, et al. High KLF4 level in normal tissue predicts poor survival in colorectal cancer patients. World J Surg Oncol. 2014;12:232. doi:10.1186/1477-7819-12-232.
63. Tili E, Croce CM, Michaille J-J. miR-155: on the crosstalk between inflammation and cancer. Int Rev Immunol. 2009;28:264–84. doi:10.1080/08830180903093796.
64. Chae YS, Kim JG, Lee SJ, Kang BW, Lee YJ, Park JY, et al. A miR-146a polymorphism (rs2910164) predicts risk of and survival from colorectal cancer. Anticancer Res. 2013;33:3233–9.
65. Koukos G, Polytarchou C, Kaplan JL, Morley-Fletcher A, Gras-Miralles B, Kokkotou E, et al. MicroRNA-124 regulates STAT3 expression and is down-regulated in colon tissues of pediatric patients with ulcerative colitis. Gastroenterology. 2013;145:842–52.e2. doi:10.1053/j.gastro.2013.07.001.
66. Lujambio A, Ropero S, Ballestar E, Fraga MF, Cerrato C, Setién F, et al. Genetic unmasking of an epigenetically silenced microRNA in human cancer cells. Cancer Res. 2007;67:1424–9. doi:10.1158/0008-5472.CAN-06-4218.
67. Ueda Y, Ando T, Nanjo S, Ushijima T, Sugiyama T. DNA methylation of microRNA-124a is a potential risk marker of colitis-associated cancer in patients with ulcerative colitis. Dig Dis Sci. 2014;59:2444–51. doi:10.1007/s10620-014-3193-4.
68. Chen Q, Wang H, Liu Y, Song Y, Lai L, Han Q, et al. Inducible microRNA-223 down-regulation promotes TLR-triggered IL-6 and IL-1β production in macrophages by targeting STAT3. PLoS One. 2012;7:e42971. doi:10.1371/journal.pone.0042971.
69. Rokavec M, Öner MG, Li H, Jackstadt R, Jiang L, Lodygin D, et al. IL-6R/STAT3/miR-34a feedback loop promotes EMT-mediated colorectal cancer invasion and metastasis. J Clin Invest. 2014;124:1853–67. doi:10.1172/JCI73531.
70. Mathé E, Nguyen GH, Funamizu N, He P, Moake M, Croce CM, et al. Inflammation regulates microRNA expression in cooperation with p53 and nitric oxide. Int J Cancer. 2012;131:760–5. doi:10.1002/ijc.26403.
71. Li W, Han W, Ma Y, Cui L, Tian Y, Zhou Z, et al. P53-dependent miRNAs mediate nitric oxide-induced apoptosis in colonic carcinogenesis. Free Radic Biol Med. 2015;85:105–13. doi:10.1016/j.freeradbiomed.2015.04.016.
72. Sachdeva M, Zhu S, Wu F, Wu H, Walia V, Kumar S, et al. p53 represses c-Myc through induction of the tumor suppressor miR-145. Proc Natl Acad Sci U S A. 2009;106:3207–12. doi:10.1073/pnas.0808042106.
73. Masciarelli S, Fontemaggi G, Di Agostino S, Donzelli S, Carcarino E, Strano S, et al. Gain-of-function mutant p53 downregulates miR-223 contributing to chemoresistance of cultured tumor cells. Oncogene. 2014;33:1601–8. doi:10.1038/onc.2013.106.
74. He L, He X, Lim LP, de Stanchina E, Xuan Z, Liang Y, et al. A microRNA component of the p53 tumour suppressor network. Nature. 2007;447:1130–4. doi:10.1038/nature05939.

75. Wang M, Zhang P, Li Y, Liu G, Zhou B, Zhan L, et al. The quantitative analysis by stem-loop real-time PCR revealed the microRNA-34a, microRNA-155 and microRNA-200c overexpression in human colorectal cancer. Med Oncol. 2012;29:3113–8. doi:10.1007/s12032-012-0241-9.
76. Hiyoshi Y, Schetter AJ, Okayama H, Inamura K, Anami K, Nguyen GH, et al. Increased microRNA-34b and -34c predominantly expressed in stromal tissues is associated with poor prognosis in human colon cancer. PLoS One. 2015;10:e0124899. doi:10.1371/journal.pone.0124899.
77. Arndt GM, Dossey L, Cullen LM, Lai A, Druker R, Eisbacher M, et al. Characterization of global microRNA expression reveals oncogenic potential of miR-145 in metastatic colorectal cancer. BMC Cancer. 2009;9:374. doi:10.1186/1471-2407-9-374.
78. Monzo M, Navarro A, Bandres E, Artells R, Moreno I, Gel B, et al. Overlapping expression of microRNAs in human embryonic colon and colorectal cancer. Cell Res. 2008;18:823–33. doi:10.1038/cr.2008.81.
79. Akao Y, Nakagawa Y, Hirata I, Iio A, Itoh T, Kojima K, et al. Role of anti-oncomirs miR-143 and −145 in human colorectal tumors. Cancer Gene Ther. 2010;17:398–408. doi:10.1038/cgt.2009.88.
80. Roy S, Levi E, Majumdar APN, Sarkar FH. Expression of miR-34 is lost in colon cancer which can be re-expressed by a novel agent CDF. J Hematol Oncol. 2012;5:58. doi:10.1186/1756-8722-5-58.
81. O'Donnell KA, Wentzel EA, Zeller KI, Dang CV, Mendell JT. c-Myc-regulated microRNAs modulate E2F1 expression. Nature. 2005;435:839–43. doi:10.1038/nature03677.
82. Pickering MT, Stadler BM, Kowalik TF. miR-17 and miR-20a temper an E2F1-induced G1 checkpoint to regulate cell cycle progression. Oncogene. 2008;28:140–5. doi:10.1038/onc.2008.372.
83. Davidson-Moncada J, Papavasiliou FN, Tam W. MicroRNAs of the immune system: roles in inflammation and cancer. Ann N Y Acad Sci. 2010;1183:183–94. doi:10.1111/j.1749-6632.2009.05121.x.
84. Zhu H, Dougherty U, Robinson V, Mustafi R, Pekow J, Kupfer S, et al. EGFR signals downregulate tumor suppressors miR-143 and miR-145 in Western diet-promoted murine colon cancer: role of G1 regulators. Mol Cancer Res. 2011;9:960–75. doi:10.1158/1541-7786.MCR-10-0531.
85. Christensen LL, Holm A, Rantala J, Kallioniemi O, Rasmussen MH, Ostenfeld MS, et al. Functional screening identifies miRNAs influencing apoptosis and proliferation in colorectal cancer. PLoS One. 2014;9:e96767. doi:10.1371/journal.pone.0096767.
86. Gregorieff A, Liu Y, Inanlou MR, Khomchuk Y, Wrana JL. Yap-dependent reprogramming of Lgr5+ stem cells drives intestinal regeneration and cancer. Nature. 2015;526:715–8. doi:10.1038/nature15382.
87. Liang L, Zhu J, Zaorsky NG, Deng Y, Wu X, Liu Y, et al. MicroRNA-223 enhances radiation sensitivity of U87MG cells in vitro and in vivo by targeting ataxia telangiectasia mutated. Int J Radiat Oncol. 2014;88:955–60. doi:10.1016/j.ijrobp.2013.12.036.
88. Tabet F, Vickers KC, Cuesta Torres LF, Wiese CB, Shoucri BM, Lambert G, et al. HDL-transferred microRNA-223 regulates ICAM-1 expression in endothelial cells. Nat Commun. 2014;5:3292. doi:10.1038/ncomms4292.
89. Ismail N, Wang Y, Dakhlallah D, Moldovan L, Agarwal K, Batte K, et al. Macrophage microvesicles induce macrophage differentiation and miR-223 transfer. Blood. 2013;121:984–95. doi:10.1182/blood-2011-08-374793.
90. Liu Q, Zhang M, Jiang X, Zhang Z, Dai L, Min S, et al. miR-223 suppresses differentiation of tumor-induced $CD11b^+$ $Gr1^+$ myeloid-derived suppressor cells from bone marrow cells. Int J Cancer. 2011;129:2662–73. doi:10.1002/ijc.25921.
91. Toh B, Wang X, Keeble J, Sim WJ, Khoo K, Wong W-C, et al. Mesenchymal transition and dissemination of cancer cells is driven by myeloid-derived suppressor cells infiltrating the primary tumor. PLoS Biol. 2011;9:e1001162. doi:10.1371/journal.pbio.1001162.
92. Knudsen KN, Nielsen BS, Lindebjerg J, Hansen TF, Holst R, Sørensen FB. MicroRNA-17 is the most up-regulated member of the miR-17-92 cluster during early colon cancer evolution. PLoS One. 2015;10:e0140503. doi:10.1371/journal.pone.0140503.
93. Ng EKO, Tsang WP, Ng SSM, Jin HC, Yu J, Li JJ, et al. MicroRNA-143 targets DNA methyltransferases 3A in colorectal cancer. Br J Cancer. 2009;101:699–706. doi:10.1038/sj.bjc.6605195.
94. Gaedcke J, Grade M, Camps J, Sokilde R, Kaczkowski B, Schetter AJ, et al. The rectal cancer microRNAome - microRNA expression in rectal cancer and matched normal mucosa. Clin Cancer Res. 2012;18:4919–30. doi:10.1158/1078-0432.CCR-12-0016.
95. Motoyama K, Inoue H, Takatsuno Y, Tanaka F, Mimori K, Uetake H, et al. Over- and under-expressed microRNAs in human colorectal cancer. Int J Oncol. 2009;34:1069–75. doi:10.3892/ijo.
96. Volinia S, Calin GA, Liu C, Ambs S, Cimmino A, Petrocca F, et al. A microRNA expression signature of human solid tumors defines cancer gene targets. Proc Natl Acad Sci U S A. 2006;103:2257–61. doi:10.1073/pnas.0510565103.
97. Liu M, Chen H. The role of microRNAs in colorectal cancer. J Genet Genomics. 2010;37:347–58. doi:10.1016/S1673-8527(09)60053-9.
98. Pekow JR, Dougherty U, Mustafi R, Zhu H, Kocherginsky M, Rubin DT, et al. miR-143 and miR-145 are downregulated in ulcerative colitis: putative regulators of inflammation and protoonco-

genes. Inflamm Bowel Dis. 2012;18:94–100. doi:10.1002/ibd.21742.
99. Faltejskova P, Svoboda M, Srutova K, Mlcochova J, Besse A, Nekvindova J, et al. Identification and functional screening of microRNAs highly deregulated in colorectal cancer. J Cell Mol Med. 2012;16:2655–66. doi:10.1111/j.1582-4934.2012.01579.x.
100. Schepeler T, Reinert JT, Ostenfeld MS, Christensen LL, Silahtaroglu AN, Dyrskjøt L, et al. Diagnostic and prognostic microRNAs in stage II colon cancer. Cancer Res. 2008;68:6416–24. doi:10.1158/0008-5472.CAN-07-6110.
101. Fasseu M, Tréton X, Guichard C, Pedruzzi E, Cazals-Hatem D, Richard C, et al. Identification of restricted subsets of mature microRNA abnormally expressed in inactive colonic mucosa of patients with inflammatory bowel disease. Dalmasso G, editor. PLoS One. 2010;5:e13160. doi:10.1371/journal.pone.0013160.
102. Pizzini S, Bisognin A, Mandruzzato S, Biasiolo M, Facciolli A, Perilli L, et al. Impact of microRNAs on regulatory networks and pathways in human colorectal carcinogenesis and development of metastasis. BMC Genomics. 2013;14:589. doi:10.1186/1471-2164-14-589.
103. Takagi T, Naito Y, Mizushima K, Hirata I, Yagi N, Tomatsuri N, et al. Increased expression of microRNA in the inflamed colonic mucosa of patients with active ulcerative colitis. J Gastroenterol Hepatol. 2010;25 Suppl 1:S129–33. doi:10.1111/j.1440-1746.2009.06216.x.
104. Qu Y-L, Wang H-F, Sun Z-Q, Tang Y, Han X-N, Yu X-B, et al. Up-regulated miR-155-5p promotes cell proliferation, invasion and metastasis in colorectal carcinoma. Int J Clin Exp Pathol. 2015;8:6988–94.
105. Li W, Han W, Zhao X, Wang H. Changes of expression of miR-155 in colitis-associated colonic carcinogenesis. Zhonghua Zhong Liu Za Zhi. 2014;36:257–62.
106. Earle JSL, Luthra R, Romans A, Abraham R, Ensor J, Yao H, et al. Association of microRNA expression with microsatellite instability status in colorectal adenocarcinoma. J Mol Diagn. 2010;12:433–40. doi:10.2353/jmoldx.2010.090154.
107. Wu F, Zikusoka M, Trindade A, Dassopoulos T, Harris ML, Bayless TM, et al. MicroRNAs are differentially expressed in ulcerative colitis and alter expression of macrophage inflammatory peptide-2 alpha. Gastroenterology. 2008;135:1624–35.e24. doi:10.1053/j.gastro.2008.07.068.
108. Wu F, Zhang S, Dassopoulos T, Harris ML, Bayless TM, Meltzer SJ, et al. Identification of microRNAs associated with ileal and colonic Crohn's disease. Inflamm Bowel Dis. 2010;16:1729–38. doi:10.1002/ibd.21267.
109. Nielsen BS, Jørgensen S, Fog JU, Søkilde R, Christensen IJ, Hansen U, et al. High levels of microRNA-21 in the stroma of colorectal cancers predict short disease-free survival in stage II colon cancer patients. Clin Exp Metastasis. 2011;28:27–38. doi:10.1007/s10585-010-9355-7.
110. Lanza G, Ferracin M, Gafà R, Veronese A, Spizzo R, Pichiorri F, et al. mRNA/microRNA gene expression profile in microsatellite unstable colorectal cancer. Mol Cancer. 2007;6:54. doi:10.1186/1476-4598-6-54.

Interplay Between Transcription Factors and MicroRNAs Regulating Epithelial-Mesenchymal Transitions in Colorectal Cancer

4

Markus Kaller and Heiko Hermeking

Abstract

The epithelial-mesenchymal-transition (EMT) represents a morphogenetic program involved in developmental processes such as gastrulation and neural crest formation. The EMT program is co-opted by epithelial tumor cells and endows them with features necessary for spreading to distant sites, such as invasion, migration, apoptosis resistance and stemness. Thereby, EMT facilitates metastasis formation and therapy resistance. A growing number of transcription factors has been implicated in the regulation of EMT. These include EMT-inducing transcription factors (EMT-TFs), the most prominent being SNAIL, SLUG, ZEB1, ZEB2 and TWIST, and negative regulators of EMT, such as p53. Furthermore, a growing number of microRNAs, such as members of the miR-200 and miR-34 family, have been characterized as negative regulators of EMT. EMT-TFs and microRNAs, such as ZEB1/2 and miR-200 or SNAIL and miR-34, are often engaged in double-negative feedback loops forming bistable switches controlling the transitions from epithelial to the mesenchymal cell states. Within this chapter, we will provide a comprehensive overview over the transcription factors and microRNAs that have been implicated in the regulation of EMT in colorectal cancer. Furthermore, we will highlight the regulatory connections between EMT-TFs and miRNAs to illustrate common principles of their interaction that regulate EMTs.

M. Kaller
Experimental and Molecular Pathology, Institute of Pathology, Ludwig-Maximilians-University Munich, Thalkirchner Strasse 36, 80337 Munich, Germany

H. Hermeking (✉)
Experimental and Molecular Pathology, Institute of Pathology, Ludwig-Maximilians-University Munich, Thalkirchner Strasse 36, 80337 Munich, Germany

German Cancer Consortium (DKTK), D-69120 Heidelberg, Germany

German Cancer Research Center (DKFZ), D-69120 Heidelberg, Germany
e-mail: heiko.hermeking@med.uni-muenchen.de

O. Slaby, G.A. Calin (eds.), *Non-coding RNAs in Colorectal Cancer*, Advances in Experimental Medicine and Biology 937, DOI 10.1007/978-3-319-42059-2_4

Keywords
miR-34 • p53 • c-MYC • Metastasis • EMT

4.1 Introduction

The ability to metastasize represents an important feature of cancer cells [1]. Whereas surgical resection and adjuvant therapy can cure spatially restricted primary tumors, metastatic disease is largely incurable because of its systemic nature and the resistance of disseminated tumor cells to existing therapeutic agents (reviewed in [2]). This explains why > 90 % of mortality from cancer is attributable to metastases, and not to the primary tumors from which these malignant lesions arise [3, 4].

The term "epithelial-mesenchymal transition" describes a cellular transdifferentiation program that is employed during embryogenic developmental stages such as gastrulation and neural crest formation, as well as (patho-)physiological processes, such as wound healing or fibrosis [5–7]. Importantly, EMT is aberrantly co-opted by epithelial tumor cells to acquire features considered to be necessary for dissemination from the primary tumor, e.g. increased migratory and invasive capacity. Therefore, EMT is critically involved in the active intravasation of primary tumor cells into the bloodstream and thought to be one of several mechanisms regulating the early stages of the invasion-metastasis cascade [5–7]. Furthermore, circulating tumor cells (CTCs) that have disseminated from the primary tumor very often display mesenchymal characteristics that allow preferential survival within the bloodstream and, therefore, seeding to distant organs [8]. The reduced proliferation rate and increased resistance to apoptosis observed in cells that have undergone an EMT also contributes to resistance to treatment with chemotherapeutic drugs or radiation. Moreover, transition to a mesenchymal state has been shown to confer stem cell characteristics, such as the ability for self-renewal and increased tumor-initiating capacity [9]. As a consequence, the presence of tumor cells that have transitioned into a mesenchymal cell state in primary and/or secondary tumors is associated with tumor recurrence after therapy and decreased patient survival [5, 7]. Therefore, a thorough understanding of the molecular networks that govern the transitions between epithelial and mesenchymal cell states may become highly relevant for therapeutic strategies in the future.

4.2 A Conceptual Framework for the Interactions Between Transcription Factors and microRNAs in the Regulation of EMT

The EMT process is activated by a variety of intracellular signaling pathways that integrate extracellular stimuli from the tumor microenvironment, such as those mediated by TGF-β, BMP, Wnt, Notch and Hedgehog and various growth factors, such as EGF, FGF and PDGF, that also control EMT during embryonic development. In addition, other forms of extracellular conditions or signals, such as hypoxia and inflammatory stimuli, have been shown to induce EMT [5].

The effectors of these EMT-inducing signals, such as β-Catenin/TCF/LEF (WNT signaling), SMADs (TGF-beta signaling), HIF1α (hypoxia), and NFkB, STAT3 (inflammation) function as direct transcriptional activators of one or several EMT-inducing transcription factors (EMT-TFs) and are therefore tightly linked to the regulatory network that orchestrates cellular reprogramming during EMT (Fig. 4.1). Hence, a set of EMT-TFs, which includes SNAI1 (SNAIL), SNAI2 (SLUG), ZEB1, ZEB2 and TWIST1, are aberrantly activated during cancer-associated EMTs and coordinate the execution of the EMT program [5–7].

The transcriptional repression of *E-Cadherin* and additional adherens junction and tight junction proteins by direct binding of EMT-TFs to the respective gene promoters is critical for the induction of EMT and a defining feature of EMT-TFs [7]. Moreover, additional transcription factors have been identified that, when ectopically expressed in epithelial cancer cell lines, lead to downregulation of *E-Cadherin* und upregulation of mesenchymal marker proteins, such as *Vimentin*. These include E47 [10], TCF4 [11–13], LEF1 [11, 14], Goosecoid (GSC) [15], FOXC2 [16], TFAP4 [17] and ZNF281 [18]. However, at least for some of these additional TFs, downregulation of *E-Cadherin* seems to be mediated by indirect mechanisms.

Conversely, a growing number of microRNAs, such as members of the miR-200 and miR-34 microRNA families, have been identified as negative regulators of EMT, most notably via regulatory feedback loops with individual EMT-TFs [19]. Strikingly, many of the microRNAs engaged in double-negative feedback loops with EMT-TFs are transcriptionally activated by the p53 tumor suppressor protein (see below), indicating that activation of p53 critically regulates the balance of these regulatory feedback-loops and the maintenance of the epithelial cell state. In addition, p53 mediates post-translational inhibition of EMT-TFs, such as SNAIL and SLUG, by MDM2–mediated protein degradation [20, 21]. Consequently, the frequent inactivation of p53 during tumorigenesis presumably contributes to deregulated activation of these EMT-TF networks and removes the barriers for reprogramming of epithelial into mesenchymal cells, resulting in augmented EMT and increased propensity of primary tumors to form metastases [22, 23].

Furthermore, several transcription factors have been identified that enforce the epithelial cell state and thereby antagonize EMT-TF function, most notably by direct transcriptional repression of EMT-TFs. Moreover, these epithelial-specific TFs are directly repressed by EMT-TFs. The ELF3 transcription factor was initially identified as an epithelial-specific marker for terminal differentiation of keratinocytes [24, 25]. ELF3 levels are reduced in SNAIL1/2-mediated EMT [26]. Interestingly, ELF3 expression is elevated in colorectal tumors and is associated with poor patient prognosis [27]. The ELF5 transcription factor suppresses EMT by direct transcriptional repression of SLUG. Furthermore, ectopic expression of ELF5 in mes-

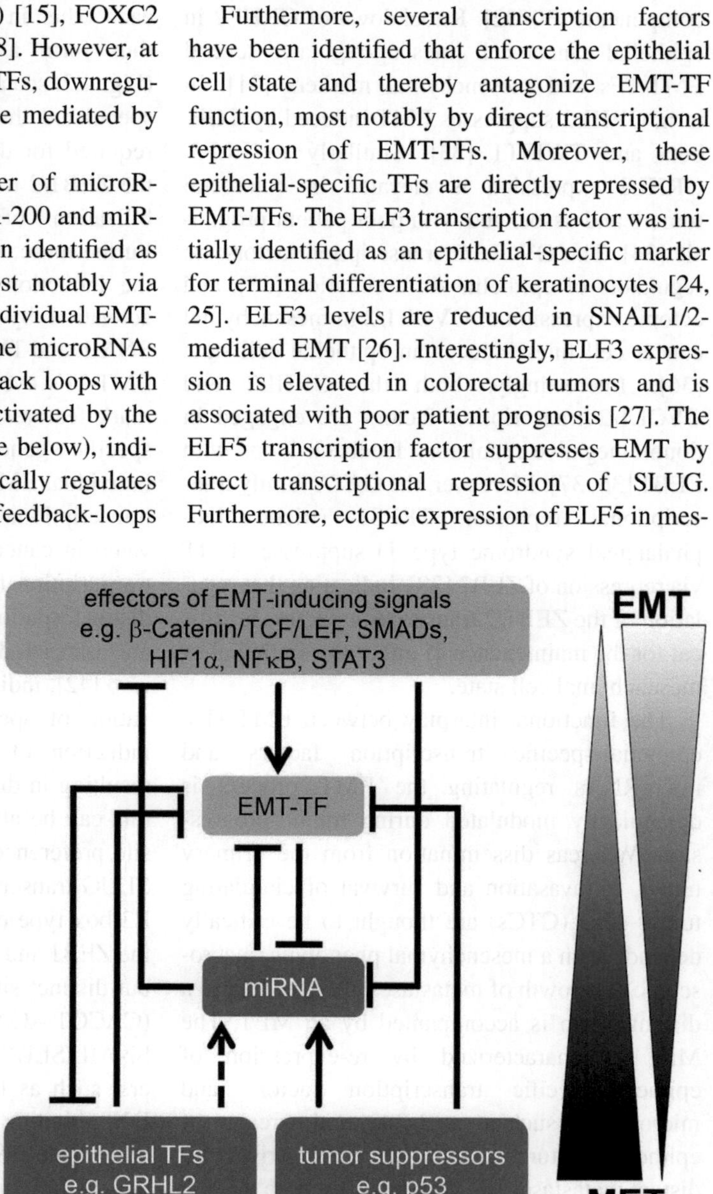

Fig. 4.1 A conceptual framework for the interactions between transcription factors and microRNAs in the regulation of epithelial and mesenchymal cell states

enchymal cancer cell lines induces a reversal of the mesenchymal phenotype, a so-called mesenchymal-epithelial-transition (MET) [28]. However, its role in colorectal cancer progression has not been determined yet. The GRHL2 transcription factor is a member of the evolutionarily conserved Grainyhead transcription factor family that plays critical roles in regulating epithelial cell differentiation [29]. GRHL2 is a positive regulator of epithelial-specific genes, such as *E-Cadherin* and other adherens and tight junction components [30–32]. Knockdown of GRHL2 in epithelial cancer cell lines up-regulates several EMT-TFs and mesenchymal markers [31]. In turn, GRHL2 suppresses EMT induced by TGF-beta and TWIST1 [33]. Similarly to ELF3, GRHL2 expression is elevated in colorectal tumors and associated with poor patient prognosis [34]. The OVOL1/2 transcription factors are regulators of epithelial differentiation [35], and ectopic expression of OVOL1/2 in mesenchymal cancer cell lines induces an epithelial cell state [36]. Interestingly, both the GRHL2 and OVOL1/2 transcription factors are engaged in double-negative regulatory feedback loops with ZEB1 [36, 37]. Moreover, the GATA family transcriptional repressor TRPS1 (tricho-rhino-phalangeal syndrome type 1) suppresses EMT via repression of ZEB2 [38], indicating that regulation of the ZEB1/2 transcription factors is critical for the maintenance of either the epithelial or mesenchymal cell state.

The functional interplay between EMT-TFs, epithelial-specific transcription factors and microRNAs regulating the EMT process is dynamically modulated during tumor progression. Whereas dissemination from the primary tumor, intravasation and survival of circulating tumor cells (CTCs) are thought to be critically dependent on a mesenchymal phenotype, macroscopic outgrowth of metastases after seeding to a distant organ is accompanied by an MET. The MET is characterized by re-expression of epithelial-specific transcription factors and microRNAs, such as miR-200, and a regain of epithelial features, which are characteristic for distant metastases [8, 39]. However, both the cell intrinsic and microenvironmental factors regulating the context-dependent balance of epithelial and mesenchymal cell states and the underlying gene expression networks, especially those regulating MET and metastatic outgrowth, are still incompletely understood.

4.3 The Regulatory Network of EMT-TFs in CRC

Remarkably, many EMT-TFs positively regulate each other in a highly cooperative manner and thus form a hierarchical regulatory network (Fig. 4.2, see also Table 4.1). Indeed, it has been speculated that EMT-TFs, such as SNAIL, are required for the initial stages of EMT, whereas the ZEB1/2 transcription factors function in stabilization of the mesenchymal state [7]. Furthermore, specific temporal requirements for the induction and maintenance of EMT upon activation by TGF-β have been described for SNAIL and TWIST1 [40]. Although activation of EMT-TFs eventually converges on shared hallmark features, e.g. the repression of epithelial-specific genes, such as *CDH1/E-Cadherin*, individual EMT-TFs nevertheless induce distinct transcriptional profiles when ectopically activated in cancer cell lines [41]. In addition, the transcription factors SNAIL and SLUG display distinct spatiotemporal expression patterns and are associated with distinct EMT signatures *in vivo* [42], indicating that context-dependent activation of specific EMT-TFs may lead to the induction of different transcriptional profiles resulting in distinct cellular phenotypes. In part, this can be attributed to different DNA binding site preferences. The highly related SNAIL and SLUG transcription factors recognize consensus E2-box type elements C/A(CAGGTG), whereas the ZEB1 and ZEB2 EMT-TFs recognize similar, but distinct sites composed of bipartite E-boxes (CACCT---CACCTG), which overlap with SNAIL/SLUG binding sites on certain promoters, such as the *CDH1* promoter [7], but their DNA binding patterns may differ on a genome-wide scale [39]. Moreover, even though SNAIL and SLUG bind to similar E-box motifs, their genome-wide binding patterns have recently

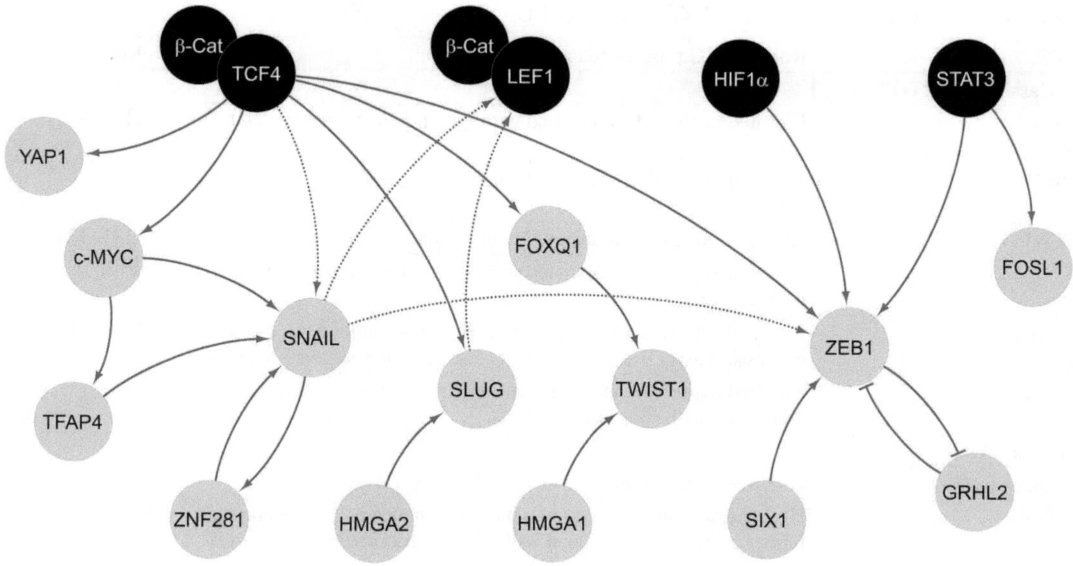

Fig. 4.2 EMT-TF transcriptional network in CRC. Experimentally validated transcriptional interactions between transcription factors involved in EMT in CRC, as inferred from literature. Presumably indirect interactions are displayed as dotted lines. For description, see main text

been shown to display extensive differences in murine mammary tumors [42].

Apart from the described EMT-TFs, a growing number of additional transcription factors have been identified that can induce an EMT in cancer cell lines from various tumor entities (see Table 4.1). Not all of these TFs have formally been shown to be involved in or sufficient to induce EMT in cell culture or mouse models of CRC yet, and their function(s) within the transcriptional network of EMT-TFs in CRC are not well understood. However, elevated expression of many of these is observed in CRC patient samples and is associated with lymphnode and distant metastasis, as well as decreased patient survival, highlighting the importance of deregulation of these factors for colorectal cancer progression (Table 4.1).

Nevertheless, a core transcriptional network of EMT-TFs operative in CRC can be inferred from the available literature (Fig. 4.2). We characterized a feed-forward loop composed of c-MYC, TFAP4 and SNAIL that induces EMT in colorectal cancer [17]. TFAP4/AP4 is a direct transcriptional target of c-MYC and mediates c-MYC-induced EMT in CRC cell lines by inducing an EMT signature [17]. Furthermore, elevated AP4 expression is associated with metastasis and poor survival of colorectal cancer and was necessary for metastasis formation in a xenograft model [17]. Moreover, we found that the c-MYC-associated zinc-finger protein ZNF281 induces EMT and is required for c-MYC-induced EMT in CRC cell lines. ZNF281 presumably achieves this by forming a positive feedback loop with SNAIL [18].

Both SNAIL and SLUG induce LEF1 expression in CRC cell lines [11]; however, since SNAIL mediated induction of LEF1 can be abrogated by ectopic expression of *E-Cadherin*, this effect may not be mediated via direct transcriptional activation [43]. Furthermore, β-Catenin/LEF1 is sufficient to induce EMT in colorectal cancer cell lines [11]. Elevated expression of LEF1 in colorectal tumors is associated with lymph node metastases, distant metastasis, and shorter overall survival of CRC patients [44, 45].

WNT/β-Catenin-mediated inhibition of GSK3β, the protein kinase which is regulating the turnover of the SNAIL protein, leads to increased SNAIL protein levels and activity [46]. In contrast, SLUG (but not SNAIL) is

Table 4.1 Transcription factors with roles in EMT and CRC

TF	Role in CRC progression and metastasis	References
"classical" EMT-TFs :		
SNAIL	Elevated expression associated with lymph node metastasis and poor overall survival	[135–137]
SLUG	Elevated expression associated with distant metastasis and shorther survival	[138]
TWIST1	Elevated expression associated with lymph node metastasis	[137, 139, 140]
	Elevated expression associated with shorter overall survival and disease-free survival	
ZEB1	Elevated expression in CRC compared to normal mucosa; elevated expression associated with liver metastasis and poor overall survival	[69–71]
ZEB2	Elevated expression at the tumor invasion front and in liver metastases associated with shorter survival	[141]
Other TFs :		
Brachyury/T	Higher expression associated with shorter survival	[142]
FOSL1	Higher expression associated with lymph node and liver metastasis; higher expression associated with shorter recurrence-free survival	[72, 73]
FOXC2	Elevated expression correlated with TNM stages; elevated expression associated with decreased overall and disease-free survival	[143–145]
FOXF2	Decreased expression in primary tumors compared to normal colon epithelium	[146]
FOXM1	Higher expression in CRC compared to normal mucosa; higher expression associated with lymph node metastasis, liver metastasis, and advanced TNM stage	[147, 148]
FOXQ1	Higher expression in CRC compared to normal mucosa	[149]
HMGA1	Higher expression in CRC compared to normal mucosa	[54]
HMGA2	Higher expression correlates with distant metastasis and poor survival	[150]
LEF1	Higher expression associated with lymph node metastases, distant metastasis, advanced TNM (tumor-node-metastasis) stage, and shorter overall survival	[44, 45]
	Higher expression associated with shorter overall survival	
c-MYC	Frequent amplification and elevated expression in CRC compared to normal colon epithelium;	[86, 151–155]
	Copy number gain associated with poor patient prognosis	
NFATC1	NFATC1-driven transcriptional program associated with shorter survival	[156]
PROX1	Higher expression associated with shorter survival	[157]
PRRX1	Higher expression associated with metastasis and poor prognosis	[158]
SATB1	Higher expression associated with poorly differentiated tumors, higher invasion depth, distant metastasis, advanced TNM stage and poorer prognosis	[159]
SIX1	High expression associated with decreased overall survival	[160]
SOX2	Higher expression correlated with lymph node and distant metastases	[161, 162]
SOX4	Higher expression associated with shorter recurrence-free and overall survival	[163, 164]
SOX9	Upregulated in CRC compared to normal mucosa; high expression asociated with shorter overall survival copy number gain detected in some primary colorectal cancers	[165, 166]
STAT3	Activation of STAT3 associated with mesenchymal gene expression signature and poor patient prognosis	[65, 126, 167]

(continued)

Table 4.1 (continued)

TF	Role in CRC progression and metastasis	References
TBX3	Higher expression associated with tumor size, poor differentiation, invasion, lymph node metastasis, TNM stage and poorer overall survival and disease-free survival	[168]
TCF4	Higher expression associated with shorter overall survival	[169]
TFAP4	Elevated expression associated with lymphnode and distant metastasis, higher tumor grade and shorter survival	[17]
TAZ	Elevated expression associated with decreased survival	[62, 63]
YAP1	Elevated expression associated with decreased survival	[61, 62]
ZNF703 / Zeppo	Elevated expression in CRC compared to normal mucosa; elevated expression associated with poor prognosis	[170]
epithelial-specific TFs :		
ELF3	Upregulation associated with poor survival; Upregulation associated with lymph node metastases	[27, 171]
GRHL2	Higher expression in CRC tissues; higher expression associated with tumor size and TNM stage, overall survival and recurrence-free survival	[34]
KLF4	Low expression associated with lymph node and distant metastasis, metastasis recurrence and poor survival	[172, 173]

transcriptionally induced by β-Catenin/TCF4 in CRC cell lines [47, 48]. Apart from inducing EMT-associated morphological changes and enhancing invasive and migratory capacities, SLUG mediates resistance to 5-fluorouracil-induced cell death in colon cancer cells [49]. SLUG expression is also regulated by HMGA2 in colon cancer cells [50]. HMGA2 is a critical downstream mediator of TGF-β signaling and also induces other EMT-TFs, such as SNAIL, albeit not in CRC models [51–53]. The related HMGA1 transcription factor induces TWIST1 expression in CRC cell lines and has been shown to be required for metastasis formation in xenograft mouse models of CRC [54].

The FOXQ1 transcription factor is transcriptionally induced by β-Catenin/TCF4 and is highly expressed in mesenchymal CRC cell lines [55]. Moreover, FOXQ1 directly induces TWIST1 and regulates migration and invasion of CRC cells [56]. Ectopic expression of FOXQ1 in breast cancer cell lines induces an EMT phenotype, which is mediated by direct repression of E-Cadherin, and promotes lung metastasis in mouse xenograft models [57].

The Hippo tumor suppressor pathway is involved in organ size control and stem-cell self renewal by inhibition of the transcriptional co-activators YAP1 and TAZ [58]. Conversely, YAP1 is directly activated by WNT/β-Catenin signaling in colorectal cancer cells and augments anchorage-independent cell growth in soft agar [59]. Moreover, YAP1 has recently been shown to cooperate with KRAS signaling in CRC cell lines and interacts with the AP-1 transcription factor FOS to regulate EMT-associated genes, such as SLUG and Vimentin [60]. Elevated expression of both YAP1 and TAZ is associated with decreased patient survival in CRC, indicating that aberrant activation of these transcription factors contributes to tumor progression [61–63].

The EMT-TF ZEB1 is transcriptionally induced by several EMT-inducing signaling pathways in CRC, such as WNT/β-Catenin [13], hypoxia [64] and STAT3 signaling [65]. Furthermore, ZEB1 is transcriptionaly activated by SNAIL, albeit in a presumably indirect manner [66]. The SIX1 transcription factor induces EMT in CRC cell lines, at least in part by activation of ZEB1 [67], which has also been demonstrated in breast cancer cell lines [37]. Moreover, inverse expression of the epithelial-specific transcription factor GRHL2 and ZEB1 in CRC cell lines indicates the existence of a double negative feedback-loop, which may regulate the transitions between epithelial and mesenchymal states [68]. However, so far this has only been demonstrated in breast cancer cell lines [37]. Elevated expression in CRC compared to normal mucosa has been shown for ZEB1, which is correlated with liver metastasis and poor overall survival [69–71].

During inflammatory signaling IL6-mediated activation of STAT3 leads to transcriptional induction of FOSL1/FRA-1 [72]. FOSL1 expression in CRC cell lines induces an EMT expression signature [73]. Moreover, strong FOSL1 expression is detected at the invasion front of colorectal tumors [73]. Therefore, FOSL1 may be an important mediator of inflammation-mediated tumor progression.

In summary, these findings indicate that numerous EMT-TFs are involved in cellular reprogramming during EMT in CRC. However, even though additional regulatory linkages between these and other transcription factors may be inferred from experimental data from other tumor entities, a comprehensive understanding of the transcription factors involved in CRC-associated EMTs and their interplay with each other is still missing.

4.4 The Regulatory Network of Interactions Between EMT-TFs and MicroRNAs in CRC

MicroRNAs (miRNAs) represent a class of small ~22-nucleotide-long non-coding RNAs that mediate posttranscriptional gene repression by inhibition of translation initiation and/or mRNA degradation via association of a ~7 nucleotide stretch, called seed-sequence, in their 5′-portion with a complementary sequence usually located in the 3′-UTR of the target mRNA. [74]. Since target mRNA recognition is primarily determined by the relatively short seed region, dozens or even hundreds of target mRNAs that harbor the complementary seed-matching sequence can be regulated by a single miRNA [75]. The majority of human protein coding mRNAs harbors evolutionarily conserved miRNA binding sites. Therefore, the genome-wide impact of miRNAs on gene expression is predicted to be widespread [76]. Moreover, most 3′-UTRs of protein-coding mRNAs harbor one or several binding sites for different miRNAs, thereby receiving multiple regulatory inputs which determine their overall mRNA stability and protein translation rates in a combinatorial manner [77–79]. Indeed, different miRNAs may exert their function via cooperative cotargeting of a shared set of mRNAs and/or mRNAs of functionally related proteins, i.e. via targeting multiple mRNAs encoding for proteins of the same cellular pathway [80]. Cooperative cotargeting by different miRNAs may be particularly relevant to understand miRNA function since microRNA-mediated repression has been reported to be rather modest for the majority of individual mRNA targets [81, 82].

A growing number of miRNAs have been identified as critical regulators of EMT, most prominently by direct regulatory interactions with mRNAs enconding EMT-TFs [83]. We performed a comprehensive survey of experimentally validated regulatory interactions between miRNAs, EMT-TFs, epithelial-specific TFs and effectors of EMT-inducing signals (i.e. β-Catenin/TCF/LEF), SMADs, HIF1α, NFkB and STAT3) based on the currently available literature and generated a chromosome-based interaction map, which illustrates the extensive regulatory network of miRNAs and transcription factors involved in the regulation of EMT (Fig. 4.3, also see Fig. 4.1). Since microRNA-mediated regulation of many of the more recently described EMT-TFs has not been analyzed in detail so far, posttranscriptional microRNA-mediated control of EMT-TFs is presumably even more pervasive.

Although not all microRNAs and transcription factors shown are likely to be involved in EMT in CRC, a substantial number of regulatory interactions is presumably conserved across tumor entities, but has not formally been validated in cell culture or mouse models of CRC. Therefore, the miRNA-TF network operative in CRC-associated EMT is probably far more complex than illustrated here. Conversely, the role of many miRNAs and transcription factors within the CRC-specific subnetwork and

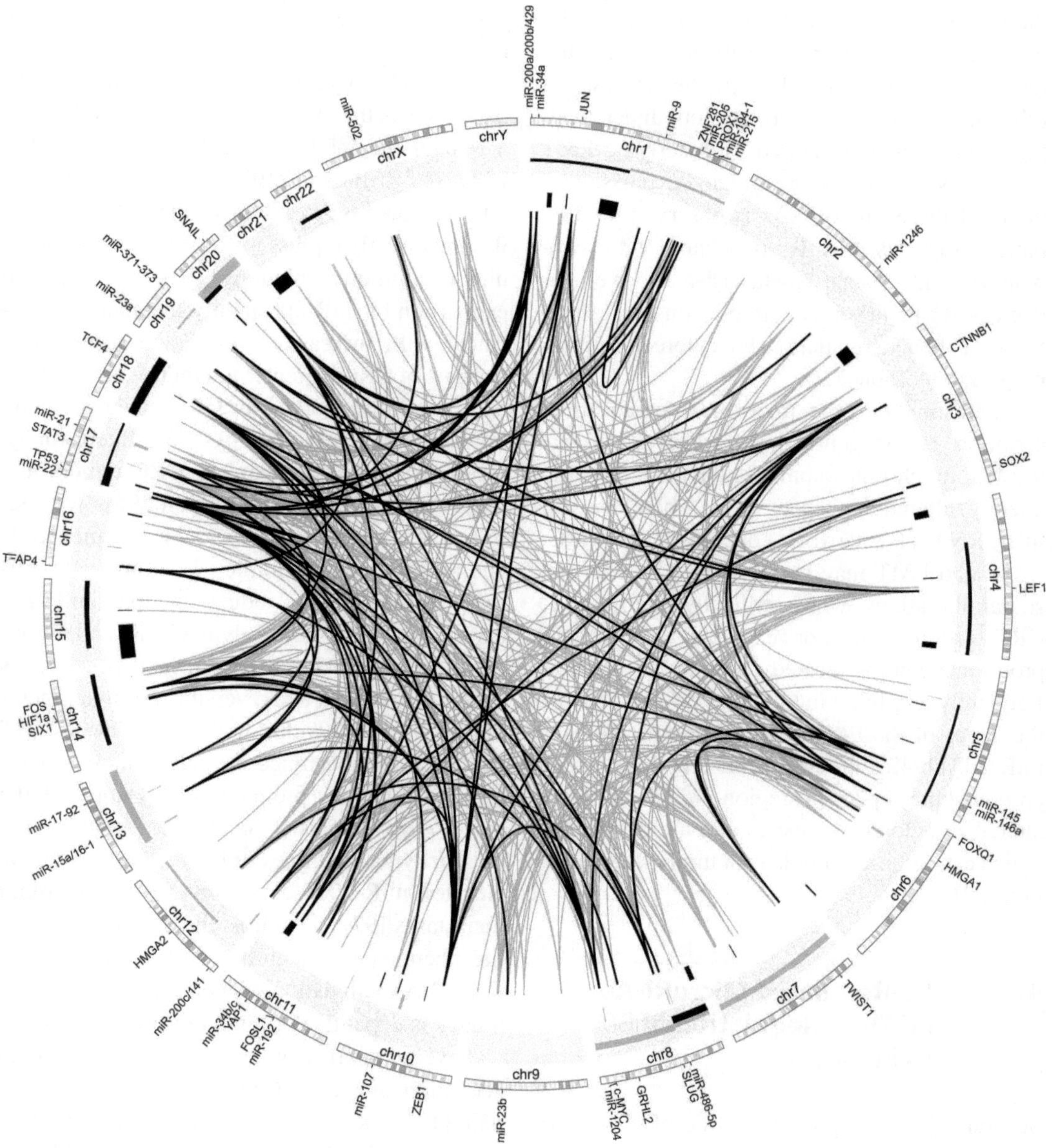

Fig. 4.3 The regulatory network of interactions between EMT-TFs and microRNAs in CRC. Experimentally validated regulatory connections between microRNAs and transcription factors involved in EMT from various tumor entities, as inferred from literature, are displayed as grey lines. Experimentally validated transcriptional regulation of miRNAs by transcription factors involved in EMT in CRC and experimentally validated transcriptional regulation between different TFs involved in EMT in CRC, as shown in Fig. 4.2, are displayed as black lines. The names of TFs and miRNAs involved in EMT in CRC are indicated. The two inner rings highlight chromosomal regions commonly altered in CRC . The first inner ring displays alterations of chromosomal arms occurring with statistical significance (amplifications are shown in grey, deletions in black). Alteration frequencies are displayed as bar height. The second inner ring displays focal copy number number alterations (CNAs) occurring with statistical significance (amplifications are shown in grey, deletions in black). The frequencies of chromosome arm alterations and the genomic coordinates of focal CNAs were obtained from [86]. The figure was generated with Circos [134]

their role in CRC progression is still not fully understood, and systematic validation of regulatory interactions derived from the analysis of other tumor entities may broaden our understanding of the regulatory networks underlying EMT in CRC. Nevertheless, deregulated expression of many of these microRNAs is observed in CRC patient samples and is associated with lymphnode and distant metastasis, as well as decreased patient survival, implicating the importance of their deregulation for colorectal cancer progression (Table 4.2).

Interestingly, some microRNA and/or transcription factor genes involved in EMT are located in chromosomal regions that are frequently altered in cancer [84]. Therefore, the microRNA-TF network operative in CRC-associated EMT may be modulated by chromosomal alterations, which critically contribute to CRC progression. For example, the c-MYC proto-oncogene is located on 8q24.21, which is frequently amplified in CRC [85, 86]. Moreover, the genes of the EMT-suppressive miR-34a and miR-200a/b/429 microRNAs are located on chromosome 1p36, a region which is often altered in colorectal cancers and the deletion of which can serve as a marker for tumor dissemination [87].

4.5 p53-Regulated MicroRNAs and Their Regulation of EMT-TFs in CRC

Several EMT-suppressing microRNAs are engaged in double-negative feedback loops with individual EMT-TFs in CRC and other tumor entities, thereby forming bistable switches that regulate the transitions from epithelial to mesenchymal cell states and *vice versa.* Strikingly, the microRNAs engaged in double-negative feedback loops with EMT-TFs are transcriptionally activated by the p53 tumor suppressor protein, indicating that activation of p53 interferes with these regulatory feedback-loops by shifting the equilibrium towards the miRNAs, which favor the epithelial state (Fig. 4.4). At least in the early stages of the invasion-metastasis cascade this may result in tumor suppression by preventing EMT.

The p53-inducible miR-200 microRNA family served as the inital example of a miRNA with a role in the inhibition of EMT [88–90]. The EMT-TFs ZEB1 and ZEB2 are posttranscriptionally regulated by members of the miR-200 family, and directly repress miR-200 transcription by direct promoter binding [91–94]. Elevated expression of miR-200 family members is associated with increased overall and disease-free survival of CRC patients, and represents a beneficial prognostic marker for CRC patients receiving adjuvant chemotherapy [95]. Interestingly, the invasion front of primary CRC tumors displays low miR-200c expression [96], whereas liver metastatic tissues have higher miR-200c and miR-141 expression levels compared to primary tumors [97], indicating that miR-200 expression levels may be dynamically and reversibly modulated during the invasion-metastasis cascade. Transient and reversible promoter methylation of *miR-200* during TGF-β induced EMT has been demonstrated in cell culture models [98]. In addition, promoter methylation of the *miR-200c/141* cluster occurs in permanently mesenchymal cell lines [99]. However, given the elevated miR-200c expression levels in distant metastases [97], as well as the primarily epithelial phenotype of metastases [100], it is likely that DNA methylation of the *miR-200* gene promoters is dynamic during tumor progression.

Several reports have shown the transcriptional activation of the miR-34 microRNA family by p53 [101–106], and members of the miR-34 microRNA family have been firmly established to act as tumor suppressors in various cancer types, at least in part, by mediating the tumor suppressive function of p53 [22, 23]. Several TFs involved in the regulation of EMT have been shown to be targeted by members of the miR-34 microRNA family. For example, the SNAIL transcription factor has been shown to be a direct target of the miR-34a/b/c [107, 108], and conversely represses miR-34a transcription [108]. In addition, c-MYC has been shown to be regulated by miR-34b and –c in colorectal cancer cells [109]. Moreover, the zinc-finger protein ZNF281, which

Table 4.2 MiRNAs with roles in EMT and CRC

microRNA	Role in CRC progression and metastasis	References
EMT suppressive miRNAs		
p53-induced :		
miR-200a/b/c	Elevated expression of miR-200 family members is associated with increased overall and disease-free survival	[95–97, 174]
miR-141		
miR-429	Beneficial prognostic marker for CRC patients receiveing adjuvant chemotherapy	
	Low expression at invasion fronts of primary CRC tumors	
	Elevated expression in liver metastatic tissues compared to primary tumors	
miR-34a/b/c	Downregulation of miR-34a and miR-34b/c expression and high frequency of silencing by CpG methylation is strongly associated with distant metastasis and poor survival	[113, 115, 117]
miR-145	Downregulation in CRC, downregulation associated with metastasis	[119, 120]
	Decreased expression at the tumor invasion front	
miR-15/16	Downregulation associated with advanced TNM stage, poor histological grade, positive lymph node metastasis, shorter overall and disease-free survival	[124]
miR-192/194/215	Downregulated in CRC tissues, downregulation associated with increased tumor size	[120, 175–179]
	Downregulation associated with advanced tumor stage	
miR-205	Downregulation in a subgroup of colorectal tumors correlating with lymphnode metastasis	[180]
Other miRNAs **:**		
miR-101	Low miR-101 expression in CR tumors; miR-101 expression associated with tumor grade	[181, 182]
miR-124	Downregulated in ulcerative colitis; Downregulation in CRC associated with shorter overall and disease-free survival	[183–185]
miR-137	Downregulated in CRC tissues compared to normal colonic mucosa	[186–188]
	Frequently downregulated by promoter methylation in primary tumors	
miR-138	Downregulation associated with lymph node metastasis, distant metastasis and poor prognosis	[189]
miR-203	Decreased levels in serum of CRC patients	[190, 191]
	Decreased expression in CRC tumor tissues correlated with tumor stage	
miR-204-5p	Downregulated in CRC compared to normal mucosa, downregulation associated with poor patient prognosis	[192]
miR-30 family	miR-30a : downregulated in metastatic tissues; miR-30b : downregulated in primary CRC; downregulated in liver metastases compared to primary tumors	[193–195]
EMT promoting miRNAs		
miR-17-92 cluster :		
hsa-mir-17	High expression associated with shorter overall survival	[196]
hsa-mir-18a	Elevated expresssion in CRC compared normal mucosa,	[197]
	Elevated plasma levels in CRC patients	
hsa-mir-19a	High expression associated with lymph node metastasis	[198]
hsa-mir-20a	High expression associated with lymph node metastases and distant metastases	[199]

(continued)

Table 4.2 (continued)

microRNA	Role in CRC progression and metastasis	References
hsa-mir-92a-1	High expression in carcinoma	[200]
miR-103/107	High expression associated with lymph node and distant metastasis, metastasis recurrence and poor survival	[201]
miR-10b	High expression associated with distant metastasis	[202–204]
	High expression with lymph node metastasis and higher tumor grade	
miR-155	High expression associated with lymph node metastases and lower overall and progression-free survival	[205, 206]
	High expression associated with tumor grade, TNM staging and distant metastasis	
miR-21	Upregulated in CRC and colitis-associated colon cancer;	[205, 207, 208]
	High expression associated with metastasis	
	High expression associated with advanced tumor stage	
miR-9	High expression associated with distant metastasis	[209]
miR-96-182-183 cluster :		
miR-96	Low expression associated with distant metastasis and poor survival	[210]
miR-182	High expression associated with lymph node metastases; elevated plasma levels in patients with CRC	[146, 211–213]
	High expression associated with tumor invasion, positive regional lymph node status, and advanced TNM stage	
miR-183	High expression associated with poor survival	[214, 215]
	Elevated plasma levels associated with shorter disease-free survival and lower overall survival	

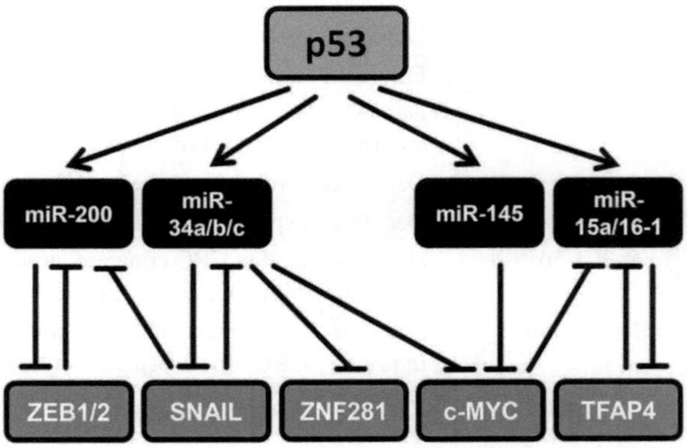

Fig. 4.4 p53-regulated microRNAs mediate inhibition of EMT-TFs in CRC. Regulatory loops between p53-induced microRNAs and transcription factors involved in EMT in CRC. For clarity, transcriptional links between individual EMT-TFs are not shown

is required for c-MYC-induced EMT, represents a direct target of miR-34a [18]. Ectopic expression of miR-34 in CRC cell lines suppresses EMT, migration, invasion and stemness *in vitro* [108, 110] and inhibits metastasis formation in xenograft models [111, 112]. Furthermore, downregulation of expression and a high frequency of silencing by CpG methylation of the *miR-34a* and *miR-34b/c* genes has been shown in CRC and several other tumor entities. Notably, *miR-34a* and *miR-34b/c* silencing has been associated with distant metastasis and poor patient survival [113–117].

The p53-induced miR-145 represses the c-MYC oncoprotein, and mediates cell cycle arrest when ectopically expressed in CRC cell lines [118]. Downregulation of miR-145 is observed in CRC compared to normal colonic mucosa, which is associated with metastasis. Furthermore, decreased expression of miR-145 at the tumor invasion front has been described, which was, however, not significantly correlated with clinical parameters [119, 120].

TFAP4 is directly regulated by p53-induced miR-15a/16-1, and has been shown to transcriptionally repress miR-15a/16-1 expression by binding to the promoter of the *miR-15a/16-1* host gene *DLEU2* [121]. Moreover, MYC represses miR-15a/16-1 expression by direct binding to the *DLEU2* promoter [122]. The microRNAs of the miR-15a/16-1 cluster are known to act as tumor suppressors in various tumor types and target factors with oncogenic potential, such as Bcl-2 and CDK4 [123]. In line with these experimental findings, downregulation of miR-15a/16-1 in primary CRCs is associated with advanced TNM stage, poor histological grade, positive lymph node metastasis and unfavorable overall survival and disease-free survival [124].

4.6 Diverse Regulatory Motifs Between TFs and MicroRNAs in CRC

MicroRNAs which are highly interconnected with the EMT-TF transcriptional network presumably exert their function not only by regulating individual transcription factors, but rather by affecting transcriptional cascades. For example, p53-induced microRNAs repress numerous EMT-TFs. Thereby, they prevent spurious activation of transcriptional cascades involved in EMT and act as a gatekeeper of the epithelial cell state. It has been suggested that miRNAs may contribute to the robustness of cellular states by reinforcing feed-forward and feed-back-loops [125]. Indeed, the EMT-TF-microRNA regulatory networks contain common regulatory motifs between pairs of transcription factors and individual microRNAs which function in feed-forward regulation of EMT-TF expression (Fig. 4.5).

Several EMT-TFs are engaged in regulatory circuits with miR-34 (Fig. 4.5a–c). SNAIL induces ZNF281 expression in a feed-forward-loop by direct transcriptional activation of ZNF281 and repression of miR-34, which is a negative regulator of ZNF281 [18]. Moreover, miR-34a and miR-34b/c have been shown to form a double-negative-feedback loop with SNAIL [108] (Fig. 4.5a). Furthermore, SNAIL directly induces LEF1 transcription [43], while regulating miR-34 levels via a double-negative-feedback loop, whereas LEF1 is a direct target of miR-34a [82, 107] (Fig. 4.5b). IL6-mediated activation of STAT3 via the IL6 receptor (IL6R) is sustained by another feedback loop involving miR-34-mediated repression of IL6R, direct transcriptional repression of miR-34a by STAT3 [126] (Fig. 4.5c). IL6-mediated activation of STAT3 leads to activation of FOSL1 presumably

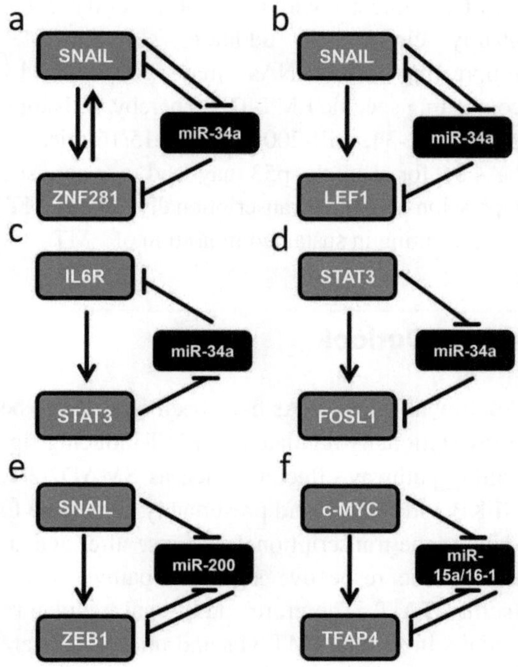

Fig. 4.5 Regulatory circuits composed of EMT-TFs and miRNAs. Regulatory connections between EMT-suppressive miRNAs, such as miR-34, miR-200 and miR-15/16 and transcriptionally linked EMT-TFs are displayed. For details, see main text

by a feed-forward loop that involves direct transcriptional induction of FOSL1 [72], which is a direct miR-34 target [127], and concomitant STAT3-mediated repression of miR-34 [126] (Fig. 4.5d).

Furthermore, the miR-200 microRNA family is engaged in a feed-forward loop regulating the expression of ZEB1. SNAIL induces transcription of ZEB1 [66], and concomitantly represses *miR-200* genes [108], which form double-negative feedback loops with ZEB1 [91, 92], thereby augmenting SNAIL induced expression of ZEB1 (Fig. 4.5e).

Another example is the c-MYC mediated induction of TFAP4 [17, 128], which is accompanied by a c-MYC-mediated repression of miR-15/16 [122]. TFAP4 and miR-15/16 form a double negative feedback loop [121], which presumably results in a robust induction of TFAP4 after activation of c-MYC (Fig. 4.5f).

Since microRNA-mediated repression has been reported to be rather modest for most targets [81, 82], the regulatory effect of a single microRNA may be augmented within the context of these regulatory circuits. In addition, several EMT-suppressing microRNAs may cooperate by cotargeting specific EMT-TFs. Thereby, activation of the miR-34, miR-200 and miR-15/16 microRNAs by, for example, p53 may lead to synergistic repression of several transcriptionally linked EMT-TFs, resulting in sustained inhibition of EMT.

4.7 Outlook

Additional microRNAs have been reported to be transcriptionally regulated by EMT-inducing signalling pathway effectors, such as SMAD2/3/4, NFKB and HIF1α, and presumably contribute to shaping the transcriptional response after activation of the respective signaling pathways and facility EMT. Therefore, additional regulatory circuits involving EMT-TFs and microRNAs are likely to be discovered in the future, some of which may be critically relevant for CRC progression and potentially provide new avenues for prognostication and treatment of colorectal cancer.

The clinical relevance of EMT for the process of tumor metastasis has recently been challenged by several reports showing that conditional ablation of individual EMT-TFs, such as SNAIL or TWIST1 or ectopic expression of miR-200 in murine models of mammary or pancreatic cancer does not inhibit invasion and metastases formation [129, 130]. However, similar studies will be necessary to validate these observations in models of other tumor entities, including CRC. In addition, due to the redundancy in the EMT-regulating networks, it is possible that other EMT-TFs substitute for the activity of the EMT-TFs inactivated in these studies. Notably, the relevance of EMT for conferring resistance to chemotherapy was even further substantiated by these studies [129, 130]. Therefore, current chemotherapeutic treatments may be more effective in combination with a therapeutic inhibition of EMT. Recently, microRNA replacement-based therapies have entered clinical trials. E.g. miR-34a based therapies are tested for treatment of unresectable forms of liver cancer or liver metastases [131, 132]. Interestingly, miR-34a may also harness the immune response towards tumors by inhibiting the checkpoint inhibitor PD-L1 besides promoting MET [133]. Therefore, a thorough understanding of the interactions between microRNAs and EMT-TFs that govern the transitions between epithelial and mesenchymal cell states may become highly relevant for therapeutic strategies in the near future.

References

1. Hanahan D, Weinberg RA. Hallmarks of cancer: the next generation. Cell. 2011;144(5):646–74.
2. Valastyan S, Weinberg RA. Tumor metastasis: molecular insights and evolving paradigms. Cell. 2011;147(2):275–92.
3. Gupta GP, Massague J. Cancer metastasis: building a framework. Cell. 2006;127(4):679–95.
4. Steeg PS. Tumor metastasis: mechanistic insights and clinical challenges. Nat Med. 2006;12(8):895–904.
5. Thiery JP, Acloque H, Huang RY, Nieto MA. Epithelial-mesenchymal transitions in development and disease. Cell. 2009;139(5):871–90.
6. Yang J, Weinberg RA. Epithelial-mesenchymal transition: at the crossroads of development and tumor metastasis. Dev Cell. 2008;14(6):818–29.

7. Peinado H, Olmeda D, Cano A. Snail, Zeb and bHLH factors in tumour progression: an alliance against the epithelial phenotype? Nat Rev Cancer. 2007;7(6):415–28.
8. Joosse SA, Gorges TM, Pantel K. Biology, detection, and clinical implications of circulating tumor cells. EMBO molecular medicine. 2015;7(1):1–11.
9. Mani SA, Guo W, Liao MJ, Eaton EN, Ayyanan A, Zhou AY, et al. The epithelial-mesenchymal transition generates cells with properties of stem cells. Cell. 2008;133(4):704–15.
10. Perez-Moreno MA, Locascio A, Rodrigo I, Dhondt G, Portillo F, Nieto MA, et al. A new role for E12/E47 in the repression of E-cadherin expression and epithelial-mesenchymal transitions. J Biol Chem. 2001;276(29):27424–31.
11. Medici D, Hay ED, Olsen BR. Snail and Slug promote epithelial-mesenchymal transition through beta-catenin-T-cell factor-4-dependent expression of transforming growth factor-beta3. Mol Biol Cell. 2008;19(11):4875–87.
12. Sobrado VR, Moreno-Bueno G, Cubillo E, Holt LJ, Nieto MA, Portillo F, et al. The class I bHLH factors E2-2A and E2-2B regulate EMT. J Cell Sci. 2009;122(Pt 7):1014–24.
13. Sanchez-Tillo E, de Barrios O, Siles L, Cuatrecasas M, Castells A, Postigo A. beta-catenin/TCF4 complex induces the epithelial-to-mesenchymal transition (EMT)-activator ZEB1 to regulate tumor invasiveness. Proc Natl Acad Sci U S A. 2011;108(48):19204–9.
14. Medici D, Hay ED, Goodenough DA. Cooperation between snail and LEF-1 transcription factors is essential for TGF-beta1-induced epithelial-mesenchymal transition. Mol Biol Cell. 2006;17(4):1871–9.
15. Hartwell KA, Muir B, Reinhardt F, Carpenter AE, Sgroi DC, Weinberg RA. The Spemann organizer gene, Goosecoid, promotes tumor metastasis. Proc Natl Acad Sci U S A. 2006;103(50):18969–74.
16. Mani SA, Yang J, Brooks M, Schwaninger G, Zhou A, Miura N, et al. Mesenchyme Forkhead 1 (FOXC2) plays a key role in metastasis and is associated with aggressive basal-like breast cancers. Proc Natl Acad Sci U S A. 2007;104(24):10069–74.
17. Jackstadt R, Roh S, Neumann J, Jung P, Hoffmann R, Horst D, et al. AP4 is a mediator of epithelial-mesenchymal transition and metastasis in colorectal cancer. J Exp Med. 2013;210(7):1331–50.
18. Hahn S, Jackstadt R, Siemens H, Hunten S, Hermeking H. SNAIL and miR-34a feed-forward regulation of ZNF281/ZBP99 promotes epithelial-mesenchymal transition. EMBO J. 2013;32(23):3079–95.
19. Brabletz T. MiR-34 and SNAIL: another double-negative feedback loop controlling cellular plasticity/EMT governed by p53. Cell Cycle. 2012;11(2):215–6.
20. Wang SP, Wang WL, Chang YL, Wu CT, Chao YC, Kao SH, et al. p53 controls cancer cell invasion by inducing the MDM2-mediated degradation of Slug. Nat Cell Biol. 2009;11(6):694–704.
21. Lim SO, Kim H, Jung G. p53 inhibits tumor cell invasion via the degradation of snail protein in hepatocellular carcinoma. FEBS Lett. 2010;584(11):2231–6.
22. Hermeking H. MicroRNAs in the p53 network: micromanagement of tumour suppression. Nat Rev Cancer. 2012;12(9):613–26.
23. Rokavec M, Li H, Jiang L, Hermeking H. The p53/miR-34 axis in development and disease. J Mol Cell Biol. 2014;6(3):214–30.
24. Andreoli JM, Jang SI, Chung E, Coticchia CM, Steinert PM, Markova NG. The expression of a novel, epithelium-specific ets transcription factor is restricted to the most differentiated layers in the epidermis. Nucleic Acids Res. 1997;25(21):4287–95.
25. Oettgen P, Alani RM, Barcinski MA, Brown L, Akbarali Y, Boltax J, et al. Isolation and characterization of a novel epithelium-specific transcription factor, ESE-1, a member of the ets family. Mol Cell Biol. 1997;17(8):4419–33.
26. Lyons JG, Patel V, Roue NC, Fok SY, Soon LL, Halliday GM, et al. Snail up-regulates proinflammatory mediators and inhibits differentiation in oral keratinocytes. Cancer Res. 2008;68(12):4525–30.
27. Wang JL, Chen ZF, Chen HM, Wang MY, Kong X, Wang YC, et al. Elf3 drives beta-catenin transactivation and associates with poor prognosis in colorectal cancer. Cell Death Dis. 2014;5:e1263.
28. Chakrabarti R, Hwang J, Andres Blanco M, Wei Y, Lukacisin M, Romano RA, et al. Elf5 inhibits the epithelial-mesenchymal transition in mammary gland development and breast cancer metastasis by transcriptionally repressing Snail2. Nat Cell Biol. 2012;14(11):1212–22.
29. Wang S, Samakovlis C. Grainy head and its target genes in epithelial morphogenesis and wound healing. Curr Top Dev Biol. 2012;98:35–63.
30. Senga K, Mostov KE, Mitaka T, Miyajima A, Tanimizu N. Grainyhead-like 2 regulates epithelial morphogenesis by establishing functional tight junctions through the organization of a molecular network among claudin3, claudin4, and Rab25. Mol Biol Cell. 2012;23(15):2845–55.
31. Xiang X, Deng Z, Zhuang X, Ju S, Mu J, Jiang H, et al. Grhl2 determines the epithelial phenotype of breast cancers and promotes tumor progression. PLoS One. 2012;7(12):e50781.
32. Werth M, Walentin K, Aue A, Schonheit J, Wuebken A, Pode-Shakked N, et al. The transcription factor grainyhead-like 2 regulates the molecular composition of the epithelial apical junctional complex. Development. 2010;137(22):3835–45.
33. Cieply B, Riley P, Pifer PM, Widmeyer J, Addison JB, Ivanov AV, et al. Suppression of the epithelial-

mesenchymal transition by Grainyhead-like-2. Cancer Res. 2012;72(9):2440–53.
34. Quan Y, Xu M, Cui P, Ye M, Zhuang B, Min Z. Grainyhead-like 2 promotes tumor growth and is associated with poor prognosis in colorectal cancer. J Cancer. 2015;6(4):342–50.
35. Li S, Yang J. Ovol proteins: guardians against EMT during epithelial differentiation. Dev Cell. 2014;29(1):1–2.
36. Roca H, Hernandez J, Weidner S, McEachin RC, Fuller D, Sud S, et al. Transcription factors OVOL1 and OVOL2 induce the mesenchymal to epithelial transition in human cancer. PLoS One. 2013;8(10):e76773.
37. Cieply B, Farris J, Denvir J, Ford HL, Frisch SM. Epithelial-mesenchymal transition and tumor suppression are controlled by a reciprocal feedback loop between ZEB1 and Grainyhead-like-2. Cancer Res. 2013;73(20):6299–309.
38. Stinson S, Lackner MR, Adai AT, Yu N, Kim HJ, O'Brien C, et al. TRPS1 targeting by miR-221/222 promotes the epithelial-to-mesenchymal transition in breast cancer. Sci Signal. 2011;4(177):ra41.
39. De Craene B, Berx G. Regulatory networks defining EMT during cancer initiation and progression. Nat Rev Cancer. 2013;13(2):97–110.
40. Tran DD, Corsa CA, Biswas H, Aft RL, Longmore GD. Temporal and spatial cooperation of Snail1 and Twist1 during epithelial-mesenchymal transition predicts for human breast cancer recurrence. Mol Cancer Res: MCR. 2011;9(12):1644–57.
41. Moreno-Bueno G, Cubillo E, Sarrio D, Peinado H, Rodriguez-Pinilla SM, Villa S, et al. Genetic profiling of epithelial cells expressing E-cadherin repressors reveals a distinct role for Snail, Slug, and E47 factors in epithelial-mesenchymal transition. Cancer Res. 2006;66(19):9543–56.
42. Ye X, Tam WL, Shibue T, Kaygusuz Y, Reinhardt F, Ng Eaton E, et al. Distinct EMT programs control normal mammary stem cells and tumour-initiating cells. Nature. 2015;525(7568):256–60.
43. Solanas G, Porta-de-la-Riva M, Agusti C, Casagolda D, Sanchez-Aguilera F, Larriba MJ, et al. E-cadherin controls beta-catenin and NF-kappaB transcriptional activity in mesenchymal gene expression. J Cell Sci. 2008;121(Pt 13):2224–34.
44. Wang WJ, Yao Y, Jiang LL, Hu TH, Ma JQ, Ruan ZP, et al. Increased LEF1 expression and decreased Notch2 expression are strong predictors of poor outcomes in colorectal cancer patients. Dis Markers. 2013;35(5):395–405.
45. Lin AY, Chua MS, Choi YL, Yeh W, Kim YH, Azzi R, et al. Comparative profiling of primary colorectal carcinomas and liver metastases identifies LEF1 as a prognostic biomarker. PLoS One. 2011;6(2):e16636.
46. Yook JI, Li XY, Ota I, Hu C, Kim HS, Kim NH, et al. A Wnt-Axin2-GSK3beta cascade regulates Snail1 activity in breast cancer cells. Nat Cell Biol. 2006;8(12):1398–406.
47. Hong CF, Chou YT, Lin YS, Wu CW. MAD2B, a novel TCF4-binding protein, modulates TCF4-mediated epithelial-mesenchymal transdifferentiation. J Biol Chem. 2009;284(29):19613–22.
48. Conacci-Sorrell M, Simcha I, Ben-Yedidia T, Blechman J, Savagner P, Ben-Ze'ev A. Autoregulation of E-cadherin expression by cadherin-cadherin interactions: the roles of beta-catenin signaling, Slug, and MAPK. J Cell Biol. 2003;163(4):847–57.
49. Findlay VJ, Wang C, Nogueira LM, Hurst K, Quirk D, Ethier SP, et al. SNAI2 modulates colorectal cancer 5-fluorouracil sensitivity through miR145 repression. Mol Cancer Ther. 2014;13(11):2713–26.
50. Li Y, Zhao Z, Xu C, Zhou Z, Zhu Z, You T. HMGA2 induces transcription factor Slug expression to promote epithelial-to-mesenchymal transition and contributes to colon cancer progression. Cancer Lett. 2014;355(1):130–40.
51. Thuault S, Tan EJ, Peinado H, Cano A, Heldin CH, Moustakas A. HMGA2 and Smads co-regulate SNAIL1 expression during induction of epithelial-to-mesenchymal transition. J Biol Chem. 2008;283(48):33437–46.
52. Thuault S, Valcourt U, Petersen M, Manfioletti G, Heldin CH, Moustakas A. Transforming growth factor-beta employs HMGA2 to elicit epithelial-mesenchymal transition. J Cell Biol. 2006;174(2):175–83.
53. Watanabe S, Ueda Y, Akaboshi S, Hino Y, Sekita Y, Nakao M. HMGA2 maintains oncogenic RAS-induced epithelial-mesenchymal transition in human pancreatic cancer cells. Am J Pathol. 2009;174(3):854–68.
54. Belton A, Gabrovsky A, Bae YK, Reeves R, Iacobuzio-Donahue C, Huso DL, et al. HMGA1 induces intestinal polyposis in transgenic mice and drives tumor progression and stem cell properties in colon cancer cells. PLoS One. 2012;7(1):e30034.
55. Christensen J, Bentz S, Sengstag T, Shastri VP, Anderle P. FOXQ1, a novel target of the Wnt pathway and a new marker for activation of Wnt signaling in solid tumors. PLoS One. 2013;8(3):e60051.
56. Abba M, Patil N, Rasheed K, Nelson LD, Mudduluru G, Leupold JH, et al. Unraveling the role of FOXQ1 in colorectal cancer metastasis. Mol Cancer Res: MCR. 2013;11(9):1017–28.
57. Zhang H, Meng F, Liu G, Zhang B, Zhu J, Wu F, et al. Forkhead transcription factor foxq1 promotes epithelial-mesenchymal transition and breast cancer metastasis. Cancer Res. 2011;71(4):1292–301.
58. Zhao B, Tumaneng K, Guan KL. The Hippo pathway in organ size control, tissue regeneration and stem cell self-renewal. Nat Cell Biol. 2011;13(8):877–83.
59. Konsavage Jr WM, Kyler SL, Rennoll SA, Jin G, Yochum GS. Wnt/beta-catenin signaling regulates Yes-associated protein (YAP) gene expression in

colorectal carcinoma cells. J Biol Chem. 2012;287(15):11730–9.
60. Shao DD, Xue W, Krall EB, Bhutkar A, Piccioni F, Wang X, et al. KRAS and YAP1 converge to regulate EMT and tumor survival. Cell. 2014;158(1):171–84.
61. Lee KW, Lee SS, Kim SB, Sohn BH, Lee HS, Jang HJ, et al. Significant association of oncogene YAP1 with poor prognosis and cetuximab resistance in colorectal cancer patients. Clin Cancer Res. 2015;21(2):357–64.
62. Wang L, Shi S, Guo Z, Zhang X, Han S, Yang A, et al. Overexpression of YAP and TAZ is an independent predictor of prognosis in colorectal cancer and related to the proliferation and metastasis of colon cancer cells. PLoS One. 2013;8(6):e65539.
63. Yuen HF, McCrudden CM, Huang YH, Tham JM, Zhang X, Zeng Q, et al. TAZ expression as a prognostic indicator in colorectal cancer. PLoS One. 2013;8(1):e54211.
64. Zhang W, Shi X, Peng Y, Wu M, Zhang P, Xie R, et al. HIF-1alpha promotes epithelial-mesenchymal transition and metastasis through direct regulation of ZEB1 in colorectal cancer. PLoS One. 2015;10(6):e0129603.
65. Xiong H, Hong J, Du W, Lin YW, Ren LL, Wang YC, et al. Roles of STAT3 and ZEB1 proteins in E-cadherin down-regulation and human colorectal cancer epithelial-mesenchymal transition. J Biol Chem. 2012;287(8):5819–32.
66. Guaita S, Puig I, Franci C, Garrido M, Dominguez D, Batlle E, et al. Snail induction of epithelial to mesenchymal transition in tumor cells is accompanied by MUC1 repression and ZEB1 expression. J Biol Chem. 2002;277(42):39209–16.
67. Ono H, Imoto I, Kozaki K, Tsuda H, Matsui T, Kurasawa Y, et al. SIX1 promotes epithelial-mesenchymal transition in colorectal cancer through ZEB1 activation. Oncogene. 2012;31(47):4923–34.
68. Quan Y, Jin R, Huang A, Zhao H, Feng B, Zang L, et al. Downregulation of GRHL2 inhibits the proliferation of colorectal cancer cells by targeting ZEB1. Cancer Biol Ther. 2014;15(7):878–87.
69. Spaderna S, Schmalhofer O, Wahlbuhl M, Dimmler A, Bauer K, Sultan A, et al. The transcriptional repressor ZEB1 promotes metastasis and loss of cell polarity in cancer. Cancer Res. 2008;68(2):537–44.
70. Spaderna S, Schmalhofer O, Hlubek F, Berx G, Eger A, Merkel S, et al. A transient, EMT-linked loss of basement membranes indicates metastasis and poor survival in colorectal cancer. Gastroenterology. 2006;131(3):830–40.
71. Zhang GJ, Zhou T, Tian HP, Liu ZL, Xia SS. High expression of ZEB1 correlates with liver metastasis and poor prognosis in colorectal cancer. Oncol Lett. 2013;5(2):564–8.
72. Liu H, Ren G, Wang T, Chen Y, Gong C, Bai Y, et al. Aberrantly expressed Fra-1 by IL-6/STAT3 transactivation promotes colorectal cancer aggressiveness through epithelial-mesenchymal transition. Carcinogenesis. 2015;36(4):459–68.
73. Diesch J, Sanij E, Gilan O, Love C, Tran H, Fleming NI, et al. Widespread FRA1-dependent control of mesenchymal transdifferentiation programs in colorectal cancer cells. PLoS One. 2014;9(3):e88950.
74. Fabian MR, Sonenberg N, Filipowicz W. Regulation of mRNA translation and stability by microRNAs. Annu Rev Biochem. 2010;79:351–79.
75. Bartel DP. MicroRNAs: target recognition and regulatory functions. Cell. 2009;136(2):215–33.
76. Friedman RC, Farh KK, Burge CB, Bartel DP. Most mammalian mRNAs are conserved targets of microRNAs. Genome Res. 2009;19(1):92–105.
77. Balaga O, Friedman Y, Linial M. Toward a combinatorial nature of microRNA regulation in human cells. Nucleic Acids Res. 2012;40(19):9404–16.
78. Friedman Y, Balaga O, Linial M. Working together: combinatorial regulation by microRNAs. Adv Exp Med Biol. 2013;774:317–37.
79. Gurtan AM, Sharp PA. The role of miRNAs in regulating gene expression networks. J Mol Biol. 2013;425(19):3582–600.
80. Tsang JS, Ebert MS, van Oudenaarden A. Genome-wide dissection of microRNA functions and cotargeting networks using gene set signatures. Mol Cell. 2010;38(1):140–53.
81. Selbach M, Schwanhausser B, Thierfelder N, Fang Z, Khanin R, Rajewsky N. Widespread changes in protein synthesis induced by microRNAs. Nature. 2008;455(7209):58–63.
82. Kaller M, Liffers ST, Oeljeklaus S, Kuhlmann K, Roh S, Hoffmann R, et al. Genome-wide characterization of miR-34a induced changes in protein and mRNA expression by a combined pulsed SILAC and microarray analysis. Mol Cell Proteomics. 2011;10(8):M111 010462.
83. Diaz-Lopez A, Moreno-Bueno G, Cano A. Role of microRNA in epithelial to mesenchymal transition and metastasis and clinical perspectives. Cancer Manag Res. 2014;6:205–16.
84. Calin GA, Sevignani C, Dumitru CD, Hyslop T, Noch E, Yendamuri S, et al. Human microRNA genes are frequently located at fragile sites and genomic regions involved in cancers. Proc Natl Acad Sci U S A. 2004;101(9):2999–3004.
85. Buffart TE, Coffa J, Hermsen MA, Carvalho B, van der Sijp JR, Ylstra B, et al. DNA copy number changes at 8q11-24 in metastasized colorectal cancer. Cell Oncol. 2005;27(1):57–65.
86. Network TCGA. Comprehensive molecular characterization of human colon and rectal cancer. Nature. 2012;487(7407):330–7.
87. Mayrhofer M, Kultima HG, Birgisson H, Sundstrom M, Mathot L, Edlund K, et al. 1p36 deletion is a marker for tumour dissemination in microsatellite

stable stage II-III colon cancer. BMC Cancer. 2014;14:872.
88. Mongroo PS, Rustgi AK. The role of the miR-200 family in epithelial-mesenchymal transition. Cancer Biol Ther. 2010;10(3):219–22.
89. Chang CJ, Chao CH, Xia W, Yang JY, Xiong Y, Li CW, et al. p53 regulates epithelial-mesenchymal transition and stem cell properties through modulating miRNAs. Nat Cell Biol. 2011;13(3):317–23.
90. Kim T, Veronese A, Pichiorri F, Lee TJ, Jeon YJ, Volinia S, et al. p53 regulates epithelial-mesenchymal transition through microRNAs targeting ZEB1 and ZEB2. J Exp Med. 2011;208(5):875–83.
91. Bracken CP, Gregory PA, Kolesnikoff N, Bert AG, Wang J, Shannon MF, et al. A double-negative feedback loop between ZEB1-SIP1 and the microRNA-200 family regulates epithelial-mesenchymal transition. Cancer Res. 2008;68(19):7846–54.
92. Burk U, Schubert J, Wellner U, Schmalhofer O, Vincan E, Spaderna S, et al. A reciprocal repression between ZEB1 and members of the miR-200 family promotes EMT and invasion in cancer cells. EMBO Rep. 2008;9(6):582–9.
93. Wellner U, Schubert J, Burk UC, Schmalhofer O, Zhu F, Sonntag A, et al. The EMT-activator ZEB1 promotes tumorigenicity by repressing stemness-inhibiting microRNAs. Nat Cell Biol. 2009;11(12):1487–95.
94. Park SM, Gaur AB, Lengyel E, Peter ME. The miR-200 family determines the epithelial phenotype of cancer cells by targeting the E-cadherin repressors ZEB1 and ZEB2. Genes Dev. 2008;22(7):894–907.
95. Diaz T, Tejero R, Moreno I, Ferrer G, Cordeiro A, Artells R, et al. Role of miR-200 family members in survival of colorectal cancer patients treated with fluoropyrimidines. J Surg Oncol. 2014;109(7):676–83.
96. Paterson EL, Kazenwadel J, Bert AG, Khew-Goodall Y, Ruszkiewicz A, Goodall GJ. Down-regulation of the miRNA-200 family at the invasive front of colorectal cancers with degraded basement membrane indicates EMT is involved in cancer progression. Neoplasia. 2013;15(2):180–91.
97. Hur K, Toiyama Y, Takahashi M, Balaguer F, Nagasaka T, Koike J, et al. MicroRNA-200c modulates epithelial-to-mesenchymal transition (EMT) in human colorectal cancer metastasis. Gut. 2013;62(9):1315–26.
98. Davalos V, Moutinho C, Villanueva A, Boque R, Silva P, Carneiro F, et al. Dynamic epigenetic regulation of the microRNA-200 family mediates epithelial and mesenchymal transitions in human tumorigenesis. Oncogene. 2012;31(16):2062–74.
99. Neves R, Scheel C, Weinhold S, Honisch E, Iwaniuk KM, Trompeter HI, et al. Role of DNA methylation in miR-200c/141 cluster silencing in invasive breast cancer cells. BMC Res Notes. 2010;3:219.
100. Brabletz T, Jung A, Spaderna S, Hlubek F, Kirchner T. Opinion: migrating cancer stem cells – an integrated concept of malignant tumour progression. Nat Rev Cancer. 2005;5(9):744–9.
101. Tarasov V, Jung P, Verdoodt B, Lodygin D, Epanchintsev A, Menssen A, et al. Differential regulation of microRNAs by p53 revealed by massively parallel sequencing: miR-34a is a p53 target that induces apoptosis and G1-arrest. Cell Cycle. 2007;6(13):1586–93.
102. Chang TC, Wentzel EA, Kent OA, Ramachandran K, Mullendore M, Lee KH, et al. Transactivation of miR-34a by p53 broadly influences gene expression and promotes apoptosis. Mol Cell. 2007;26(5):745–52.
103. He L, He X, Lim LP, de Stanchina E, Xuan Z, Liang Y, et al. A microRNA component of the p53 tumour suppressor network. Nature. 2007;447(7148):1130–4.
104. Bommer GT, Gerin I, Feng Y, Kaczorowski AJ, Kuick R, Love RE, et al. p53-mediated activation of miRNA34 candidate tumor-suppressor genes. Curr Biol. 2007;17(15):1298–307.
105. Raver-Shapira N, Marciano E, Meiri E, Spector Y, Rosenfeld N, Moskovits N, et al. Transcriptional activation of miR-34a contributes to p53-mediated apoptosis. Mol Cell. 2007;26(5):731–43.
106. Corney DC, Flesken-Nikitin A, Godwin AK, Wang W, Nikitin AY. MicroRNA-34b and MicroRNA-34c are targets of p53 and cooperate in control of cell proliferation and adhesion-independent growth. Cancer Res. 2007;67(18):8433–8.
107. Kim NH, Kim HS, Li XY, Lee I, Choi HS, Kang SE, et al. A p53/miRNA-34 axis regulates Snail1-dependent cancer cell epithelial-mesenchymal transition. J Cell Biol. 2011;195(3):417–33.
108. Siemens H, Jackstadt R, Hunten S, Kaller M, Menssen A, Gotz U, et al. miR-34 and SNAIL form a double-negative feedback loop to regulate epithelial-mesenchymal transitions. Cell Cycle. 2011;10(24):4256–71.
109. Kress TR, Cannell IG, Brenkman AB, Samans B, Gaestel M, Roepman P, et al. The MK5/PRAK kinase and Myc form a negative feedback loop that is disrupted during colorectal tumorigenesis. Mol Cell. 2011;41(4):445–57.
110. Siemens H, Jackstadt R, Kaller M, Hermeking H. Repression of c-Kit by p53 is mediated by miR-34 and is associated with reduced chemoresistance, migration and stemness. Oncotarget. 2013;4:1399–415.
111. Yang S, Li Y, Gao J, Zhang T, Li S, Luo A, et al. MicroRNA-34 suppresses breast cancer invasion and metastasis by directly targeting Fra-1. Oncogene. 2012.
112. Yan K, Gao J, Yang T, Ma Q, Qiu X, Fan Q, et al. MicroRNA-34a inhibits the proliferation and metastasis of osteosarcoma cells both in vitro and in vivo. PLoS One. 2012;7(3):e33778.

113. Lodygin D, Tarasov V, Epanchintsev A, Berking C, Knyazeva T, Korner H, et al. Inactivation of miR-34a by aberrant CpG methylation in multiple types of cancer. Cell Cycle. 2008;7(16):2591–600.
114. Lujambio A, Calin GA, Villanueva A, Ropero S, Sanchez-Cespedes M, Blanco D, et al. A microRNA DNA methylation signature for human cancer metastasis. Proc Natl Acad Sci U S A. 2008;105(36):13556–61.
115. Siemens H, Neumann J, Jackstadt R, Mansmann U, Horst D, Kirchner T, et al. Detection of miR-34a promoter methylation in combination with elevated expression of c-Met and beta-catenin predicts distant metastasis of colon cancer. Clin Cancer Res. 2013;19(3):710–20.
116. Thorstensen L, Qvist H, Heim S, Liefers GJ, Nesland JM, Giercksky KE, et al. Evaluation of 1p losses in primary carcinomas, local recurrences and peripheral metastases from colorectal cancer patients. Neoplasia. 2000;2(6):514–22.
117. Vogt M, Munding J, Gruner M, Liffers ST, Verdoodt B, Hauk J, et al. Frequent concomitant inactivation of miR-34a and miR-34b/c by CpG methylation in colorectal, pancreatic, mammary, ovarian, urothelial, and renal cell carcinomas and soft tissue sarcomas. Virchows Arch. 2011;458:313–22.
118. Sachdeva M, Zhu S, Wu F, Wu H, Walia V, Kumar S, et al. p53 represses c-MYC through induction of the tumor suppressor miR-145. Proc Natl Acad Sci U S A. 2009;106(9):3207–12.
119. Feng Y, Zhu J, Ou C, Deng Z, Chen M, Huang W, et al. MicroRNA-145 inhibits tumour growth and metastasis in colorectal cancer by targeting fascin-1. Br J Cancer. 2014;110(9):2300–9.
120. Kahlert C, Klupp F, Brand K, Lasitschka F, Diederichs S, Kirchberg J, et al. Invasion front-specific expression and prognostic significance of microRNA in colorectal liver metastases. Cancer Sci. 2011;102(10):1799–807.
121. Shi L, Jackstadt R, Siemens H, Li H, Kirchner T, Hermeking H. p53-induced miR-15a/16-1 and AP4 form a double-negative feedback loop to regulate epithelial-mesenchymal transition and metastasis in colorectal cancer. Cancer Res. 2014;74(2):532–42.
122. Xue G, Yan HL, Zhang Y, Hao LQ, Zhu XT, Mei Q, et al. c-Myc-mediated repression of miR-15-16 in hypoxia is induced by increased HIF-2alpha and promotes tumor angiogenesis and metastasis by upregulating FGF2. Oncogene. 2015;34(11):1393–406.
123. Aqeilan RI, Calin GA, Croce CM. miR-15a and miR-16-1 in cancer: discovery, function and future perspectives. Cell Death Differ. 2010;17(2):215–20.
124. Xiao G, Tang H, Wei W, Li J, Ji L, Ge J. Aberrant expression of MicroRNA-15a and MicroRNA-16 synergistically associates with tumor progression and prognosis in patients with colorectal cancer. Gastroenterol Res Pract. 2014;2014:364549.
125. Ebert MS, Sharp PA. Roles for microRNAs in conferring robustness to biological processes. Cell. 2012;149(3):515–24.
126. Rokavec M, Oner MG, Li H, Jackstadt R, Jiang L, Lodygin D, et al. IL-6R/STAT3/miR-34a feedback loop promotes EMT-mediated colorectal cancer invasion and metastasis. J Clin Invest. 2014;124(4):1853–67.
127. Wu J, Wu G, Lv L, Ren YF, Zhang XJ, Xue YF, et al. MicroRNA-34a inhibits migration and invasion of colon cancer cells via targeting to Fra-1. Carcinogenesis. 2012;33(3):519–28.
128. Jung P, Menssen A, Mayr D, Hermeking H. AP4 encodes a c-MYC-inducible repressor of p21. Proc Natl Acad Sci U S A. 2008;105(39):15046–51.
129. Zheng X, Carstens JL, Kim J, Scheible M, Kaye J, Sugimoto H, et al. Epithelial-to-mesenchymal transition is dispensable for metastasis but induces chemoresistance in pancreatic cancer. Nature. 2015;527:525–30.
130. Fischer KR, Durrans A, Lee S, Sheng J, Li F, Wong ST, et al. Epithelial-to-mesenchymal transition is not required for lung metastasis but contributes to chemoresistance. Nature. 2015;527:472–6.
131. Bader AG. miR-34 - a microRNA replacement therapy is headed to the clinic. Front Genet. 2012;3:120.
132. Agostini M, Knight RA. miR-34: from bench to bedside. Oncotarget. 2014;5(4):872–81.
133. Cortez MA, Ivan C, Valdecanas D, Wang X, Peltier HJ, Ye Y, et al. PDL1 regulation by p53 via miR-34. J Nat Cancer Inst. 2016;108(1).
134. Krzywinski M, Schein J, Birol I, Connors J, Gascoyne R, Horsman D, et al. Circos: an information aesthetic for comparative genomics. Genome Res. 2009;19(9):1639–45.
135. Roy HK, Smyrk TC, Koetsier J, Victor TA, Wali RK. The transcriptional repressor SNAIL is overexpressed in human colon cancer. Dig Dis Sci. 2005;50(1):42–6.
136. Fan XJ, Wan XB, Yang ZL, Fu XH, Huang Y, Chen DK, et al. Snail promotes lymph node metastasis and Twist enhances tumor deposit formation through epithelial-mesenchymal transition in colorectal cancer. Hum Pathol. 2013;44(2):173–80.
137. Kim YH, Kim G, Kwon CI, Kim JW, Park PW, Hahm KB. TWIST1 and SNAI1 as markers of poor prognosis in human colorectal cancer are associated with the expression of ALDH1 and TGF-beta1. Oncol Rep. 2014;31(3):1380–8.
138. Shioiri M, Shida T, Koda K, Oda K, Seike K, Nishimura M, et al. Slug expression is an independent prognostic parameter for poor survival in colorectal carcinoma patients. Br J Cancer. 2006;94(12):1816–22.
139. Gomez I, Pena C, Herrera M, Munoz C, Larriba MJ, Garcia V, et al. TWIST1 is expressed in colorectal carcinomas and predicts patient survival. PLoS One. 2011;6(3):e18023.

140. Okada T, Suehiro Y, Ueno K, Mitomori S, Kaneko S, Nishioka M, et al. TWIST1 hypermethylation is observed frequently in colorectal tumors and its overexpression is associated with unfavorable outcomes in patients with colorectal cancer. Genes, chromosomes & cancer. 2010;49(5):452–62.
141. Kahlert C, Lahes S, Radhakrishnan P, Dutta S, Mogler C, Herpel E, et al. Overexpression of ZEB2 at the invasion front of colorectal cancer is an independent prognostic marker and regulates tumor invasion in vitro. Clin Cancer Res. 2011;17(24):7654–63.
142. Kilic N, Feldhaus S, Kilic E, Tennstedt P, Wicklein D, Wasielewski R, et al. Brachyury expression predicts poor prognosis at early stages of colorectal cancer. Eur J Cancer. 2011;47(7):1080–5.
143. Cui YM, Jiao HL, Ye YP, Chen CM, Wang JX, Tang N, et al. FOXC2 promotes colorectal cancer metastasis by directly targeting MET. Oncogene. 2015;34(33):4379–90.
144. Watanabe T, Kobunai T, Yamamoto Y, Matsuda K, Ishihara S, Nozawa K, et al. Gene expression of mesenchyme forkhead 1 (FOXC2) significantly correlates with the degree of lymph node metastasis in colorectal cancer. Int Surg. 2011;96(3):207–16.
145. Li Q, Wu J, Wei P, Xu Y, Zhuo C, Wang Y, et al. Overexpression of forkhead Box C2 promotes tumor metastasis and indicates poor prognosis in colon cancer via regulating epithelial-mesenchymal transition. Am J Cancer Res. 2015;5(6):2022–34.
146. Zhang Y, Wang X, Wang Z, Tang H, Fan H, Guo Q. miR-182 promotes cell growth and invasion by targeting forkhead box F2 transcription factor in colorectal cancer. Oncol Rep. 2015;33(5):2592–8.
147. Chu XY, Zhu ZM, Chen LB, Wang JH, Su QS, Yang JR, et al. FOXM1 expression correlates with tumor invasion and a poor prognosis of colorectal cancer. Acta Histochem. 2012;114(8):755–62.
148. Zhang HG, Xu XW, Shi XP, Han BW, Li ZH, Ren WH, et al. Overexpression of forkhead box protein M1 (FOXM1) plays a critical role in colorectal cancer. Clin Transl Oncology. 2015;18:527–32.
149. Kaneda H, Arao T, Tanaka K, Tamura D, Aomatsu K, Kudo K, et al. FOXQ1 is overexpressed in colorectal cancer and enhances tumorigenicity and tumor growth. Cancer Res. 2010;70(5):2053–63.
150. Wang X, Liu X, Li AY, Chen L, Lai L, Lin HH, et al. Overexpression of HMGA2 promotes metastasis and impacts survival of colorectal cancers. Clin Cancer Res. 2011;17(8):2570–80.
151. He TC, Sparks AB, Rago C, Hermeking H, Zawel L, da Costa LT, et al. Identification of c-MYC as a target of the APC pathway. Science. 1998;281(5382):1509–12.
152. Lee KS, Kwak Y, Nam KH, Kim DW, Kang SB, Choe G, et al. c-MYC copy-number gain is an independent prognostic factor in patients with colorectal cancer. PLoS One. 2015;10(10):e0139727.
153. Takatsuno Y, Mimori K, Yamamoto K, Sato T, Niida A, Inoue H, et al. The rs6983267 SNP is associated with MYC transcription efficiency, which promotes progression and worsens prognosis of colorectal cancer. Ann Surg Oncol. 2013;20(4):1395–402.
154. Rochlitz CF, Herrmann R, de Kant E. Overexpression and amplification of c-myc during progression of human colorectal cancer. Oncology. 1996;53(6):448–54.
155. Sato K, Miyahara M, Saito T, Kobayashi M. c-myc mRNA overexpression is associated with lymph node metastasis in colorectal cancer. Eur J Cancer. 1994;30A(8):1113–7.
156. Tripathi MK, Deane NG, Zhu J, An H, Mima S, Wang X, et al. Nuclear factor of activated T-cell activity is associated with metastatic capacity in colon cancer. Cancer Res. 2014;74(23):6947–57.
157. Skog M, Bono P, Lundin M, Lundin J, Louhimo J, Linder N, et al. Expression and prognostic value of transcription factor PROX1 in colorectal cancer. Br J Cancer. 2011;105(9):1346–51.
158. Takahashi Y, Sawada G, Kurashige J, Uchi R, Matsumura T, Ueo H, et al. Paired related homoeobox 1, a new EMT inducer, is involved in metastasis and poor prognosis in colorectal cancer. Br J Cancer. 2013;109(2):307–11.
159. Zhang Y, Tian X, Ji H, Guan X, Xu W, Dong B, et al. Expression of SATB1 promotes the growth and metastasis of colorectal cancer. PLoS One. 2014;9(6):e100413.
160. Kahlert C, Lerbs T, Pecqueux M, Herpel E, Hoffmeister M, Jansen L, et al. Overexpression of SIX1 is an independent prognostic marker in stage I-III colorectal cancer. Int J Cancer. 2015;137(9):2104–13.
161. Neumann J, Bahr F, Horst D, Kriegl L, Engel J, Luque RM, et al. SOX2 expression correlates with lymph-node metastases and distant spread in right-sided colon cancer. BMC Cancer. 2011;11:518.
162. Han X, Fang X, Lou X, Hua D, Ding W, Foltz G, et al. Silencing SOX2 induced mesenchymal-epithelial transition and its expression predicts liver and lymph node metastasis of CRC patients. PLoS One. 2012;7(8):e41335.
163. Andersen CL, Christensen LL, Thorsen K, Schepeler T, Sorensen FB, Verspaget HW, et al. Dysregulation of the transcription factors SOX4, CBFB and SMARCC1 correlates with outcome of colorectal cancer. Br J Cancer. 2009;100(3):511–23.
164. Chen J, Ju HL, Yuan XY, Wang TJ, Lai BQ. SOX4 is a potential prognostic factor in human cancers: a systematic review and meta-analysis. Clin Trans Oncol. 2015;18(1):65–72.
165. Matheu A, Collado M, Wise C, Manterola L, Cekaite L, Tye AJ, et al. Oncogenicity of the developmental transcription factor Sox9. Cancer Res. 2012;72(5):1301–15.
166. Lu B, Fang Y, Xu J, Wang L, Xu F, Xu E, et al. Analysis of SOX9 expression in colorectal cancer. Am J Clin Pathol. 2008;130(6):897–904.

167. Kusaba T, Nakayama T, Yamazumi K, Yakata Y, Yoshizaki A, Inoue K, et al. Activation of STAT3 is a marker of poor prognosis in human colorectal cancer. Oncol Rep. 2006;15(6):1445–51.
168. Shan ZZ, Yan XB, Yan LL, Tian Y, Meng QC, Qiu WW, et al. Overexpression of Tbx3 is correlated with Epithelial-Mesenchymal Transition phenotype and predicts poor prognosis of colorectal cancer. Am J Cancer Res. 2015;5(1):344–53.
169. Kriegl L, Horst D, Reiche JA, Engel J, Kirchner T, Jung A. LEF-1 and TCF4 expression correlate inversely with survival in colorectal cancer. J Transl Med. 2010;8:123.
170. Ma F, Bi L, Yang G, Zhang M, Liu C, Zhao Y, et al. ZNF703 promotes tumor cell proliferation and invasion and predicts poor prognosis in patients with colorectal cancer. Oncol Rep. 2014;32(3):1071–7.
171. Nakarai C, Osawa K, Matsubara N, Ikeuchi H, Yamano T, Okamura S, et al. Significance of ELF3 mRNA expression for detection of lymph node metastases of colorectal cancer. Anticancer Res. 2012;32(9):3753–8.
172. Wei D, Kanai M, Huang S, Xie K. Emerging role of KLF4 in human gastrointestinal cancer. Carcinogenesis. 2006;27(1):23–31.
173. Cui J, Shi M, Quan M, Xie K. Regulation of EMT by KLF4 in gastrointestinal cancer. Curr Cancer Drug Targets. 2013;13(9):986–95.
174. Toiyama Y, Hur K, Tanaka K, Inoue Y, Kusunoki M, Boland CR, et al. Serum miR-200c is a novel prognostic and metastasis-predictive biomarker in patients with colorectal cancer. Ann Surg. 2014;259(4):735–43.
175. Chiang Y, Song Y, Wang Z, Liu Z, Gao P, Liang J, et al. microRNA-192, −194 and −215 are frequently downregulated in colorectal cancer. Exp Ther Med. 2012;3(3):560–6.
176. Geng L, Chaudhuri A, Talmon G, Wisecarver JL, Are C, Brattain M, et al. MicroRNA-192 suppresses liver metastasis of colon cancer. Oncogene. 2014;33(46):5332–40.
177. Basati G, Razavi AE, Pakzad I, Malayeri FA. Circulating levels of the miRNAs, miR-194, and miR-29b, as clinically useful biomarkers for colorectal cancer. Tumour Biol. 2015;37(2):1781–8.
178. Karaayvaz M, Pal T, Song B, Zhang C, Georgakopoulos P, Mehmood S, et al. Prognostic significance of miR-215 in colon cancer. Clin Colorectal Cancer. 2011;10(4):340–7.
179. Slattery ML, Herrick JS, Mullany LE, Valeri N, Stevens J, Caan BJ, et al. An evaluation and replication of miRNAs with disease stage and colorectal cancer-specific mortality. Int J Cancer. 2015;137(2):428–38.
180. Orang AV, Safaralizadeh R, Hosseinpour Feizi MA, Somi MH. Diagnostic and prognostic value of miR-205 in colorectal cancer. Asian Pac J Cancer Prev. 2014;15(9):4033–7.
181. Strillacci A, Valerii MC, Sansone P, Caggiano C, Sgromo A, Vittori L, et al. Loss of miR-101 expression promotes Wnt/beta-catenin signalling pathway activation and malignancy in colon cancer cells. J Pathol. 2013;229(3):379–89.
182. Strillacci A, Griffoni C, Sansone P, Paterini P, Piazzi G, Lazzarini G, et al. MiR-101 downregulation is involved in cyclooxygenase-2 overexpression in human colon cancer cells. Exp Cell Res. 2009;315(8):1439–47.
183. Koukos G, Polytarchou C, Kaplan JL, Morley-Fletcher A, Gras-Miralles B, Kokkotou E, et al. MicroRNA-124 regulates STAT3 expression and is down-regulated in colon tissues of pediatric patients with ulcerative colitis. Gastroenterology. 2013;145(4):842–52 e2.
184. Wang MJ, Li Y, Wang R, Wang C, Yu YY, Yang L, et al. Downregulation of microRNA-124 is an independent prognostic factor in patients with colorectal cancer. Int J Color Dis. 2013;28(2):183–9.
185. Jinushi T, Shibayama Y, Kinoshita I, Oizumi S, Jinushi M, Aota T, et al. Low expression levels of microRNA-124-5p correlated with poor prognosis in colorectal cancer via targeting of SMC4. Cancer Med. 2014;3(6):1544–52.
186. Bandres E, Agirre X, Bitarte N, Ramirez N, Zarate R, Roman-Gomez J, et al. Epigenetic regulation of microRNA expression in colorectal cancer. Int J Cancer . 2009;125(11):2737–43.
187. Steponaitiene R, Kupcinskas J, Langner C, Balaguer F, Venclauskas L, Pauzas H, et al. Epigenetic silencing of miR-137 is a frequent event in gastric carcinogenesis. Mol Carcinog. 2015.
188. Balaguer F, Link A, Lozano JJ, Cuatrecasas M, Nagasaka T, Boland CR, et al. Epigenetic silencing of miR-137 is an early event in colorectal carcinogenesis. Cancer Res. 2010;70(16):6609–18.
189. Long L, Huang G, Zhu H, Guo Y, Liu Y, Huo J. Down-regulation of miR-138 promotes colorectal cancer metastasis via directly targeting TWIST2. J Transl Med. 2013;11:275.
190. Wang J, Huang SK, Zhao M, Yang M, Zhong JL, Gu YY, et al. Identification of a circulating microRNA signature for colorectal cancer detection. PLoS One. 2014;9(4):e87451.
191. Chiang Y, Song Y, Wang Z, Chen Y, Yue Z, Xu H, et al. Aberrant expression of miR-203 and its clinical significance in gastric and colorectal cancers. J Gastrointest Surg. 2011;15(1):63–70.
192. Yin Y, Zhang B, Wang W, Fei B, Quan C, Zhang J, et al. miR-204-5p inhibits proliferation and invasion and enhances chemotherapeutic sensitivity of colorectal cancer cells by downregulating RAB22A. Clin Cancer Res. 2014;20(23):6187–99.
193. Zhong M, Bian Z, Wu Z. miR-30a suppresses cell migration and invasion through downregulation of PIK3CD in colorectal carcinoma. Cell Physiol Biochem. 2013;31(2–3):209–18.

194. Zhang Q, Tang Q, Qin D, Yu L, Huang R, Lv G, et al. Role of microRNA 30a targeting insulin receptor substrate 2 in colorectal tumorigenesis. Mol Cell Biol. 2015;35(6):988–1000.
195. Zhao H, Xu Z, Qin H, Gao Z, Gao L. miR-30b regulates migration and invasion of human colorectal cancer via SIX1. Biochem J. 2014;460(1):117–25.
196. Ma Y, Zhang P, Wang F, Zhang H, Yang Y, Shi C, et al. Elevated oncofoetal miR-17-5p expression regulates colorectal cancer progression by repressing its target gene P130. Nat Commun. 2012;3:1291.
197. Zhang GJ, Zhou T, Liu ZL, Tian HP, Xia SS. Plasma miR-200c and miR-18a as potential biomarkers for the detection of colorectal carcinoma. Mol Clin Oncol. 2013;1(2):379–84.
198. Huang L, Wang X, Wen C, Yang X, Song M, Chen J, et al. Hsa-miR-19a is associated with lymph metastasis and mediates the TNF-alpha induced epithelial-to-mesenchymal transition in colorectal cancer. Sci Rep. 2015;5:13350.
199. Zhang GJ, Li Y, Zhou H, Xiao HX, Zhou T. miR20a is an independent prognostic factor in colorectal cancer and is involved in cell metastasis. Mol Med Rep. 2014;10(1):283–91.
200. Tsuchida A, Ohno S, Wu W, Borjigin N, Fujita K, Aoki T, et al. miR-92 is a key oncogenic component of the miR-17-92 cluster in colon cancer. Cancer Sci. 2011;102(12):2264–71.
201. Chen HY, Lin YM, Chung HC, Lang YD, Lin CJ, Huang J, et al. miR-103/107 promote metastasis of colorectal cancer by targeting the metastasis suppressors DAPK and KLF4. Cancer Res. 2012;72(14):3631–41.
202. Hur K, Toiyama Y, Schetter AJ, Okugawa Y, Harris CC, Boland CR, et al. Identification of a metastasis-specific MicroRNA signature in human colorectal cancer. J Nat Cancer Inst. 2015;107(3).
203. Wang YF, Li Z, Zhao XH, Zuo XM, Zhang Y, Xiao YH, et al. MicroRNA-10b is upregulated and has an invasive role in colorectal cancer through enhanced Rhoc expression. Oncol Rep. 2015;33(3):1275–83.
204. Chang KH, Miller N, Kheirelseid EA, Lemetre C, Ball GR, Smith MJ, et al. MicroRNA signature analysis in colorectal cancer: identification of expression profiles in stage II tumors associated with aggressive disease. Int J Color Dis. 2011;26(11):1415–22.
205. Shibuya H, Iinuma H, Shimada R, Horiuchi A, Watanabe T. Clinicopathological and prognostic value of microRNA-21 and microRNA-155 in colorectal cancer. Oncology. 2010;79(3–4):313–20.
206. Qu YL, Wang HF, Sun ZQ, Tang Y, Han XN, Yu XB, et al. Up-regulated miR-155-5p promotes cell proliferation, invasion and metastasis in colorectal carcinoma. Int J Clin Exp Pathol. 2015;8(6):6988–94.
207. Ferraro A, Kontos CK, Boni T, Bantounas I, Siakouli D, Kosmidou V, et al. Epigenetic regulation of miR-21 in colorectal cancer: ITGB4 as a novel miR-21 target and a three-gene network (miR-21-ITGBeta4-PDCD4) as predictor of metastatic tumor potential. Epigenetics. 2014;9(1):129–41.
208. Xiong Y, Zhang YY, Wu YY, Wang XD, Wan LH, Li L, et al. Correlation of over-expressions of miR-21 and Notch-1 in human colorectal cancer with clinical stages. Life Sci. 2014;106(1–2):19–24.
209. Zhu L, Chen H, Zhou D, Li D, Bai R, Zheng S, et al. MicroRNA-9 up-regulation is involved in colorectal cancer metastasis via promoting cell motility. Med Oncol. 2012;29(2):1037–43.
210. Ress AL, Stiegelbauer V, Winter E, Schwarzenbacher D, Kiesslich T, Lax S, et al. MiR-96-5p influences cellular growth and is associated with poor survival in colorectal cancer patients. Mol Carcinog. 2015;54(11):1442–50.
211. Yang MH, Yu J, Jiang DM, Li WL, Wang S, Ding YQ. microRNA-182 targets special AT-rich sequence-binding protein 2 to promote colorectal cancer proliferation and metastasis. J Transl Med. 2014;12:109.
212. Perilli L, Vicentini C, Agostini M, Pizzini S, Pizzi M, D'Angelo E, et al. Circulating miR-182 is a biomarker of colorectal adenocarcinoma progression. Oncotarget. 2014;5(16):6611–9.
213. Rapti SM, Kontos CK, Papadopoulos IN, Scorilas A. Enhanced miR-182 transcription is a predictor of poor overall survival in colorectal adenocarcinoma patients. Clin Chem Lab Med. 2014;52(8):1217–27.
214. Zhou T, Zhang GJ, Zhou H, Xiao HX, Li Y. Overexpression of microRNA-183 in human colorectal cancer and its clinical significance. Eur J Gastroenterol Hepatol. 2014;26(2):229–33.
215. Yuan D, Li K, Zhu K, Yan R, Dang C. Plasma miR-183 predicts recurrence and prognosis in patients with colorectal cancer. Cancer Biol Ther. 2015;16(2):268–75.

Non-coding RNAs Functioning in Colorectal Cancer Stem Cells

5

Daniele Fanale, Nadia Barraco, Angela Listì, Viviana Bazan, and Antonio Russo

Abstract

In recent years, the hypothesis of the presence of tumor-initiating cancer stem cells (CSCs) has received a considerable support. This model suggested the existence of CSCs which, thanks to their self-renewal properties, are able to drive the expansion and the maintenance of malignant cell populations with invasive and metastatic potential in cancer. Increasing evidence showed the ability of such cells to acquire self-renewal, multipotency, angiogenic potential, immune evasion, symmetrical and asymmetrical divisions which, along with the presence of several DNA repair mechanisms, further enhance their oncogenic potential making them highly resistant to common anticancer treatments. The main signaling pathways involved in the homeostasis of colorectal (CRC) stem cells are the Wnt, Notch, Sonic Hedgehog, and Bone Morfogenic Protein (BMP) pathways, which are mostly responsible for all the features that have been widely referred to stem cells. The same pathways have been identified in colorectal cancer stem cells (CRCSCs), conferring a more aggressive phenotype compared to non-stem CRC cells. Recently, several evidences suggested that non-coding RNAs (ncRNAs) may play a crucial role in the regulation of different biological mechanisms in CRC, by modulating the expression of critical stem cell transcription factors that have been found active in CSCs. In this chapter, we will discuss the involvement of ncRNAs, especially microRNAs (miRNAs) and long non-coding RNAs (lncRNAs), in stemness acquisition and maintenance by CRCSCs, through the regulation

D. Fanale • N. Barraco • A. Listì • V. Bazan
A. Russo (✉)
Department of Surgical, Oncological and Oral Sciences, Section of Medical Oncology, University of Palermo, Via del Vespro 129, 90127 Palermo, Italy
e-mail: fandan@libero.it; barraconadia@gmail.com; alisti@live.it; viviana.bazan@unipa.it; antonio.russo@usa.net

O. Slaby, G.A. Calin (eds.), *Non-coding RNAs in Colorectal Cancer*, Advances in Experimental Medicine and Biology 937, DOI 10.1007/978-3-319-42059-2_5

of pathways modulating the CSC phenotype and growth, carcinogenesis, differentiation, and epithelial to mesenchymal transition (EMT).

Keywords
Cancer stem cells • Colorectal cancer • Differentiation • Epithelial-mesenchymal transition • MicroRNAs • Non-coding RNAs • Self-renewal • Signaling pathways • Stemness • Tumorigenicity

5.1 Introduction

Colorectal cancer (CRC) is one of the most common malignant tumors worldwide representing the second leading cause of cancer-related death in the Western Europe countries and the third in the United States [1, 2]. Although, in the last few years, early detection methods and new therapeutic strategies have been implemented in order to prolong survival and improve life quality of patients, however, the development resistance mechanisms to chemo- and radiotherapy is one of the major issues for the clinical management of CRC patients, leading to tumor recurrence and, consequently, poor prognosis [3, 4]. Numerous evidence revealed that a possible mechanism by which CRC cells can evade common therapeutic treatment (chemo- and radiotherapy) is the maintenance of a cancer stem cell (CSC) phenotype via the regulation of pathways modulating the carcinogenesis, differentiation, epithelial to mesenchymal transition (EMT), and CSC growth [5].

Recent findings suggested that non-coding RNAs (ncRNAs), such as microRNAs (miRNAs) and long non-coding RNAs (lncRNAs), could be involved in stemness acquisition and maintenance of colorectal CSCs (CRCSCs), though the role of these molecules has yet to be clarified [6]. In this chapter, we will discuss the involvement of ncRNAs, especially miRNAs and lncRNAs, in regulation of pathways characterizing the CSC phenotype in CRCSCs.

5.2 Relevance of CSCs in Tumors

To date, there are two main hypotheses about tumor-triggering growth and progression. The so-called "stochastic hypothesis" suggests that each cell may be responsible for the tumor transformation by random acquisition of genetic and epigenetic alterations. The second theory states that only few cells are involved in the tumorigenesis process, and probably CSCs may play a crucial role in such process. CSCs refer to a subpopulation of SCs localized in the tumor microenvironment. Tumorigenic potential of CSCs was first suggested in 2006 by Jordan et al. [7], who precisely argued the role played by this small group of cells in supporting the tumor growth. Thanks to their self-renewal properties, CSCs drive the expansion and the maintenance of malignant cell populations with invasive and metastatic potential in cancer tissue [8]. One of the first pieces in the mosaic of the CSC model focuses on the heterogeneity of the proliferative capacity of the cells found in the tumor microenvironment. Such heterogeneity is the result of the production by the multipotent CSCs (highly proliferative) of a wide variety of progenitors (averagely proliferative) and differentiated (non proliferative) cells [9]. CSCs are in close contact with a special microenvironment, called stem cell niche, which confers them the potential to self-renew. The niche consists of cellular and extracellular components. Cross-talking between microenvironment, cytokines and growth factors is necessary to beat the physiological rhythm of both adult epithelial cells and SC maintenance and growth [10].

Furthermore, CSCs may undergo asymmetric and symmetric divisions [11]. The asymmetric division occurs when the mitotic spindle rotates at 90° driven by a morphogen gradient. This causes the retention of one daughter cell into the niche, while the other one will differentiate into a mature cell. The progeny in contact with the stem cell niche retains the self-renewal ability, while the other undergoes differentiation [12]. Differently, symmetric division refers to the generation of two daughter cells with the same differentiation potential of mother cell. This division seems to be necessary to increase the expansion rate of each tumor cell population [11].

It has been discussed for a long time about the potential origin of CSCs in order to understand how and why these cells can acquire genetic and epigenetic mutations, which usually characterize the tumor CSCs but are absent in non-CSCs. Some studies suggested that CSCs can originate from the somatic cell, which regains stem-like properties due to acquired mutations, mainly the capacity of renewal and proliferation. Another theory hypothesized that CSCs arise from progenitors which acquired genetic or epigenetic alterations. Indeed, several studies showed that tumors appear to arise from adult stem cells which present anomalies in the ability to accomplish asymmetric mitotic divisions [13]. During normal ageing, stem cells accumulate damage and subsequent stress-dependent changes, such as, for example, de-repression of the *INK4a/ARF* (*CDKN2a*) locus or telomere shortening [14]. This leads to the increasing abundance of senescent cells within differentiated tissues. Incipient tumors, arising directly from SCs or from more committed cells, undergo a rapid proliferation. These pre-malignant tumor cells rapidly accumulate damage, in part owing to the presence of oncogenes, leading to a higher proportion of cancer cells which become senescent. Tumor progression to full malignancy is favored when tumor cells acquire mutations that impair the senescence program (e.g., mutations in *Tp53* or *CDKN2a*). In the same contest, it has been proposed also the concept of "tumor-initiating CSCs", which refers to the ability of such cells to initiate a tumor if transplanted in xenograft models [15].

CSCs, like all SCs, are able to acquire self-renewal, multipotency, angiogenic potential and immune evasion. In addition, the presence of DNA repair mechanisms, activity of detoxification enzymes and not least the ability of symmetrical and asymmetrical divisions further enhance their oncogenic potential. Therefore, the occurrence of acquired genetic and epigenetic alterations confers to CSCs a more aggressive cancer phenotype compared to somatic cancer cells, making them highly resistant to common anticancer treatments [16], as well as determining a potential source of metastatic spread in different sites of the body [17] and in different tumor types, including breast cancer, colorectal cancer, pancreatic cancer, small cell lung cancer, etc [18]. One of the main limitations which have delayed the research about the CSCs was just how identify them in a tumor tissue. The first study dating back to 20 years ago about the isolation of the CSCs in leukemia was published in 1997 by Bonnet and Dick [19]. They isolated subpopulations of leukemic cells from acute myeloid leukemie expressing CD34 but not CD38. CD34 identifies hematopoietic SCs and bone marrow progenitor cells, whereas CD38 is expressed during the differentiation of a subset of these cells [19]. Subsequently, several other studies identified CSCs in brain, breast, ovary, colon, pancreas and prostate cancers, melanoma and multiple myeloma.

5.3 Colorectal Cancer Stem Cells (CRCSCs)

Colorectal stem cells, like all SCs, represent a reserve of undifferentiated, multipotent and self-renewable cells, useful in the homeostasis process. These cells follow an asymmetrical division pattern, giving rise to two daughter cells: one will remain undifferentiated, while the other will become an intestinal progenitor cell [20]. Under physiological conditions, intestinal homeostasis appears to be the result of the interaction between stem cells, progenitor cells and microenvironment. In particular, cross-talking between mesenchymal and epithelial cells induces the morphogen gradients, required by intestinal subepithelial myofibroblast cells, which are responsible for the balance between proliferation, differentiation, migration, and renewal [21]. The homeostasis of colorectal SCs is under the control of the Wingless/Int (Wnt), Notch and Sonic Hedgehog, Bone Morfogenic Protein (BMP) signaling pathways, which are mainly responsible for all the features that have been widely referred to all SCs [22].

The identification of CRCSCs has witnessed a strong development in recent years [23, 24]. Today the search for surface markers represents one of the most performed approaches for the identification of stem cells within the tumor. Other experimental studies have been performed to isolate CSCs by other specific features, such as the long-term preservation of labeled DNA, morphological traits, epigenetic modifications, or differential gene expression. CRCSCs or CRC stem-like cells (CRCSLCs) have been shown to express surface markers, such as CD44, CD166, CD133 (or Promonin-1) and ESA (epithelial-specific antigen, also known as EpCAM). More recently, Lgr5, Musashi-1 and aldehyde-dehydrogenase 1 (ALDH-1) have been added to the list of stem cell markers for CRC [25, 26].

The research of surface biomarkers remains the most promising approach under current investigation, although several limitations and controversies regarding the identification and localization of such biomarkers, including the lack of widely accepted specific molecular markers and the low sensitivity of the different used techniques, were detected.

5.3.1 Signaling Pathways Involved in CRCSCs

Alterations in signaling pathways regulating the homeostasis of colorectal SCs, including Wnt network, Notch and Sonic Hedgehog, Bone Morfogenic Protein (BMP) signaling, have been reported in CRCSCs, driving their malignant behavior.

Wnt belongs to a multigene family whose members appear to be conserved across species. The different members encode cytokines triggering intracellular responses. Several experiments showed that Wnt pathway plays a key role in intestinal crypt development, maintenance, and proliferation, and is under control of Sonic Hedgehog signaling pathway. In particular, it plays a crucial role in the regulation of self-renewal and proliferation of CRCSCs. Indeed, Wnt knockout in xenograft models was associated with the loss of ability to develop the colon crypts [27]. Progenitors at the bottom of the crypt accumulate intracellular β-catenine which by TCF/LEF activation induces the transcription of genes responsible for SC features. This hypothesis was investigated through *in vitro* studies which aimed to observe the growth and formation spherical colonies under anchorage-independent conditions in serum-free cultures. Experimental evidence revealed the expansion of colon CSCs by generating "colonospheres" from colon cancer cell lines, confirming the pivotal role of Wnt/β-catenin signaling pathway in the regulation of epithelial stem cell self-renewal [28, 29] and dysregulation of this signaling pathway in colon carcinogenesis [30]. The Notch signaling is another pathway involved in stemness maintenance, by preventing differentiation and inducing proliferation [31, 32]. In addition, Notch may contribute to tumorigenesis by inhibiting differentiation, promoting survival or accelerating proliferation [33]. Potentially oncogenic targets of Notch-1 include cyclins D1 and D3 [34, 35], cyclin A, SKP2, phosphatidylinositol 3-kinase, AKT, ERBB2, nuclear factor-nB, and nuclear factor-nB2, h-catenin, signal transducers and activators of transcription-3, and hypoxia-inducible factor-1a [36]. Deregulated expression of Notch receptors, ligands, and targets has been observed in several solid tumor [37]. Wnt and Notch signaling are also involved in the process of EMT, tumor invasion and metastasis [38–42].

EMT, originally described during the embryogenesis, is characterized by the loss of epithelial cell features and gain of mesenchymal-like phenotype. Subsequently, it has been well studied and defined in cancer cells where it plays an important role in tumor progression, metastasis and drug-resistance [43]. CRCSCs are characterized by the expression of several surface markers which reflect the EMT process. Cancer cell with acquired mesenchymal-like phenotype shows the ability to cross endothelial barriers and invade blood and lymphatic circulations to reach new tissues on which to take root [44]. A large number of pathways regulated by factors, such as TGFβ, Wnt, NF-kB, Notch, integrins, and tyrosine-kinase receptors (EGF, FGF, HGF, PDGF, IGF), have been associated with EMT. Functional interaction between these pathways might result in signal amplification and induce EMT and metastasis [45]. The molecular mechanisms involved in EMT cause inhibition of E-cadherins and overexpression of surface mesenchymal markers, such as vimentin and fibronectin, by zinc-finger transcriptional factors, such as SNAIL, SLUG, TWIST, ZEB1, SIP1, and E47 [46]. Recent findings showed that EMT is not only governed by such signaling pathways, but also by ncRNAs which seem to play a key role in this transition process [47]. Another important molecular mechanism involved in stem cell regulation is modulated by signaling pathway of BMP, which is a member of the transforming growth factor-β (TGF-β) superfamily. BMP signaling promotes differentiation of normal colon stem cells, while its inactivation confers increased tumorigenesis. BMPs have been associated to the regulation of cancer pathogenesis and metastasis, possibly due to its ability to counteract TGF-β-induced SMAD3-dependent EMT. This protein, also, is expressed in non-CSCs, but not in CSCs, and it would induce differentiation, apoptosis and sensitivity to chemotherapy in the CSC population of human CRC cells, suggesting a potential role as therapeutic agent against CRCSCs [48].

Another set of genes in CRCSCs, especially *SOX2* and *Oct4*, seems to inhibit other genes that start differentiation and thus maintain the self-renewal ability of stem cells. The levels of *Oct4* and *SOX2* mRNA in peripheral blood of patients with metastatic CRC were found to be higher than in healthy controls [49].

5.4 Non-coding RNAs Involved in CRCSCs

Recently, several studies have suggested that a large remarkably class of ncRNAs, including miRNAs and lncRNAs, seems to have a potential role in the genetic and epigenetic regulatory networks [50]. These biological molecules are regulatory ncRNAs that are included in share of genome that is unable to codify for proteins. NcRNAs may play a critical role in regulating the induction of induced pluripotent stem cells (iPSCs), maintenance and differentiation of CRCSCs [51]. In particular, miRNAs and lncRNAs are involved in different biological mechanisms of CRC. They represent more than 98 % of whole genome, originally known as "junk DNA", appearing to lack a protein-coding potential essential in normal cellular physiology and disease. Several studies have also described their involvement in various cellular functions and development processes [52].

5.4.1 Role of miRNAs and Their Relevance in CRCSCs

Several evidences suggested the involvement of miRNAs in maintaining stemness of CRCSCs through the regulation of pathways that modulate the CSC phenotype, carcinogenesis, differentiation, EMT, and CSC growth [53] (Table 5.1).

Monzo et al. [54] have hypothesized that overexpression of miR-17-5p, a member of the miR-17-92 cluster detected in crypt progenitor compartment, could be involved in cell differentiation, proliferation, and homeostatic self-renewal of colon tissue, indicating a potential role in stemness regulation. In fact, suppression of miR-17-5p determined a reduction of proliferation and clonogenicity in DLD1 CRC cells [54].

Further, Yu et al. [55] have reported that miR-21 silencing induces differentiation of CSCs/cancer stem-like cells (CSLCs)-enriched chemoresistant HCT-116 and HT-29 cells, by decreasing the ability to form colonospheres *in vitro*, reducing the T-cell factor/lymphoid enhancer

Table 5.1 MiRNAs functioning in CRCSCs

miRNA	Expression	Targets	Effects	References
miR-17-5p	↑	*E2F1*	Differentiation, proliferation, self-renewal	[54]
miR-21	↑	*PDCD4, TGFβR2, miR-145*	Growth, differentiation, stemness, chemoresistance	[55, 56]
miR-23a	↑	*MTSS1*	Migration, cell motility, invasion and metastasis	[57]
miR-27a	↑	*FBXW7*	Proliferation, metastasis	[57, 60]
miR-34a	↑	*Notch, c-Kit*	Differentiation, inhibition of stemness, chemosensitivity	[61]
miR-34a	↓	*Notch, c-Kit*	Self-renewal, stemness, symmetric division	[62]
miR-93	↓	*HDAC8, TLE4*	Proliferation and colony formation	[71]
miR-106b	↑	*PTEN, p21*	Radioresistance, apoptosis inhibition, proliferation, DNA damage, tumour-initiating cell capacity, self-renewal, stemness	[72]
miR-124	↓	*PRRX1*	Radioresistance, EMT, stemness, self-renewal	[73]
miR-125a/b	↓	*Mcl1, ALDH1A3*	Paclitaxel resistance, tumor growth, stemness, CSC survival, apoptosis inhibition	[74]
miR-140	↑	*HDAC4*	Chemoresistance	[87]
miR-145	↓	*SOX2, CD44, Nanog, β-catenin, Oct4, miR-21*	Proliferation, stemness, differentiation, chemoresistance	[56]
miR-146a	↑	*Numb*	Symmetric division, stemness, cetuximab resistance	[69, 70]
miR-183	↓	*Bmi1*	Stemness, EMT, invasion and metastasis	[79]
miR-200a	↓	*Bmi1, ZEB1, ZEB2*	Stemness, EMT, invasion and metastasis	[80–84]
miR-200c	↓	*SOX2, KLF4, Bmi1, ZEB1, ZEB2*	Stemness, EMT, proliferation, migration, invasion and metastasis	[79–82, 85]
miR-203	↓	*Bmi1, SOX2, KLF4*	Stemness, self-renewal, EMT, tumorigenicity, invasion and metastasis	[79, 86]
miR-215	↑	*DTL*	Chemoresistance	[88]
miR-215	↓	*Bmi1*	Stemness, self-renewal, clonogenicity, inhibition of differentiation	[89]
miR-302b	↑	N/S	Stemness, colony-forming ability, self-renewal, invasion, migration	[90]
miR-328	↓	*ABCG2, MMP16*	Maintenance of CSLC phenotype, self-renewal, differentiation, invasion, chemoresistance	[91]
miR-449b	↓	*CCND1, E2F3*	Proliferation, self-renewal	[92]
miR-451	↓	*MIF, ABCB1*	Stemness, self-renewal, tumorigenicity, irinotecan resistance	[93]

↑Up-regulated and ↓down-regulated miRNAs in CRCSCs. *EMT* Epithelial-mesenchymal transition, *CSC* cancer stem cell, *CSLC* cancer stem-like cell, N/S target not specified factor (TCF/LEF) activity, increasing the expression of pro-apoptotic *PDCD4* target gene, and consequently enhancing cancer cell chemosensitivity to combined therapeutic regimens containing 5-fluorouracil and oxaliplatin (FOLFOX). Conversely, overexpression of miR-21 has been shown to induce stemness in CRCs by down-regulating also the expression of TGFβR2 (transforming growth factor beta receptor 2), which is involved in cell differentiation, resulting in activation of Wnt/β-catenin signaling pathway. Furthermore, following the miR-21 up-regulation, CRCSCs exhibited increased levels of β-catenin, cyclin-D and c-Myc, and greater TCF/LEF activity [55]. Recently, a work carried out by same researchers group revealed that miR-21, miR-145 and their networks cooperate in modulating growth and/or differentiation of CRCSCs, and acquisition of chemoresistance [56]. Indeed, CSCs/CSLCs-enriched CRC cells showed increased miR-21 expression and decreased miR-145 expression, suggesting a role for these miRNAs in inducing stemness and increasing proliferation of CRCSCs. Whereas ectopic expression of miR-145 or miR-21 silencing in CRC cells has been shown to induce differentiation and inhibit stemness and growth of CRC xenografts in SCID mice, by reducing the expression of CD44 (colon CSC marker), β-catenin (stem cell growth regulator), SOX2 (a miR-145 target), Nanog, and Oct4, instead miR-21 up-regulation caused opposed effects. Moreover, *in vitro* experiments demonstrated that miR-145 negatively modulates the expression of miR-21 and vice versa. In addition, KRAS seems to be involved in this process, since *KRAS*-deficient chemoresistant CRC cells showed increased miR-145 expression levels, suppression of miR-21, and inhibition of the negative regulative loop between miR-21 and miR-145 [56].

In vivo and *in vitro* evidence showed that miR-23a and miR-27a were overexpressed in mouse intestinal adenocarcinomas, primary tumors from stage I/II CRC patients, as well as in human CRC cell lines and CRCSCs, triggering mechanisms modulating the transition from indolent to invasive CRC [57]. Up-regulation of miR-23a promotes migration, invasion and metastasis of CRC cells and stem cells, by directly targeting the *MTSS1* (Metastasis Suppressor 1) gene and down-regulating the expression of the encoded protein, which activates SRC signaling pathway and determines filopodia formation via interaction with cortactin [57, 58]; whereas up-regulation of miR-27a induces proliferation and prevents secretory lineage differentiation in CRCSCs, by enhancing Jun, Myc and Notch signaling pathways via direct inhibition of ubiquitin ligase F-box protein FBXW7 [57, 59]. In particular, miR-27a-induced down-regulation of *FBXW7* inhibits proteasome-dependent degradation of the transcription factors *JUN* and *Myc*, and up-regulates Notch signaling components [60].

Recently, Siemens et al. [61] have demonstrated that p53-induced up-regulation of tumor suppressor miR-34a mediated repression of *c-Kit* by p53 through down-regulation of *c-Kit* mRNA, causing an increased CRC cell sensitivity to 5-fluorouracil, and leading to a decrease in Erk signaling and transformation induced by c-Kit down-regulation, and inhibition of stem cell factor (SCF)-induced invasion/migration. Moreover, ectopic expression of miR-34a in CRC cells inhibited the ability to form colonospheres following exposure to SCF, reducing the expression of several stemness markers such as CD44, Lgr5 and BMI-1 [61]. Concomitantly, Bu and colleagues [62] demonstrated that miR-34a significantly affects cell fate of early stage dividing colon CSCs, regulating a bimodal switch through suppression of Notch signaling to promote progeny differentiation. Reduced expression levels of miR-34a have been shown to induce self-renewal in colon CSCs via up-regulation of Notch signaling, whereas an increase in miR-34 expression was correlated with differentiation of the progeny, by prompting daughter cells to become non-CRCSCs. *In vitro* and *in vivo* experiments have showed that the choice between self-renewal *versus* differentiation is determined by losses or gains of miR-34a function, respectively, that generate a bimodal Notch signal. Furthermore, the authors have observed that early stage colon CSCs arising from well-differentiated CRCs can produce both asymmetric divisions generating a daughter colon CSC and a differentiated, non-

CSC daughter cell, and symmetric divisions generating two daughter colon CSCs [62]. Since the asymmetric cell division mechanism is generally used both by normal stem cells and CSCs in order to maintain stemness and tissue homeostasis [63–66], and Notch pathway has been shown to play a key role in regulation of asymmetric division [67, 68], miR-34a levels seem to be involved in control of either symmetric or asymmetric divisions in colon CSCs. Indeed, *in vitro* assays highlighted that high and low levels of miR-34a inhibit asymmetric divisions of colon CSCs, whereas low miR-34a levels promote symmetric divisions, resulting in a higher number of daughter colon CSCs [62]. Conversely, Hwang et al. [69] showed that Snail-induced up-regulation of miR-146a expression causes symmetric division in CRCSCs, by directly targeting *Numb* to stabilize β-catenin, which, in turn, drives the symmetrical cell division triggering a feedback mechanism to maintain Wnt activity. Therefore, the asymmetrical-to-symmetrical cell division transition in CRCSCs is induced by EMT promoted by Snail, which increases the miR-146a expression via β-catenin-TCF4 complex [70]. Impairment of the Snail-miR-146a-β-catenin signaling axis resulting in suppression of the Wnt or MEK activity reduces the symmetric division, tumorigenicity and cetuximab resistance in CRCSCs. Indeed, high expression levels of Snail and decreased Numb expression were associated with unfavorable prognosis and cetuximab resistance in CRC patients [69].

Yu and collaborators [71] performed a microarray analysis to indentify differentially expressed miRNAs in human CRCSCs (SW1116csc) compared to original cell line. They found 35 miRNAs up-regulated, including miR-29b, miR-32, miR-33a, miR-192, miR-194 and miR-215, and 11 miRNAs down-regulated, including miR-93, miR-524-3p, miR-561, miR-886-3p and miR-1231. Among these, miR-93, if ectopically expressed, has been shown to inhibit proliferation and colony formation of SW1116csc, by down-regulating also the expression of HDAC8 and TLE4 [71].

Recently, *in vitro* and *in vivo* studies revealed that miR-106b may be involved in radioresistance of CRC cells by directly targeting *PTEN* and *p21*, thus resulting in alteration of the PTEN/PI3K/AKT signaling pathway, inhibition of cell apoptosis, increase of cell proliferation trough induction of G1 to S transition, and repression of DNA damage repair. Additionally, these effects induced by miR-106b overexpression may be associated with enhanced tumour-initiating cell capacity in presence or absence of irradiation, and increased expression of stemness-related genes, such as *CD133* and *SOX2*, conferring the ability to form colonospheres and induce cell self-renewal [72]. Unlike miR-106b, up-regulation of miR-124 has been shown to promote radiosensitivity of CRC cells *in vitro* and *in vivo* through direct inhibition of PRRX1, which is a stemness regulator and EMT inducer, and concomitant decrease of the expression of stemness-related genes, such as *ABCG2*, *SOX2*, and *Oct4* [73]. Furthermore, another study from Chen et al. [74] reported that low expression levels of miR-125a/b may mediate paclitaxel resistance in CRC cells, inducing the expression of the aldehyde dehydrogenase ALDH1A3, a stem cell marker [75], and Mcl1, a member of the pro-survival Bcl-2 family [76], resulting in enhanced activation and survival of CRCSCs. On the contrary, increased cell apoptosis and decreased survival were observed in CRC cells treated with paclitaxel, following overexpression of miR-125a/b and subsequent down-regulation of the *ALDH1A3* and *Mcl1* expression [74].

Since, in recent years, EMT has been reported to be correlated with stemness and therapy resistance [77, 78], several studies investigated the role of EMT activation in tumorigenicity, and stemness acquisition and maintenance through repression of stemness-inhibiting miRNAs. Wellner et al. [79] showed that ZEB1 (zinc finger E-box binding homeobox 1), a EMT inducer and metastasis promoter, is able to suppress the expression of stemness-inhibiting miR-203 and miR-200 family members, whose targets, such KLF4 and SOX2, are stem cell factors. On the other hand, miR-200 family members may significantly induce epithelial differentiation triggering a feedback mechanism that involves EMT inhibition by repressing translation of *ZEB1* and *ZEB2* mRNAs [80–82]. Also, miR-183 has been shown to cooperate with miR-200c and miR-203 in order to inhibit the expression of stem cell factors in CSCs via inhibition of the polycomb repressor Bmi1. Therefore, ZEB1-induced EMT seems to play a pivotal role in promoting invasion and metastasis, and inducing tumour-initiating capacity in CRC cells [79]. Recently, Pichler and colleagues [83] further confirmed the regulatory function of miR-200a on EMT and its correlation with CSC phenotype in CRC, observing a reduced miR-200a expression in CSCs-enriched CRC cells and CRC patients with unfavorable prognosis. Another study reported that EMT may be activated by SIX1-induced repression of the miR-200a expression in CRC cells [84]. In a recent work, Lu et al. [85] highlighted that miR-200c modulates *SOX2* expression and vice versa through a negative feedback loop, inhibiting the PI3K-AKT signaling pathway, resulting in regulation of stemness, growth and metastasis in CRC. Low expression levels of miR-200c have been shown to enhance the proliferation, migration and invasion, and increase the expression of CRCSC markers and ability to form colonospheres *in vitro* [85]. Concerning the stemness inhibitor miR-203, Ju et al. [86] demonstrated also that Snail-induced repression of this miRNA was required for stemness maintenance in CD44^{+} CRC cells, via c-Src kinase activation promoted by the interaction between CD44 and hyaluronic acid (HA). Therefore, as a result of CD44-mediated up-regulation of Snail and consequent miR-203 down-regulation, CD44^{+} CRC cells showed CSC properties, such as increased *in vitro* colony-forming capacity and *in vivo* tumorigenicity, and a greater invasion, metastasis and differentiation potential [86].

Two independent studies from Song et al. [87, 88] suggested the involvement of high expression levels of endogenous miR-140 and miR-215 in CRC chemoresistance induced by CD133^{+high}/CD44^{+high} CRCSLCs/CRCSCs, through inhibition of cell proliferation and cell cycle G2-arrest as response mechanisms to the cellular and DNA damage caused by chemotherapy agents. MiR-140 and miR-215 have been shown to exert their functions by suppressing the expression of their target genes *HDAC4* (histone deacetylase 4) and *DTL* (denticleless protein homolog), respectively [87, 88]. Recently, Jones and colleagues [89] have proposed that the caudal-type homeobox 1 (CDX1)-miR-215 axis promotes CRCSC differentiation, by inhibiting the expression of cell cycle- and stemness-related genes downstream of CDX1, including *Bmi1*, *EFNB2*, *EGR1*, *EREG*, and *HOXA10*. In particular, expression levels of the transcription factor CDX1 were found closely correlated with those of miR-215, but not with Bmi1, since CDX1 directly activates the expression of miR-215, which, in turn, down-regulates the *Bmi1* expression. Therefore, expression of Bmi1, but not CDX1, was observed in CRCSCs, determining stemness, self-renewal, clonogenicity, inhibition of differentiation [89].

In vitro and *in vivo* studies revealed that down-regulation of the transcription factor achaete scute-like 2 (Ascl2) is involved in loss of CSC properties by HT-29 and LS174T CRC cells, through reduction of the expression of stemness-related genes, such as *CD133*, *Sox2*, *Oct4*, *Lgr5*, *Bmi1*, *c-myc*, leading to a decrease in cellular proliferation, colony-forming ability, self-renewal, invasion and migration *in vitro*, and *in vivo* tumor growth arrest. Furthermore, Ascl2 knockdown caused down-regulation of the miRNA-17, miRNA-20a and miR-302b expression in CRC cells compared to control cells. However, only ectopic miR-302b expression reversed the previous effects, restoring the stemness in Ascl2-deficient CRC cells [90].

Experimental evidence reported that decreased miR-328 expression is responsible for CSC-like characteristics, such as self-renewal, differentiation, invasiveness, chemoresistance, and tumor formation ability, detected in side population (SP) cells sorted from CRC. Conversely, miR-328 overexpression has been shown to suppress chemoresistance and invasive capacity of SP cells, by directly targeting *ABCG2* and *MMP16* genes [91].

A recent work suggested that proliferative ability of CD133^{+}/CD44^{+} SW1116 CRCSCs may be inhibited by miR-449b up-regulation, via repression of CCND1 and E2F3 expression [92].

Bitarte et al. [93] showed that low expression levels of miR-451 were associated with chemoresistance of CRCSCs to irinotecan-based treatments. Also, the authors found that down-regulation of miR-451 induces increased expression of its target gene macrophage migration inhibitory factor (*MIF*), responsible for the expression of cyclooxygenase-2 (COX-2), which mediates, in turn, the activation of Wnt pathway, promoting growth, tumorigenicity, self-renewal in CRCSCs [93]. Therefore, miR-451 could modulate CSC phenotype by inhibiting the Wnt signaling, since the Wnt pathway activation is crucial for the stemness maintenance in CRCSCs [94]. Conversely, miR-451 up-regulation has been shown to induce chemosensitivity to irinotecan, by suppressing the expression of the ATP-binding cassette drug transporter ABCB1 [93].

Lastly, Zhang and collaborators [95] identified, using microarray analysis, a miRNA expression profile which appears to be involved in stemness maintenance of $CD133^+$ CRCSCs, including EMT-modulating miRNAs, such as miR-429 and miR-155, and carcinogenesis-modulating miRNAs, such as miR-185, miR-320, miR-494, miR-221, and miR-31.

5.4.2 Impact of lncRNAs in Stemness Regulation of CRCSCs

Experimental evidences suggested that dysregulation of lncRNAs promotes tumorigenesis and metastasis of several human cancers, including CRC. Furthermore, the aberrant expression of lncRNAs has been associated with poor prognosis in a variety of tumor histotypes [52]. The CRC cell phenotype is the result of genetic and epigenetic aberrations [96]. The CRC malignancy is mainly connected to a subset of CSCs. As previously described, the activation of several development pathways, especially the Wnt/β-catenin signaling, allows to maintain or acquire the stem-like traits in CRCSCs. Therefore, targeting such molecular pathways could represent a promising strategy for the anticancer treatment. LncRNAs can act in the regulation of cellular functions inhibiting the expression of critical stem cell transcription factors that have been found active in CSCs. LncRNAs take part and probably are responsible for an precise amount of these aberrations. Preclinical studies have identified more than 900 lincRNAs (long intergenic non-coding RNAs) in mouse embryonic stem cells (mESCs) and human ESCs (hESCs). These molecules may modulate three mechanisms: self-renewal control and pluripotency of ESCs, reprogramming somatic cells and differentiation of PSCs [97, 98].

LincRNA-p21 may be essential for the maintenance of CRCSC self-renewal, even if the precise molecular mechanism has not been fully understood. Indeed, such lincRNA is a direct transcription target of p53 functioning as tumor suppressor. *In vitro*, lincRNA-p21 inhibits the β-catenin signal transduction leading to the down-regulation of related-genes expression [99]. LncRNAs seem to maintain CSC pluripotency thanks to the repression of the differentiation programs or the generation of iPSCs. Indeed, *in vivo* experiments have shown that more of 100 lincRNAs in mESCs appear to be bound by SOX2, Oct4 and Nanog as well as by other ESC-specific transcription factors [100, 101]. The inhibition or dysregulation of any of these lincRNAs determined changes in expression levels of these factors, demonstrating their critical role in maintaining pluripotency of mESCs, probably due to the repression of differentiation programs. For example, Panct 1-3 were identified as modulators of mESC pluripotency based on reduced *Oct4* promoter activity, and again lincRNA-RoR was found as regulator of reprogramming in hESC. Furthermore, these lncRNAs are involved in the generation of iPSCs. Any of these lincRNAs exhibited overexpression in iPSCs versus ESCs but not comparable to that detected in somatic cells. The reason can be found in their promotion of reprogramming pluripotency. In addition, the expression of these iPSC-enriched lincRNAs is directly regulated by pluripotency transcription factors and their related pathways. The overexpression or down-regulation of lincRNAs induces high or low levels of iPSCs, respectively, and their absence is associated with dysregulation of p53, confirming the key role of lincRNAs in the induction of pluripotency [102].

It's important to report also the epigenetic changes, involving altered DNA methylation patterns, histone modifications and chromatin structure in CRCSCs. Different lncRNAs can interact with chromatin modifying polycomb complex groups (PcGs). This complex is present in several tumor types and is known to regulate the pluripotency and self-renewal of ESCs. Other data suggested that lncRNAs may recruit PcGs into specific sites of chromatin to turn-off differentiation programs in CSCs [50]. The demonstration that quiescent cells (a frequent characteristic of stem cells) show a reduction in global H3K9me3 and H4K20me3 methylation states could be a factor indicating enhanced plasticity of the epigenome. Finally, it has been demonstrated that targeting lncRNAs may cause great inhibitory effects in cancer cells, suggesting the these lncRNAs may become novel therapeutic targets [103].

5.4.3 Therapeutic Potential

The resistance of colorectal CSCs to conventional therapies makes these cells a potential target to optimize current anticancer therapies. Strategies to identify and target specific cell surface markers are currently under investigation, with a particular focus on the functional characterization of CRCSCs [104]. Recent evidence suggested that miRNAs and lncRNAs may provide potential new therapeutic approaches for patients with resistance to current therapies and drug-induced toxicity. Since ncRNAs may modulate the pathways required for the maintenance of a CSC phenotype, potential targeting of specific ncRNAs affecting therapy resistance could be useful for designing novel and targeted ncRNA-based therapeutic strategies in order to improve the clinical outcome of CRC patients [105]. Furthermore, modulation of the expression of specific stemness-related ncRNAs during therapeutic treatment could offer a new tool for the prediction of acquired resistance. To date, several studies showed the involvement of specific stemness-related ncRNAs (miR-21, miR-34a, miR-106b, miR-124, miR-125a/b, miR-140, miR-145, miR-146a, miR-215, miR-328, miR-451) in CRCSC-induced therapeutic resistance. Targeting of oncogenic ncRNAs, by means of antisense oligonucleotides, and forced expression of tumor suppressor ncRNAs may inhibit the stemness characteristics, self-renewal ability, and invasive and metastatic potential of CRCSCs, allowing to reverse the CSC phenotype, enhance sensitivity of CRC cells to therapy and prevent tumor recurrence [53].

This investigation is highly innovative and could have important clinical implications in understanding the molecular mechanisms responsible for the CSC phenotype, stemness maintenance, and chemo- and radioresistance in CRC.

5.5 Conclusions

There is much interest as regards investigation of CSCs because it offers the possibility to generate novel targets. CSCs can evade common therapeutic treatment (chemo- and radiotherapy), inducing chemo- and radioresistance and compromising therapeutic efficacy. Aberrant expression of ncRNAs has been reported in several types of human cancer, including CRC, suggesting a potential role in cancer pathogenesis, tumor initiation, progression, metastatic processes and acquisition of tumor resistance to treatment. Many studies suggested that some ncRNAs can play a key role in the regulation of CRCSCs, directly or indirectly, through the interplay of signaling pathways involved in cancer stemness (Fig. 5.1). The identification of possible ncRNA-mediated biological mechanisms that regulate the behavior of CRCSCs constitutes the future objective. Moreover, the discovery of new potential molecular mechanisms involved in stemness acquisition, maintenance and regulation of CRCSCs could be an important clinical tool to develop new ncRNA-based therapeutic strategies for CRC patients who may benefit from individualized therapies. However, despite encouraging

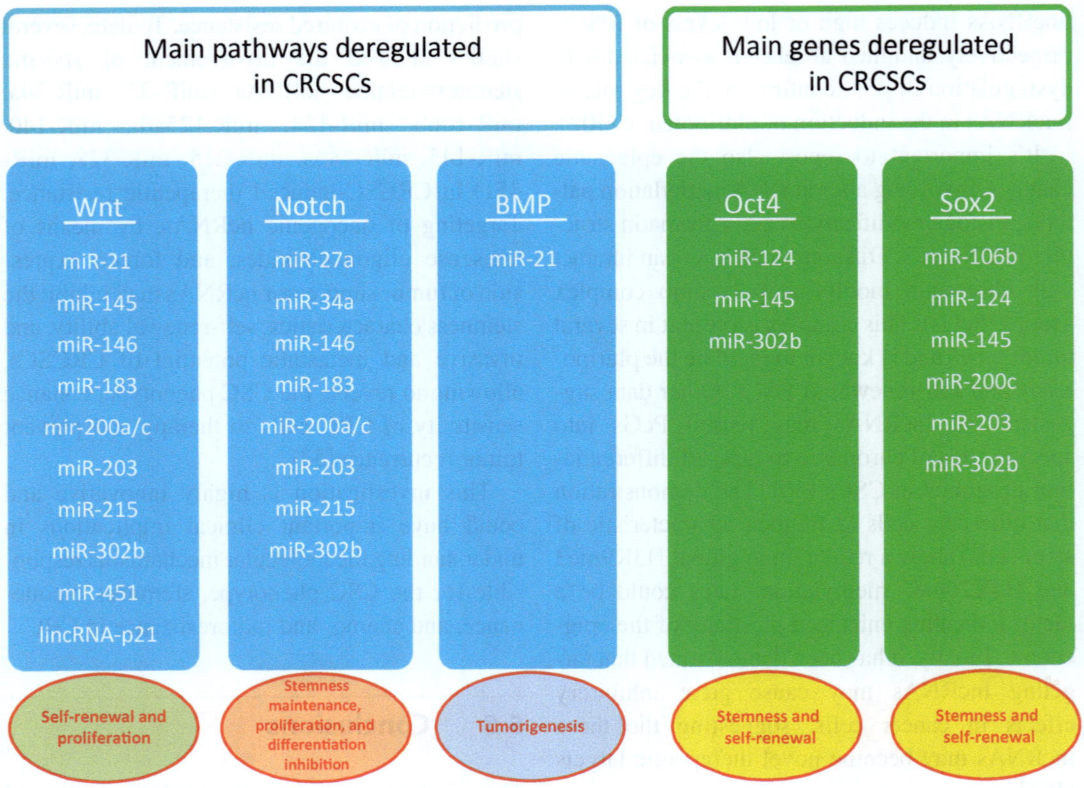

Fig. 5.1 Involvement of ncRNAs in the main stemness-related signaling pathways deregulated in CRCSCs

obtained results, the introduction of ncRNAs in clinical practice appears to be still far. Hopefully, in the near future, specific ncRNA signatures could offer new opportunities to identify and select in a specific manner CRCSCs from CRCs. Finally, the modulation of the expression of specific ncRNAs involved in stemness of CRC, using miRNA mimics or antagomiRs, could provide a new tool to reverse CSC phenotype and overcome therapy resistance.

References

1. Siegel RL, Miller KD, Jemal A. Cancer statistics, 2015. CA Cancer J Clin. 2015;65(1):5–29.
2. Torre LA, Bray F, Siegel RL, Ferlay J, Lortet-Tieulent J, Jemal A. Global cancer statistics, 2012. CA Cancer J Clin. 2015;65(2):87–108.
3. Marin JJ, Sanchez de Medina F, Castano B, Bujanda L, Romero MR, Martinez-Augustin O, et al. Chemoprevention, chemotherapy, and chemoresistance in colorectal cancer. Drug Metab Rev. 2012;44(2):148–72.
4. Skvortsova I, Debbage P, Kumar V, Skvortsov S. Radiation resistance: cancer stem cells (CSCs) and their enigmatic pro-survival signaling. Semin Cancer Biol. 2015;35:39–44.
5. Raza U, Zhang JD, Sahin O. MicroRNAs: master regulators of drug resistance, stemness, and metastasis. J Mol Med (Berl). 2014;92(4):321–36.
6. Sun X, Jiao X, Pestell TG, Fan C, Qin S, Mirabelli E, et al. MicroRNAs and cancer stem cells: the sword and the shield. Oncogene. 2014;33(42):4967–77.
7. Jordan CT, Guzman ML, Noble M. Cancer stem cells. N Engl J Med. 2006;355(12):1253–61.
8. Reya T, Morrison SJ, Clarke MF, Weissman IL. Stem cells, cancer, and cancer stem cells. Nature. 2001;414(6859):105–11.
9. Dalerba P, Cho RW, Clarke MF. Cancer stem cells: models and concepts. Annu Rev Med. 2007;58:267–84. doi:10.1146/annurev.med.58.062105.204854.
10. Medema JP, Vermeulen L. Microenvironmental regulation of stem cells in intestinal homeostasis and cancer. Nature. 2011;474(7351):318–26.
11. Vermeulen L, Sprick MR, Kemper K, Stassi G, Medema JP. Cancer stem cells–old concepts, new insights. Cell Death Differ. 2008;15(6):947–58.

12. Lechler T, Fuchs E. Asymmetric cell divisions promote stratification and differentiation of mammalian skin. Nature. 2005;437(7056):275–80.
13. Clarke MF, Dick JE, Dirks PB, Eaves CJ, Jamieson CH, Jones DL, et al. Cancer stem cells–perspectives on current status and future directions: AACR Workshop on cancer stem cells. Cancer Res. 2006;66(19):9339–44.
14. Gostjeva EV, Thilly WG. Stem cell stages and the origins of colon cancer: a multidisciplinary perspective. Stem Cell Rev. 2005;1(3):243–51.
15. Al-Hajj M, Clarke MF. Self-renewal and solid tumor stem cells. Oncogene. 2004;23(43):7274–82.
16. Visvader JE, Lindeman GJ. Cancer stem cells in solid tumours: accumulating evidence and unresolved questions. Nat Rev Cancer. 2008;8(10):755–68.
17. Soltysova A, Altanerova V, Altaner C. Cancer stem cells. Neoplasma. 2005;52(6):435–40.
18. Wang Z, Li Y, Ahmad A, Azmi AS, Kong D, Banerjee S, et al. Targeting miRNAs involved in cancer stem cell and EMT regulation: an emerging concept in overcoming drug resistance. Drug Resist Updat. 2010;13(4-5):109–18.
19. Bonnet D, Dick JE. Human acute myeloid leukemia is organized as a hierarchy that originates from a primitive hematopoietic cell. Nat Med. 1997;3(7):730–7.
20. Davies EJ, Marsh V, Clarke AR. Origin and maintenance of the intestinal cancer stem cell. Mol Carcinog. 2011;50(4):254–63.
21. Todaro M, Francipane MG, Medema JP, Stassi G. Colon cancer stem cells: promise of targeted therapy. Gastroenterology. 2010;138(6):2151–62.
22. Kosinski C, Li VS, Chan AS, Zhang J, Ho C, Tsui WY, et al. Gene expression patterns of human colon tops and basal crypts and BMP antagonists as intestinal stem cell niche factors. Proc Natl Acad Sci U S A. 2007;104(39):15418–23.
23. Lugli A, Iezzi G, Hostettler I, Muraro MG, Mele V, Tornillo L, et al. Prognostic impact of the expression of putative cancer stem cell markers CD133, CD166, CD44s, EpCAM, and ALDH1 in colorectal cancer. Br J Cancer. 2010;103(3):382–90.
24. Finkel T, Serrano M, Blasco MA. The common biology of cancer and ageing. Nature. 2007;448(7155):767–74.
25. Moitra K. Overcoming multidrug resistance in cancer stem cells. Biomed Res Int. 2015;2015:635745.
26. Cherciu I, Barbalan A, Pirici D, Margaritescu C, Saftoiu A. Stem cells, colorectal cancer and cancer stem cell markers correlations. Curr Health Sci J. 2014;40(3):153–61.
27. Fevr T, Robine S, Louvard D, Huelsken J. Wnt/beta-catenin is essential for intestinal homeostasis and maintenance of intestinal stem cells. Mol Cell Biol. 2007;27(21):7551–9.
28. Korkaya H, Paulson A, Charafe-Jauffret E, Ginestier C, Brown M, Dutcher J, et al. Regulation of mammary stem/progenitor cells by PTEN/Akt/beta-catenin signaling. PLoS Biol. 2009;7(6):e1000121.
29. Brabletz S, Schmalhofer O, Brabletz T. Gastrointestinal stem cells in development and cancer. J Pathol. 2009;217(2):307–17.
30. Morin PJ, Sparks AB, Korinek V, Barker N, Clevers H, Vogelstein B, et al. Activation of beta-catenin-Tcf signaling in colon cancer by mutations in beta-catenin or APC. Science. 1997;275(5307):1787–90.
31. Miele L. Notch signaling. Clin Cancer Res. 2006;12(4):1074–9.
32. Amini S, Fathi F, Mobalegi J, Sofimajidpour H, Ghadimi T. The expressions of stem cell markers: Oct4, Nanog, Sox2, nucleostemin, Bmi, Zfx, Tcl1, Tbx3, Dppa4, and Esrrb in bladder, colon, and prostate cancer, and certain cancer cell lines. Anat Cell Biol. 2014;47(1):1–11. doi:10.5115/acb.2014.47.1.1.
33. Miele L, Osborne B. Arbiter of differentiation and death: notch signaling meets apoptosis. J Cell Physiol. 1999;181(3):393–409.
34. Ronchini C, Capobianco AJ. Induction of cyclin D1 transcription and CDK2 activity by Notch(ic): implication for cell cycle disruption in transformation by Notch(ic). Mol Cell Biol. 2001;21(17):5925–34.
35. Chen Y, Fischer WH, Gill GN. Regulation of the ERBB-2 promoter by RBPJkappa and NOTCH. J Biol Chem. 1997;272(22):14110–4.
36. Wang Z, Li Y, Kong D, Sarkar FH. The role of Notch signaling pathway in epithelial-mesenchymal transition (EMT) during development and tumor aggressiveness. Curr Drug Targets. 2010;11(6):745–51.
37. Leethanakul C, Patel V, Gillespie J, Pallente M, Ensley JF, Koontongkaew S, et al. Distinct pattern of expression of differentiation and growth-related genes in squamous cell carcinomas of the head and neck revealed by the use of laser capture microdissection and cDNA arrays. Oncogene. 2000;19(28):3220–4.
38. Kessler M, Hoffmann K, Brinkmann V, Thieck O, Jackisch S, Toelle B, et al. The Notch and Wnt pathways regulate stemness and differentiation in human fallopian tube organoids. Nat Commun. 2015;6:8989.
39. Su J, Wu S, Wu H, Li L, Guo T. CD44 is functionally crucial for driving lung cancer stem cells metastasis through Wnt/beta-catenin-FoxM1-Twist signaling. Mol Carcinog. 2015. doi:10.1002/mc.22443.
40. Yang K, Wang X, Zhang H, Wang Z, Nan G, Li Y, et al. The evolving roles of canonical WNT signaling in stem cells and tumorigenesis: implications in targeted cancer therapies. Lab Invest. 2015.
41. Li JL, Sainson RC, Oon CE, Turley H, Leek R, Sheldon H, et al. DLL4-Notch signaling mediates tumor resistance to anti-VEGF therapy in vivo. Cancer Res. 2011;71(18):6073–83.
42. McAuliffe SM, Morgan SL, Wyant GA, Tran LT, Muto KW, Chen YS, et al. Targeting Notch, a key pathway for ovarian cancer stem cells, sensitizes tumors to platinum therapy. Proc Natl Acad Sci U S A. 2012;109(43):E2939–48.

43. Huang B, Jolly MK, Lu M, Tsarfaty I, Ben-Jacob E, Onuchic JN. Modeling the transitions between collective and solitary migration phenotypes in cancer metastasis. Sci Rep. 2015;5:17379.
44. Li P, Yang R, Gao WQ. Contributions of epithelial-mesenchymal transition and cancer stem cells to the development of castration resistance of prostate cancer. Mol Cancer. 2014;13:55.
45. Peinado H, Olmeda D, Cano A. Snail, Zeb and bHLH factors in tumour progression: an alliance against the epithelial phenotype? Nat Rev Cancer. 2007;7(6):415–28.
46. Wu Y, Zhou BP. Inflammation: a driving force speeds cancer metastasis. Cell Cycle. 2009;8(20):3267–73.
47. Garzon R, Fabbri M, Cimmino A, Calin GA, Croce CM. MicroRNA expression and function in cancer. Trends Mol Med. 2006;12(12):580–7.
48. Sakaki-Yumoto M, Katsuno Y, Derynck R. TGF-beta family signaling in stem cells. Biochim Biophys Acta. 2013;1830(2):2280–96.
49. Saigusa S, Tanaka K, Toiyama Y, Yokoe T, Okugawa Y, Ioue Y, et al. Correlation of CD133, OCT4, and SOX2 in rectal cancer and their association with distant recurrence after chemoradiotherapy. Ann Surg Oncol. 2009;16(12):3488–98.
50. Moran VA, Perera RJ, Khalil AM. Emerging functional and mechanistic paradigms of mammalian long non-coding RNAs. Nucleic Acids Res. 2012;40(14):6391–400.
51. Ohnuki M, Tanabe K, Sutou K, Teramoto I, Sawamura Y, Narita M, et al. Dynamic regulation of human endogenous retroviruses mediates factor-induced reprogramming and differentiation potential. Proc Natl Acad Sci U S A. 2014;111(34):12426–31.
52. Martin L, Chang HY. Uncovering the role of genomic "dark matter" in human disease. J Clin Invest. 2012;122(5):1589–95.
53. Caruso S, Bazan V, Rolfo C, Insalaco L, Fanale D, Bronte G, et al. MicroRNAs in colorectal cancer stem cells: new regulators of cancer stemness? Oncogenesis. 2012;1:e32.
54. Monzo M, Navarro A, Bandres E, Artells R, Moreno I, Gel B, et al. Overlapping expression of microRNAs in human embryonic colon and colorectal cancer. Cell Res. 2008;18(8):823–33.
55. Yu Y, Kanwar SS, Patel BB, Oh PS, Nautiyal J, Sarkar FH, et al. MicroRNA-21 induces stemness by downregulating transforming growth factor beta receptor 2 (TGFbetaR2) in colon cancer cells. Carcinogenesis. 2012;33(1):68–76.
56. Yu Y, Nangia-Makker P, Farhana L, S GR, Levi E, Majumdar AP. miR-21 and miR-145 cooperation in regulation of colon cancer stem cells. Mol Cancer. 2015;14:98.
57. Jahid S, Sun J, Edwards RA, Dizon D, Panarelli NC, Milsom JW, et al. miR-23a promotes the transition from indolent to invasive colorectal cancer. Cancer Discov. 2012;2(6):540–53.
58. Saarikangas J, Mattila PK, Varjosalo M, Bovellan M, Hakanen J, Calzada-Wack J, et al. Missing-in-metastasis MIM/MTSS1 promotes actin assembly at intercellular junctions and is required for integrity of kidney epithelia. J Cell Sci. 2011;124(Pt 8):1245–55.
59. Sancho R, Jandke A, Davis H, Diefenbacher ME, Tomlinson I, Behrens A. F-box and WD repeat domain-containing 7 regulates intestinal cell lineage commitment and is a haploinsufficient tumor suppressor. Gastroenterology. 2010;139(3):929–41.
60. Babaei-Jadidi R, Li N, Saadeddin A, Spencer-Dene B, Jandke A, Muhammad B, et al. FBXW7 influences murine intestinal homeostasis and cancer, targeting Notch, Jun, and DEK for degradation. J Exp Med. 2011;208(2):295–312.
61. Siemens H, Jackstadt R, Kaller M, Hermeking H. Repression of c-Kit by p53 is mediated by miR-34 and is associated with reduced chemoresistance, migration and stemness. Oncotarget. 2013;4(9):1399–415.
62. Bu P, Chen KY, Chen JH, Wang L, Walters J, Shin YJ, et al. A microRNA miR-34a-regulated bimodal switch targets Notch in colon cancer stem cells. Cell Stem Cell. 2013;12(5):602–15.
63. Neumuller RA, Knoblich JA. Dividing cellular asymmetry: asymmetric cell division and its implications for stem cells and cancer. Genes Dev. 2009;23(23):2675–99.
64. Dey-Guha I, Wolfer A, Yeh AC, J GA, Darp R, Leon E, et al. Asymmetric cancer cell division regulated by AKT. Proc Natl Acad Sci U S A. 2011;108(31):12845–50.
65. O'Brien CA, Kreso A, Ryan P, Hermans KG, Gibson L, Wang Y, et al. ID1 and ID3 regulate the self-renewal capacity of human colon cancer-initiating cells through p21. Cancer Cell. 2012;21(6):777–92.
66. Pine SR, Ryan BM, Varticovski L, Robles AI, Harris CC. Microenvironmental modulation of asymmetric cell division in human lung cancer cells. Proc Natl Acad Sci U S A. 2010;107(5):2195–200.
67. van Es JH, van Gijn ME, Riccio O, van den Born M, Vooijs M, Begthel H, et al. Notch/gamma-secretase inhibition turns proliferative cells in intestinal crypts and adenomas into goblet cells. Nature. 2005;435(7044):959–63.
68. Yoo YD, Kwon YT. Molecular mechanisms controlling asymmetric and symmetric self-renewal of cancer stem cells. J Anal Sci Technol. 2015;6(1):28.
69. Hwang WL, Jiang JK, Yang SH, Huang TS, Lan HY, Teng HW, et al. MicroRNA-146a directs the symmetric division of Snail-dominant colorectal cancer stem cells. Nat Cell Biol. 2014;16(3):268–80.
70. Lerner RG, Petritsch C. A microRNA-operated switch of asymmetric-to-symmetric cancer stem cell divisions. Nat Cell Biol. 2014;16(3):212–4.

71. Yu XF, Zou J, Bao ZJ, Dong J. miR-93 suppresses proliferation and colony formation of human colon cancer stem cells. World J Gastroenterol. 2011;17(42):4711–7. doi:10.3748/wjg.v17.i42.4711.
72. Zheng L, Zhang Y, Liu Y, Zhou M, Lu Y, Yuan L, et al. MiR-106b induces cell radioresistance via the PTEN/PI3K/AKT pathways and p21 in colorectal cancer. J Transl Med. 2015;13:252.
73. Zhang Y, Zheng L, Huang J, Gao F, Lin X, He L, et al. MiR-124 Radiosensitizes human colorectal cancer cells by targeting PRRX1. PLoS ONE. 2014;9(4):e93917.
74. Chen J, Chen Y, Chen Z. MiR-125a/b regulates the activation of cancer stem cells in paclitaxel-resistant colon cancer. Cancer Invest. 2013;31(1):17–23. doi: 10.3109/07357907.2012.743557.
75. Douville J, Beaulieu R, Balicki D. ALDH1 as a functional marker of cancer stem and progenitor cells. Stem Cells Dev. 2009;18(1):17–25.
76. Schwickart M, Huang X, Lill JR, Liu J, Ferrando R, French DM, et al. Deubiquitinase USP9X stabilizes MCL1 and promotes tumour cell survival. Nature. 2010;463(7277):103–7.
77. Mani SA, Guo W, Liao MJ, Eaton EN, Ayyanan A, Zhou AY, et al. The epithelial-mesenchymal transition generates cells with properties of stem cells. Cell. 2008;133(4):704–15.
78. Hollier BG, Evans K, Mani SA. The epithelial-to-mesenchymal transition and cancer stem cells: a coalition against cancer therapies. J Mammary Gland Biol Neoplasia. 2009;14(1):29–43.
79. Wellner U, Schubert J, Burk UC, Schmalhofer O, Zhu F, Sonntag A, et al. The EMT-activator ZEB1 promotes tumorigenicity by repressing stemness-inhibiting microRNAs. Nat Cell Biol. 2009;11(12):1487–95.
80. Burk U, Schubert J, Wellner U, Schmalhofer O, Vincan E, Spaderna S, et al. A reciprocal repression between ZEB1 and members of the miR-200 family promotes EMT and invasion in cancer cells. EMBO Rep. 2008;9(6):582–9.
81. Korpal M, Lee ES, Hu G, Kang Y. The miR-200 family inhibits epithelial-mesenchymal transition and cancer cell migration by direct targeting of E-cadherin transcriptional repressors ZEB1 and ZEB2. J Biol Chem. 2008;283(22):14910–4.
82. Park SM, Gaur AB, Lengyel E, Peter ME. The miR-200 family determines the epithelial phenotype of cancer cells by targeting the E-cadherin repressors ZEB1 and ZEB2. Genes Dev. 2008;22(7):894–907.
83. Pichler M, Ress AL, Winter E, Stiegelbauer V, Karbiener M, Schwarzenbacher D, et al. MiR-200a regulates epithelial to mesenchymal transition-related gene expression and determines prognosis in colorectal cancer patients. Br J Cancer. 2014;110(6):1614–21.
84. Ono H, Imoto I, Kozaki K, Tsuda H, Matsui T, Kurasawa Y, et al. SIX1 promotes epithelial-mesenchymal transition in colorectal cancer through ZEB1 activation. Oncogene. 2012;31(47):4923–34.
85. Lu YX, Yuan L, Xue XL, Zhou M, Liu Y, Zhang C, et al. Regulation of colorectal carcinoma stemness, growth, and metastasis by an miR-200c-Sox2-negative feedback loop mechanism. Clin Cancer Res. 2014;20(10):2631–42.
86. Ju SY, Chiou SH, Su Y. Maintenance of the stemness in CD44(+) HCT-15 and HCT-116 human colon cancer cells requires miR-203 suppression. Stem Cell Res. 2014;12(1):86–100.
87. Song B, Wang Y, Xi Y, Kudo K, Bruheim S, Botchkina GI, et al. Mechanism of chemoresistance mediated by miR-140 in human osteosarcoma and colon cancer cells. Oncogene. 2009;28(46):4065–74.
88. Song B, Wang Y, Titmus MA, Botchkina G, Formentini A, Kornmann M, et al. Molecular mechanism of chemoresistance by miR-215 in osteosarcoma and colon cancer cells. Mol Cancer. 2010;9:96.
89. Jones MF, Hara T, Francis P, Li XL, Bilke S, Zhu Y, et al. The CDX1-microRNA-215 axis regulates colorectal cancer stem cell differentiation. Proc Natl Acad Sci U S A. 2015;112(13):E1550–8.
90. Zhu R, Yang Y, Tian Y, Bai J, Zhang X, Li X, et al. Ascl2 knockdown results in tumor growth arrest by miRNA-302b-related inhibition of colon cancer progenitor cells. PLoS ONE. 2012;7(2):e32170.
91. Xu XT, Xu Q, Tong JL, Zhu MM, Nie F, Chen X, et al. MicroRNA expression profiling identifies miR-328 regulates cancer stem cell-like SP cells in colorectal cancer. Br J Cancer. 2012;106(7):1320–30.
92. Fang Y, Gu X, Li Z, Xiang J, Chen Z. miR-449b inhibits the proliferation of SW1116 colon cancer stem cells through downregulation of CCND1 and E2F3 expression. Oncol Rep. 2013;30(1):399–406.
93. Bitarte N, Bandres E, Boni V, Zarate R, Rodriguez J, Gonzalez-Huarriz M, et al. MicroRNA-451 is involved in the self-renewal, tumorigenicity, and chemoresistance of colorectal cancer stem cells. Stem Cells. 2011;29(11):1661–71.
94. Vermeulen L, De Sousa EMF, van der Heijden M, Cameron K, de Jong JH, Borovski T, et al. Wnt activity defines colon cancer stem cells and is regulated by the microenvironment. Nat Cell Biol. 2010;12(5):468–76.
95. Zhang H, Li W, Nan F, Ren F, Wang H, Xu Y, et al. MicroRNA expression profile of colon cancer stem-like cells in HT29 adenocarcinoma cell line. Biochem Biophys Res Commun. 2011;404(1):273–8.
96. Vaiopoulos AG, Kostakis ID, Koutsilieris M, Papavassiliou AG. Colorectal cancer stem cells. Stem Cells. 2012;30(3):363–71.
97. Guttman M, Amit I, Garber M, French C, Lin MF, Feldser D, et al. Chromatin signature reveals over a thousand highly conserved large non-coding RNAs in mammals. Nature. 2009;458(7235):223–7.
98. Khalil AM, Guttman M, Huarte M, Garber M, Raj A, Rivea Morales D, et al. Many human large intergenic noncoding RNAs associate with chromatin-

modifying complexes and affect gene expression. Proc Natl Acad Sci U S A. 2009;106(28):11667–72.

99. Wang G, Li Z, Zhao Q, Zhu Y, Zhao C, Li X, et al. LincRNA-p21 enhances the sensitivity of radiotherapy for human colorectal cancer by targeting the Wnt/beta-catenin signaling pathway. Oncol Rep. 2014;31(4):1839–45.
100. Boo K, Bhin J, Jeon Y, Kim J, Shin HJ, Park JE, et al. Pontin functions as an essential coactivator for Oct4-dependent lincRNA expression in mouse embryonic stem cells. Nat Commun. 2015;6:6810.
101. Sheik Mohamed J, Gaughwin PM, Lim B, Robson P, Lipovich L. Conserved long noncoding RNAs transcriptionally regulated by Oct4 and Nanog modulate pluripotency in mouse embryonic stem cells. RNA. 2010;16(2):324–37.
102. Loewer S, Cabili MN, Guttman M, Loh YH, Thomas K, Park IH, et al. Large intergenic non-coding RNA-RoR modulates reprogramming of human induced pluripotent stem cells. Nat Genet. 2010;42(12):1113–7.
103. Gui X, Li H, Li T, Pu H, Lu D. Long noncoding RNA CUDR regulates HULC and beta-catenin to govern human liver stem cell malignant differentiation. Mol Ther. 2015.
104. Dragu DL, Necula LG, Bleotu C, Diaconu CC, Chivu-Economescu M. Therapies targeting cancer stem cells: current trends and future challenges. World J Stem Cells. 2015;7(9):1185–201.
105. Fanale D, Caruso S, Bazan V, Bronte G, Di Piazza F, Rolfo C et al. MicroRNAs in colorectal cancer drug resistance: shooters become targets. J Carcinog Mutag. 2013;04(02).

6 MicroRNA Methylation in Colorectal Cancer

Sippy Kaur, Johanna E. Lotsari-Salomaa, Riitta Seppänen-Kaijansinkko, and Päivi Peltomäki

Abstract

Epigenetic alterations such as DNA methylation, histone modifications and non-coding RNA (including microRNA) associated gene silencing have been identified as a major characteristic in human cancers. These alterations may occur more frequently than genetic mutations and play a key role in silencing tumor suppressor genes or activating oncogenes, thereby affecting multiple cellular processes. In recent years, studies have shown that microRNAs, that act as posttranscriptional regulators of gene expression are frequently deregulated in colorectal cancer (CRC), via aberrant DNA methylation. Over the past decade, technological advances have revolutionized the field of epigenetics and have led to the identification of numerous epigenetically dysregulated miRNAs in CRC, which are regulated by CpG island hypermethylation and DNA hypomethylation. In addition, aberrant DNA methylation of miRNA genes holds a great promise in several clinical applications such as biomarkers for early screening, prognosis, and therapeutic applications in CRC.

Keywords

MicroRNA • Epigenetic regulation • DNA methylation • Colorectal cancer

S. Kaur (✉) • R. Seppänen-Kaijansinkko
Department of Oral and Maxillofacial Diseases, University of Helsinki and Helsinki University Hospital, Helsinki, Finland
e-mail: sippy.kaur@helsinki.fi

J.E. Lotsari-Salomaa • P. Peltomäki
Department of Medical and Clinical Genetics, University of Helsinki, Helsinki, Finland

6.1 Introduction to Epigenetic Mechanisms, Genomic Instability and Colon Cancer

Epigenetic mechanisms are hereditary and reversible changes in gene regulation without alterations in DNA. These modifications may be divided to covalent (DNA methylation and histone modifications) or non-covalent (chromatin

O. Slaby, G.A. Calin (eds.), *Non-coding RNAs in Colorectal Cancer*, Advances in Experimental Medicine and Biology 937, DOI 10.1007/978-3-319-42059-2_6

remodeling and small non-coding RNAs) subgroups. The most common epigenetic phenomenon is DNA methylation which regulates gene activity [1]. 60 % of methylation patterns are detected within CG-rich regions called CpG islands, located at the 5′ end of human gene promoters [2, 3]. In cancer, DNA methylation can be used to silence the function of tumor suppressor genes controlling many essentials functions such as proliferation, differentiation or promoting cell cycle arrest or apoptosis [4, 5]. However, it should be noticed that methylation also regulates the silencing of gene expression as part of normal development or to maintain homeostasis in a cell by inactivating X chromosome or by silencing the imprinted genes [3]. The methylation process in mammals is catalyzed by DNA methyltransferase (DNMT) enzymes, so-called maintenance methylator DNMT1 and de novo methylators DNMT3A and DNMT3B [4]. In addition to tumor suppressor gene silencing by promoter methylation, tumorigenesis may benefit from global hypomethylation in both gene-coding and non-coding regions of the genome to increase chromosomal instability (CIN) and to activate proto-oncogenes and other transposable elements [4, 5].

CIN occurs in 80 % of sporadic colon cancers (CRC)s [6]. CIN leads to missegregation of chromosomes and aneuploidy or unbalanced structural rearrangements and loss of heterozygosity (LOH) [6, 7]. Colon carcinomas are generally divided to microsatellite-stable (MSS) and microsatellite-unstable (MSI) cancers [6]. The term CpG island methylator phenotype (CIMP) is a distinct form of epigenomic instability which indicates high frequency of CpG island promoter methylation of tumor suppressor genes originally described in sporadic colon carcinomas [8]. Aging promotes methylation in normal colonic mucosa but methylation of tumor suppressor genes is seen at high frequency especially with CIMP-positive colon tumors which have specific molecular and clinical features [9]. If the *MLH1* gene is among the methylated genes in colon carcinoma, microsatellite instability (MSI) is detected [10]. Failure in proof-reading mechanisms leads to numeric errors in microsatellite repeats when deleted or inserted nucleotides are not corrected in the tumor genome. MSI is detected in 10–15 % of sporadic colon tumors [11, 12].

Micro-RNAs (miRNAs) are an abundant class of non-coding RNAs whose abnormalities accompany multistep colorectal tumorigenesis [13]. Increased expression of miR-135 suppresses the gatekeeper gene *APC*, which results in the Wnt pathway activation and early adenoma formation. Dysregulation of additional miRNAs leads to activation of EGFR signaling, inactivation of TGF-β response, loss of p53 function, and epithelial-mesenchymal transition as tumorigenesis proceeds towards adenomas with increasing dysplasia, CRC, and metastasis [13]. CRCs are composed of various subtypes based on histological characteristics and molecular features, including MSI and CIMP status. MSI can result from overexpression of miRNAs targeting mismatch repair genes, including miR-155 [14] and miR-21 [14]. Conversely, precursor genes for miRNAs (e.g., miR-1303) may contain repeat sequences that can be targets for MSI in CRC [15]. Expression profiling has demonstrated unique patterns of miRNAs depending on the MSI status of CRCs [16, 17]. While the mechanistic basis of the associations between miRNA expression and MSI mostly remains unknown, they may have important clinical significance since CRCs with vs. without MSI and CIMP have distinct clinical outcomes [18]. In addition to expression patterns, the methylation status of miRNAs has also been shown to be associated with MSI status [19, 20]. Epigenetic regulation of miRNAs and its correlations with clinicopathological parameters of CRCs will be discussed in greater detail below.

6.2 Epigenetic Regulation of Micro-RNAs

Mature miRNAs are approximately 22 nucleotide single-stranded non-coding RNAs which post-transcriptionally inhibit gene expression in a sequence-specific manner [21]. These small genes take part in controlling several essential

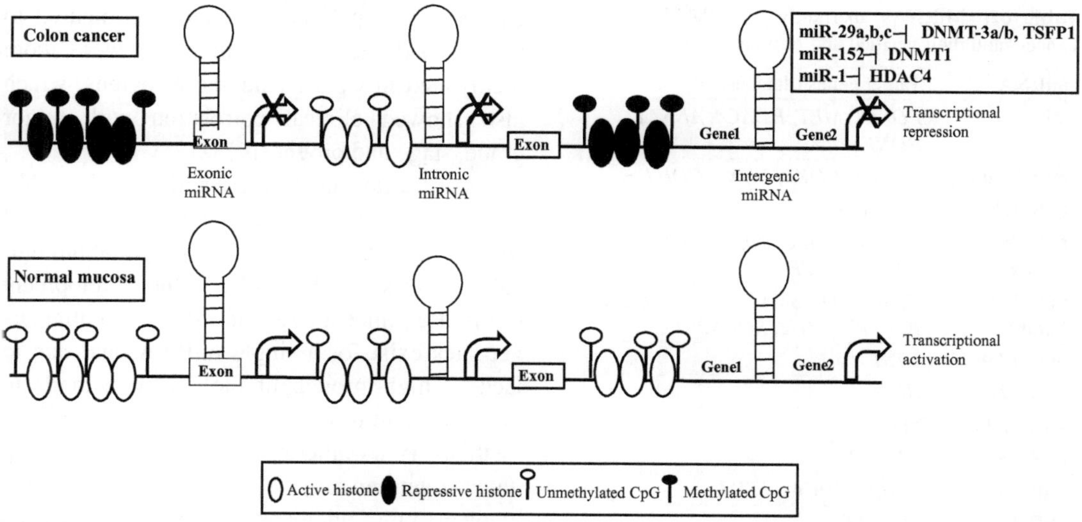

Fig. 6.1 Illustration of epigenetic alterations of miRNA genes in colorectal cancer

functions in a cell including embryonic development, cell differentiation, proliferation, and apoptosis [22, 23]. The first described miRNA, *line-4*, was identified in the genetic model organism *Caenorhabditis elegans* [24]. miRNAs can be encoded from gene-coding, non-coding, and intronic regions of the genome, and be oriented in antisense or sense direction relative to the host gene [21, 23]. Some miRNAs are clustered as polycistronic transcripts regulated together, for example, with certain developmental stages [21]. The maturation of miRNAs starts from primary miRNA (pri-miRNA) and can proceed along canonical and non-canonical pathways; the former pathway involves the enzyme Drosha to generate precursor miRNA (pre-miRNA) whereas the latter pathway ("mirtrons") uses the splicing machinery for the same purpose [23, 25].

There are two separate miRNA-related ways to control gene activity [22]. If miRNA binds complementary to its target messenger-RNAs (mRNAs) the RNA-mediated interference (RNAi) pathway will be induced leading to degradation of mRNA [22]. The initiation of translation will be inhibited when miRNAs are imperfectly binding to the 3′ untranslated regions (UTRs) of target mRNAs [22].

In addition to chromosomal abnormalities, miRNA expression is regulated by epigenetic mechanisms, such as promotor methylation or histone modifications to promote tumorigenesis as shown in Fig. 6.1 [22, 26]. Approximately half of the miRNAs are associated with CpG islands, suggesting that DNA methylation is crucially involved in the normal regulation as well as dysregulation of miRNAs. Based on the annotation of their genomic location, most of the miRNAs are located within intergenic regions (transcribed independently), although their location within exonic and intronic regions of the host gene (co-transcribed) either in the sense or antisense orientation is also observed [27]. A comprehensive characterization to map miRNA promoter sequences in order to identify the role of epigenetic mechanisms in the regulation of miRNA expression revealed that the distance of miRNA promoter sequences to the miRNA coding sequence can span up to up to 50 kb. One third of intragenic miRNAs have independent transcription sites and more than half of all miRNA transcription sites are associated with a CpG island. The probability of an independent promoter increases when the distance of the mature miRNA sequence to the host gene promoter exceeds 10 kb, whereas miRNAs in the vicinity of the host gene promoter are generally co-expressed with the host transcript. Moreover, as the protein coding genes, miRNA promoters are also associ-

Table 6.1 MiRNAs aberrantly methylated in various cancers and their verified and/or putative target genes

miRNA	Target genes (databases)[a]
miR-1-1	*FOXP1, MET, HDAC4, ANXA2, BDNF*
miR-9 family	*FGFR1, CDK6, CDX2, CDH1*
miR-10b	*MAPRE1*
miR-34 family	*MET, CDK4, CCNE2, C-MYC, CDK6, E2F3, NOTCH4*
miR-124 family	*CDK6, VIM, SMYD3, E2F6, IQGAP1, IGFBP7, EZH2, ROCK2*
miR-125b	*ETS1*
miR-127	*BCL6*
miR-129-2	*SOX4*
miR-132	*TALIN2*
miR-137	*CDK6, CDC42, LSD1*
miR-143	*MLL-AF4*
miR-148a	*TGIF2*
miR-152	*DNMT1, E2F3, MET, RICTOR*
miR-181c	*NOTCH4, KRAS*
miR-193a	*SRSF2, KIT*
miR-200 family	*ZEB1, ZEB2*
miR-203	*ABL1, ABCE1, CDK6*
miR-205	*ZEB1, ZEB2*
miR-218	*RICTOR*
miR-335	*SOX4, TNC*
miR-345	*BAG3*
miR-375	*IGF1R, PDK1*
miR-512	*MCL1*

[a]www.mirtarbase.mbc, www.targetscan.org

ated with TATA-box elements, TFIIB recognition elements or an initiator [28–30].

Advances in microarray and sequencing technologies have enabled comprehensive analysis of the epigenome and miRNA expression in cancer cells, which has led to the identification of miRNA promotors which are frequent targets of aberrant DNA methylation in cancer, followed by the identification and verification of their mRNA targets (Table 6.1). A majority of studies, including colon cancer-related investigations to be described below, are focused on methylation of CpG islands located in the promoter regions of miRNAs. In addition to CpG islands, CpG island shores, the regions located within 2 kb of CpG islands are also involved in regulation of miRNAs and protein coding genes [31]. In bladder cancer it was pointed out that as compared to CpG island methylation, miRNAs were more methylated in CpG island shore regions, which also showed clinical correlation with tumor grade, stage and prognosis [32].

Identification and further validation of miRNA targets is critical for understanding the functional role of miRNAs in the context of normal biological processes and their roles in the development of disease. Since a large number of potential target sites exist for a single miRNA, and due to lack of high-throughput biological methods to identify the miRNA targets, many computational methods (www.microrna.org/; http://mirdb.org; www.targetscan.org, www.mirtarbase.mbc, www.ma.uni-heidelberg.de/apps/zmf/mirwalk/) are freely available for predicting miRNA targets which facilitates the process of narrowing down putative targets for experimental validation.

Tumor suppressive miRNAs, whose expression is decreased in cancer, target oncogenes whereas oncogenic miRNAs, which show elevated expression, target suppressor genes. In addition to controlling growth-regulatory genes, miRNAs can control the epigenetic machinery by targeting the components responsible for DNA methylation and histone modifications. Such miRNAs are called epi-miRNAs. Several epi-miRNA identified in various cancers are listed in Table 6.2. The mir-29 family directly targets DNMT3A and DNMT3B in lung cancer [33] and represented the first epi-miRNAs identified. In prostate carcinoma cells, miR-34b can target both methyltransferase enzymes and histone deacetylases (HDACs) [34]. To sum up, miRNAs can serve as drivers of tumor-suppressive and oncogenic functions, and tumorigenesis-related dysregulation of certain miRNAs may induce aberrant methylation of specific gene promoters [30].

Besides promoter CpG island methylation which silences tumor suppressor genes, global hypomethylation is also observed in various cancers mediating overexpression of oncogenes. Hypomethylation generally occurs at repetitive sequences including LINE-1 [35] and several miRNAs are also upregulated by hypomethylation. The first hypomethylated miRNA identi-

Table 6.2 List of epi-miRNAs targeting enzymes of the epigenetic machinery in various cancers

miRNA	Epi targets	References
miR-29a, b, c	*DNMT3A, DNMT3B, TSFP1*	Fabbri et al. [33]
miR-124a	*EZH2*	Zheng et al. [45]
miR-148a, b	*DNMT3B*	Duursma et al. [104]
miR-152, miR-301	*DNMT1*	Braconi et al. [105]
miR-101	*EZH2*	Varambally et al. [106]
miR-1	*HDAC4*	Chen et al. [107]
miR-449a	*HDAC1*	Noonan et al. [108]
miR-137	*LSD1/KDM1A, KDM5, EZH2*	Althoff et al. [60], Denis et al. [61], Ren et al. [62]

fied was *let-7a-3* in lung cancer. It was methylated in normal tissue and unmethylated in cancer tissue which caused an oncogenic effect [36]. Since then many hypomethylated miRNAs in cancers have been identified [37, 38]. Therefore, DNA hyper- and hypo-methylation leading to silencing of tumor suppressor genes or activating oncogenes, respectively, indicates a mechanism used by the tumor cell to favor carcinogenesis.

6.2.1 Methylation of miRNA Genes in Colorectal Cancer

The CIMP phenotype appears early in colorectal tumorigenesis. For example, certain antagonists of Wnt signaling may be inactivated by promoter methylation already in transition from normal colonic epithelium to aberrant crypt foci, the earliest detectable lesions of the colon [39]. Since methylation of miRNA-associated CpG islands has been shown to correlate with methylation of conventional tumor suppressor genes [19], it is conceivable that methylation of many miRNAs, too, takes place early in CRC development. A systematic screen for miRNAs regulating the canonical Wnt pathway identified 38 candidate miRNAs that either repress or activate the canonical Wnt pathway [40]. The former group of miRNAs includes several known to be targeted by CpG island methylation, such as miR-1 silenced in early and advanced CRC and miR-200a altered late in tumorigenesis (see below). The following paragraphs will provide brief individual descriptions of selected CRC-associated miRNAs.

miR-124 Family The first evidence of epigenetic silencing of miR-124 was reported in CRC [41]. Since then this miRNA is known to be methylated in several other cancers. MiR-124 is associated with CpG islands and is encoded by three independent loci (miR-124-1, 2 and 3). Methylation of miR-124 family is reported in over 70 % of CRC tissues and in hematological malignancies, and also in cervical cancer [42–44]. Tumor suppressor effect of miR-124 is caused by targeting various oncogenes, such as the *CDK6* (cyclin-dependent kinase 6), which is activated as a result of miR-124 silencing, leading to phosphorylation and inactivation of the retinoblastoma protein [27]. MiR-124 also functions as epi-miRNA by suppressing *EZH2*, a key component of the polycomb repressive complex two and responsible for H3K27 methylation [45].

miR-9 Family In breast and pancreatic cancer, methylation of the CpG island in miR-9 was first identified [46, 47]. Methylation of other family members of miR-9 (miR-9-1, 2, 3) have been reported in various cancers including lymph node metastasis in CRC [48–50]. MiR-9 targets *FGFR1*, *CDK6*, *CDX2*, and *CDH1* genes [51–53] and functions as tumor suppressor miRNA.

miR-34 Family miR-34 family members (miR-34a and mir-34b/c) are direct targets of p53. MiR-34a is located on chromosome 1p36, while mir-34b/c are co-transcribed from chromosome 11q23. All members of miR-34 family are targets of CpG island methylation in CRC and various other cancers such as esophageal and oral [54, 55]. Methylation of miR-34b/c was associated with cancer metastases and invasion in several cancers. MiR-34 members act as tumor suppressors targeting *MET, CDK4, CCNE2* and *MYC* [56, 57].

miR-200 Family MiR-200 gene family contains five members which are divided into two polycistronic units: miR-200b/200a/429 located on chromosome one and miR-200c/141 located on chromosome 12. This miRNA family and miR-205 encode important epithelial-mesenchymal transition regulators and it directly targets *ZEB1* and *ZEB2* which are direct repressors of the E-cadherin [58]. In colorectal cancer it was demonstrated that CpG island DNA methylation of the miR-200 family was a dynamic process which could be shifted to hypermethylation or unmethylation depending on epithelial or mesenchymal origin of the cells. CpG islands were unmethylated in cancers displaying epithelial features and methylated in cancer cells with mesenchymal characteristics. Thus epigenetic silencing of the miR-200 family, specifically miR-200c plays an important role in metastatic behavior of CRC [59].

miR-137 This miRNA was first reported in oral cancer, and then subsequently identified in colon and other cancers. MiR-137 is embedded in a CpG island and reported to be frequently methylated in several cancers [48, 55]. Candidate genes targeted by miR-137 include *AURKA*, *CDK6* [55], and several members of the epigenetic machinery, such as the histone demethylases *LSD1/KDM1A* [60] and *KDM5* [61] and the histone methyltransferase *EZH2* [62]. miR-137 acts a tumor suppressor and is frequently silenced in colorectal adenomas through promoter hypermethylation, which suggests that silencing of miR-137 by methylation is an early event in CRC, which could have prognostic and therapeutic implications [48, 63].

Other miRNA Genes/Families CpG island methylation of miR-1 has been reported in CRC and other malignancies [30, 64, 65]. MiR-1 was reported to be frequently methylated in early and advanced CRC and it may act as a tumor suppressor. Ectopic expression of miR-1 in CRC suppressed cell growth, motility, and invasion and two novel targets identified for miR-1 are *ANXA2* and *BDNF* [30, Tables 6.1 and 6.2].

Numerous miRNA expression studies have demonstrated differential expression of miRNAs depending on MMR status [16, 17, 66, 67] miRNA methylation profiles in the context of microsatellite instability (MSI) status, even though less studied, have also been demonstrated to discriminate tumors according to their MMR status. For example, a genome-wide miRNA assay was utilized to explore the miRNA signature in CRC displaying MSS and MSI phenotypes [20]. Eight miRNAs were identified that could distinguish the MSI status of CRCs. Due to promotor methylation miR-484 expression was significantly decreased in the MSI CRC group compared with the MSS group. MiR-484 was identified as a tumor suppressor miRNA which targeted and repressed the *CD137L*, leading to decreased production of IL-8 and possibly, attenuated anti-tumor immune response in MSI-CRC cells [20].

Recently, three miRNAs (miR-129-2, miR-345, and miR-132) emerged as promising targets for subgroup-specific methylation in Finnish and Australian CRC populations [68, Fig. 6.2]. Methylation of miR-132 was linked with sporadic MSI CRC as compared to MSS CRC and moreover, distinguished sporadic MSI CRC from Lynch-CRC. In the clinical correlation analysis for methylated miR-132, MSI CRCs showed a significant association with miR-132 methylation and female gender, increased age and proximal location in the bowel. In the same study, miR-129-2 and miR-345 hypermethylation were also more frequent in MSI than MSS CRC tumors (the difference was statistically significant for miR-345) [68].

6.2.2 Methylation of Other Non-coding RNA Genes

Besides miRNAs, non-coding RNA includes several other classes of genes that are highly abundant and functionally important [69]. These are broadly divided into two groups based on their size: small noncoding RNAs, which are less than 200 nucleotides and constitute RNAs such as piRNA and small nucleolar RNA, and long non-

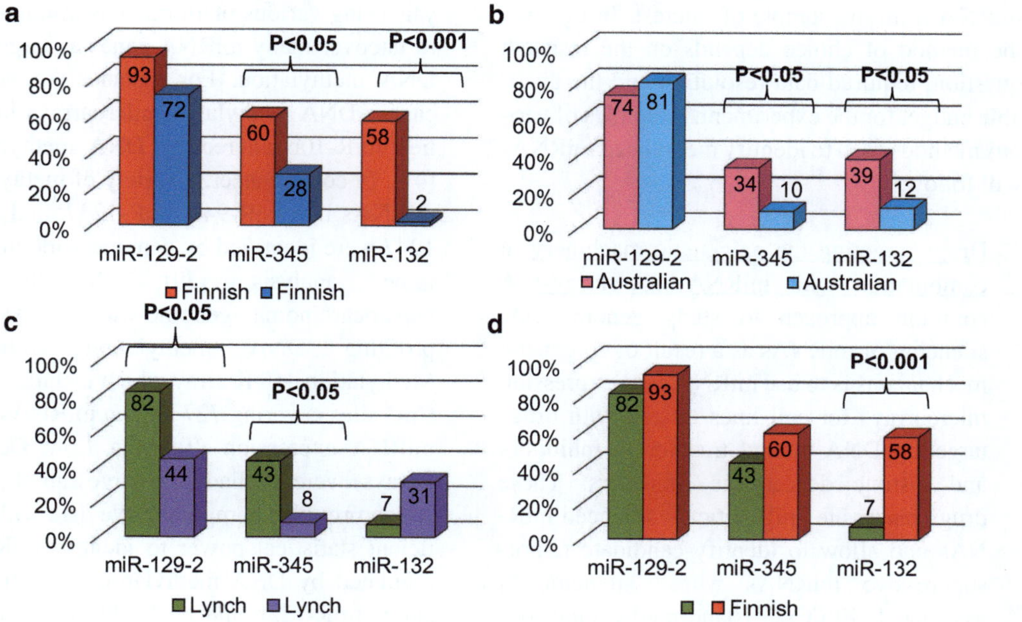

Fig. 6.2 Methylation frequency of miRNA genes in various colon cancer groups. Average hypermethylation frequencies of miRNA genes are shown between Finnish sporadic MSI vs. MSS tumors (**a**); Australian sporadic MSI vs. MSS tumors (**b**); LS colorectal vs. endometrial tumors (**c**); LS colorectal vs. Finnish sporadic MSI tumors (**d**). Y axis indicates percentage of tumors with hypermethylation

coding RNAs that are longer than 200 nucleotides.

Long non-coding (lnc) RNAs are the second most commonly studied classes of non-coding RNAs after miRNAs. Recently it was reported by Di Ruscio and colleagues [70] that lncRNA termed *ecCEBPA* arising from the *CEBPA* gene locus, by virtue of association with DNA methyltransferase 1 (DNMT1) regulates DNA methylation patterns. This study was only focused on DNMT1, and it is very likely that other DNA methyltransferase enzymes may also show similar association with lncRNA. Using RNA co-immunoprecipitation (RIP) followed by next generation RNA sequencing (RIP-seq) a subset of lncRNAs were identified that interact with DNMT1 in a colon cancer cell line. The lncRNA DACOR1 identified with this method depicted high tissue-specific expression in the normal colon, but was repressed in a panel of colon tumors [71]. Accordingly, their study showed that deregulation of DNMT1-associated lncRNAs contributes to aberrant DNA methylation and gene expression during colon tumorigenesis. Collectively, regulation of DNA methylation by lncRNAs indicates new regulatory functions for noncoding RNAs which are likely to be important in controlling gene expression during development and disease. Furthermore, in addition to DNA methylation, specific interactions between lncRNAs and several chromatin-modifying complexes have been identified and it has also been demonstrated that these interactions are required for regulating gene expression [72].

6.3 Methods to Identify Methylation of miRNA Genes

One of the most common mechanisms by which tumor suppressor genes are inactivated during tumorigenesis is via epigenetic gene silencing due to promoter CpG island hypermethylation. As described above, accumulating evidence indicates that in addition to coding genes, non-coding genes such as miRNAs are also targets of epigenetic silencing in cancer. Currently there are several methods available to identify methylated miRNA from the sample of interest. In the end, the method of choice depends on the research question, required data resolution and the available budget for the experiment. Examples of various methods used to identify methylated miRNAs will follow.

1. Drugs targeting the epigenetic machinery in combination with miRNA microarrays: A common approach to study genome-wide silencing of miRNAs as a result of epigenetic mechanisms is to use miRNA gene expression microarrays for cell lines treated with drugs targeting DNA methyl transferase inhibitors and histone deacetylase inhibitors. These drugs reactivate epigenetically silenced miRNAs and allow to identify candidate tumor-suppressive miRNAs whose silencing is associated with CpG island methylation. For example, in bladder cancer, 17 genes were upregulated by drug treatment [73]. Among these, miR-127 was shown to be embedded within a CpG island and was regulated by both DNA methylation and histone modifications. Subsequently, this miRNA was experimentally verified to target the proto-oncogene B-cell lymphoma 6 (*BCL6*). Among other miRNAs, miR-373, miR-345, and miR-133b were identified by drug screens in CRC [74–76]. Similarly, miR-345 and miR-132 were identified as novel differentially methylated miRNAs allowing the sub classification of CRC, and miR-129 methylation turned out to be a marker of progression in early endometrial tumorigenesis [68]. By unmasking epigenetically silenced miRNAs by drug treatment, a number of other methylated miRNAs have been identified in several cancers [55, 77, 78]. An alternative approach to identify epigenetically regulated miRNAs is to use RNA from genetically disrupted cell lines for profiling on microarrays. In CRC, the HCT116 cell line where the DNA methyltransferases DNMT1 and DNMT3B were genetically knocked out (DNMT KO) was utilized to identify epigenetically regulated miRNAs [41].
2. Genome-wide methylation analysis by microarrays: Genome wide DNA methylation analysis using various platforms has also enabled to uncover many miRNA genes as targets of DNA methylation. For instance, in gastric cancer DNA methylation microarrays identified miR-10b altered by DNA methylation [47]. In colon cancer, a variety of methylated miRNAs like miR-941, miR-1237, and miR-1247 were identified by genome-wide methylome analysis [79]. Recently, in hepatocarcinoma genome-wide arrays for profiling DNA methylation (Infinium Methylation 450 K array which includes 3439 CpG sites covering 727 human miRNAs) and miRNA expression (TaqMan Low Density Arrays) were applied on a large sample size, which provided comprehensive data with sufficient statistical power to identify miRNAs regulated by DNA methylation [80]. In that study, miR-125b and miR-199a were shown to be dramatically regulated by DNA hypermethylation, supporting their tumor suppressor role in the repression of downstream target oncogenes which play a key role in hepatocarcinogenesis. In breast cancer, DNA methylation of miRNA gene promoters was comprehensively evaluated by using 5-methylcytosine immunoprecipitation combined with miRNA tiling microarray hybridization [81]. One third (55/167) of miRNA promoters were targets for aberrant methylation, which surprisingly exceeds the percentage of protein coding genes targeted for aberrant DNA methylation in cancer. MiR-31, miR-130a, let-7a-3/let-7b, miR-155 and miR-34b/mir-34c were found to be silenced by aberrant DNA methylation [81].
3. Targeted methylation analysis by different methods: Methylation-specific multiplex ligation-dependent probe amplification (MS-MLPA, MRC-Holland, Netherlands) is based on selective amplification of methylated sequences that the methylation-sensitive enzyme *HhaI* is unable to cut [82]. It has been proven to be a semiquantitative, convenient and fast technique for evaluating the methylation status of multiple sequences simultaneously in all types of samples including formalin-fixed paraffin embedded (FFPE) tis-

sues [35, 68, 83]. Even 10 ng of DNA is enough for methylation analysis and no sodium bisulphite conversion is needed [82, 84]. The sensitivity limit of MS-MLPA is considered to be 10 % and lower methylation levels are usually interpreted as background [82]. The methylation statuses of multiple gene loci can be detected in the same analysis, and gene copy number variations as well as point mutations analyzed [82, 85]. The methylation analyses are limited to GCGC sites recognized by *HhaI* [85]. With paraffin-embedded tumors, the methylation analysis will be restricted to a specific block and not the whole tumor [85]. Other options for targeted methylation analysis are also available such as methylation sensitive PCR (MSP) and combined bisulphite restriction analysis (COBRA). Bisulphite conversion of DNA is needed for the methylation analysis and also the quality of DNA is critical with these methods, since the amplified genomic region will be restricted [86, 87].

4. Next generation sequencing: Next-generation sequencing (NGS) of the methyl-CpG binding domain allows an in-depth focus on the methylated regions in the genome. For instance in breast cancer, in order to clarify the association between DNA methylation and transcription of miRNAs, NGS and microarrays were used to analyze the methylation and expression of miRNAs and other genes. Though this approach epigenetic similarities and differences between miRNA and protein-coding genes were identified [88]. In pancreatic cancer, hypermethylation of miR-130b and miR-210 were discovered by a combination of several methods such as methyl capture sequencing (methylCap-seq), methylation-specific PCR (MSP), bisulfite sequencing PCR (BSP), and methylation sensitive restriction enzyme-based qPCR [89].
5. Functional screening: Tumor suppressor miRNAs silenced by DNA hypermethylation were identified by a function-based screening and a series of sequential analyses in endometrial and oral cancer [90, 91]. The screen included 327 synthetic miRNAs, of which nearly 100 showed growth suppressive effects, and half were associated with CpG islands. Integrated DNA methylation and expression analysis identified miR-218 and miR-152 as targets of DNA methylation in oral and endometrial cancer, respectively.

6.4 Clinical Application of Epigenetically Silenced miRNAs

The expression profiles of miRNAs can be useful indicators for detecting disease and predicting disease outcome, and epigenetically silenced miRNAs also hold a promise as biomarkers in cancer management [92, 93]. Methylation of tumor suppressor miR-34b/c in case of non-small cell lung cancer was associated with a high incidence of recurrence and poor overall survival [94]. Methylation of miR-124 family genes in acute lymphoblastic leukemia was identified as an independent prognostic factor for disease-free and overall survival [42]. In another study on acute lymphoblastic leukemia, 13 methylated miRNA genes (miR-9-1, miR-9-2, miR-9-3, miR-10b, miR-34b/c, miR-124-1, miR-124-2, miR-124-3, miR-132/212, miR-196b, and miR-203) were identified. At least one methylated miRNA gene was an important prognostic factor since patients with miRNA methylation displayed significantly poorer disease-free and overall survival than the unmethylated patient group [95]. In non-small cell lung cancer, methylation of several miRNA gene loci is strongly linked with larger tumor size and also poorer progression-free survival [96].

Regarding translational applications of CRC-related miRNAs specifically, epigenetic silencing of miR-124a and miR-34b/c occurred frequently in normal adjacent colonic mucosa, and likewise miR-342 methylation was detected in adenomas as well as in normal colorectal mucosa [97], indicating that methylation of these miRNAs is an early event in CRC. Moreover, methylation of miR-124a in adjacent normal colonic mucosa was associated with MSI status of CRCs, while methylation of miR-34b/c correlated with older age at diagnosis [98]. MiR-34b/c methylation was also evaluated in fecal specimens of CRC patients, which represents a noninvasive method for early CRC detection [99]. Methylation of miR-375 was associated with lymph node metastasis and worse histological type in CRC [75]. In another investigation, miR-204-5p was found to be downregulated by promoter methylation in CRC and adjacent noncancerous tissue and shown to be a marker of poor prognosis through overexpression of *RAB22*, a member of the RAS oncogene family and a direct target of miR-204-5p [100]. The authors suggested that therapeutic overexpression of miR-204-5p might be useful in CRC by targeting RAB22A and thereby improving sensitivity to chemotherapy. The associations of miR-132 and miR-345 methylation with sporadic MSI (vs. MSS) CRC [68] were discussed above. The methylation status of certain CRC-relevant protein-coding genes can already be monitored in blood plasma or stool as commercial tests [101], and it is likely that similar diagnostic or prognostic tests based on epigenetically regulated miRNAs will become available in the near future.

Some of the epigenetically silenced miRNA genes possess tumor suppressor potential, and restoration of their expression by demethylation drugs has a major potential as an alternative strategy in cancer treatment. Several experiments using *in vitro* and *in vivo* models have demonstrated the promising effect that restoring the expression of methylation-silenced tumor suppressor miRNAs can have on inhibiting cancer growth by downregulating target oncogenes [41, 57]. For example, restoration of expression of the tumor suppressive miR-34 family in pancreatic cancer led to the reduction of tumor initiating cells [102]. Analogous therapeutic opportunities exist in the case of miRNAs methylated in CRC [103].

6.5 Conclusions

Discovery of the functional role of noncoding RNA and particularly miRNAs over two decades ago has revolutionized our knowledge regarding their impact on carcinogenesis and also provided important insights into their potential use as clinically important biomarkers of disease. It is well known that miRNA expression is epigenetically regulated and methylation of miRNAs is involved in CRC pathogenesis. As DNA methylation, a reversible process, is mainly catalyzed by members of the DNMT family, DNMT inhibitors may be suitable as demethylation drugs and have a great potential as anticancer agents. More extensive studies concerning the methylation of miRNAs are required to fully elucidate the role of methylated miRNAs in cancer initiation and progression and their biomarker potential (see Fig. 6.2). The ongoing developments in research on aberrant DNA methylation and alterations in noncoding RNAs suggest that epigenetic alterations might be routinely used in the diagnosis and treatment of CRC in the future.

References

1. Sharma S, Kelly TK, Jones PA. Epigenetics in cancer. Carcinogenesis. 2010;31:27–36.
2. Medvedeva YA, Fridman MV, Oparina NJ. Intergenic, gene terminal, and intragenic CpG islands in the human genome. BMC Genomics. 2010;11.
3. Bird A. DNA methylation patterns and epigenetic memory. Genes Dev. 2002;16:6–21.
4. Li W, Chen BF. Aberrant DNA methylation in human cancers. J Huazhong Univ Sci Technolog. 2013;33:798–804.
5. Feinberg AP, Ohlsson R, Henikoff S. The epigenetic progenitor origin of human cancer. Nature Rev Genet. 2006;7:21–33.
6. Migliore L, Migheli F, Spisni R, Coppede F. Genetics, cytogenetics, and epigenetics of colorectal cancer. J Biomed Biotechnol. 2011;2011:1–19.
7. Pino MS, Chung DC. The chromosomal instability pathway in colon cancer. Gastroenterology. 2010;138:2059–72.
8. Toyota M, Ahuja N, Ohe-Toyota M, Herman JG, Baylin SB, Issa JP. CpG island methylator phenotype in colorectal cancer. Proc Natl Acad Sci U S A. 1999;96:8681–6.
9. Issa JP. Colon cancer: it's CIN or CIMP. Clin Cancer Res. 2009;19:5939–40.
10. Weisenberger DJ, Siegmund KD, Campan M, Young J, Long TI, Faasse MA, et al. CpG island methylator phenotype underlies sporadic microsatellite instability and is tightly associated with BRAF mutation in colorectal cancer. Nat Genet. 2006;38:787–93.

11. Jass JR. Classification of colorectal cancer based on correlation of clinical, morphological and molecular features. Histopathology. 2007;50:113–30.
12. de la Chapelle A. Microsatellite instability. N Eng J Med. 2003;349:209–10.
13. Fearon ER. Molecular genetics of colorectal cancer. Annu Rev Pathol: Mech Dis. 2011;6:479–507. doi:10.1146/annurev-pathol-011110-130235.
14. Valeri N, Gasparini P, Fabbri M. Modulation of mismatch repair and genomic stability by miR-155. Proc Natl Acad Sci U S A. 2010;15:6982–7.
15. El-Murr N, Abidi Z, Wanherdrick K, Svrcek M, Gaub MP, Flejou JF, et al. MiRNA genes constitute new targets for microsatellite instability in colorectal cancer. PLoS One. 2012;2:e31862.
16. Balaguer F, Moreira L, Lozano JJ, Link A, Ramirez G, Shen Y, et al. Colorectal cancers with microsatellite instability display unique miRNA profiles. Clin Cancer Res. 2011;17:6239–49.
17. Sarver AL, French AJ, Borralho PM. Human colon cancer profiles show differential microRNA expression depending on mismatch repair status and are characteristic of undifferentiated proliferative states. BMC Cancer. 2009;9:401.
18. Ogino S, Goel A. Molecular classification and correlates in colorectal cancer. J Mol Diagn. 2008;1:13–27.
19. Pavicic W, Perkio E, Kaur S, Peltomaki P. Altered methylation at microRNA-associated CpG islands in hereditary and sporadic carcinomas: a methylation-specific multiplex ligation-dependent probe amplification (MS-MLPA)-based approach. Mol Med. 2011;17:726–35.
20. Mei Q, Xue G, Li X, Wu Z, Li X, Yan H, et al. Methylation-induced loss of miR-484 in microsatellite-unstable colorectal cancer promotes both viability and IL-8 production via CD137L. J Pathol. 2015;236:165–74.
21. He L, Hannon GJ. MicroRNAs: small RNAs with a big role in gene regulation. Nat Rev Genet. 2004;5:522–31.
22. Suzuki H, Maruyama R, Yamamoto E, Kai M. Epigenetic alteration and microRNA dysregulation in cancer. Front Genet. 2013;4:258.
23. Li Z, Rana TM. Therapeutic targeting of microRNAs: current status and future challenges. Nat Rev Drug Dis. 2014;13:622–38.
24. Lee RC, Feinbaum RL, Ambros V. The C. elegans heterochronic gene lin-4 encodes small RNAs with antisense complementarity to lin-14. Cell. 1993;75:843–54.
25. Chan SP, Slack FJ. And now introducing mammalian mirtrons. Dev Cell. 2007;5:605–7.
26. Saito Y, Jones PA. Epigenetic activation of tumor suppressor microRNAs in human cancer cells. Cell Cycle. 2006;5:2220–2.
27. Weber B, Stresemann C, Brueckner B, Lyko F. Methylation of human microRNA genes in normal and neoplastic cells. Cell Cycle. 2007;6:1001–5.
28. Chien CH, Sun YM, Chang WC, Chiang Hsieh PY, Lee TY, Tsai WC, et al. Identifying transcriptional start sites of human microRNAs based on high-throughput sequencing data. Nucleic Acids Res. 2011;39:9345–56.
29. Ozsolak F, Poling LL, Wang Z, Liu H, Liu XS, Roeder RG, et al. Chromatin structure analyses identify miRNA promoters. Genes Dev. 2008;22:3172–83.
30. Suzuki H, Takatsuka S, Akashi H, Yamamoto E, Nojima M, Maruyama R, et al. Genome-wide profiling of chromatin signatures reveals epigenetic regulation of MicroRNA genes in colorectal cancer. Cancer Res. 2011;71:5646–58.
31. Irizarry RA, Ladd-Acosta C, Wen B, Wu Z, Montano C, Onyango P, et al. The human colon cancer methylome shows similar hypo- and hypermethylation at conserved tissue-specific CpG island shores. Nat Genet. 2009;41:178–86.
32. Dudziec E, Miah S, Choudhry HM, Owen HC, Blizard S, Glover M, et al. Hypermethylation of CpG islands and shores around specific microRNAs and mirtrons is associated with the phenotype and presence of bladder cancer. Clin Cancer Res. 2011;17:1287–96.
33. Fabbri M, Garzon R, Cimmino A, Liu Z, Zanesi N, Callegari E, et al. MicroRNA-29 family reverts aberrant methylation in lung cancer by targeting DNA methyltransferases 3A and 3B. Proc Natl Acad Sci U S A. 2007;104:15805–10.
34. Majid S, Dar AA, Saini S, Shahryari V, Arora S, Zaman MS, et al. miRNA-34b inhibits prostate cancer through demethylation, active chromatin modifications, and AKT pathways. Clin Cancer Res. 2013;1:73–84.
35. Pavicic W, Joensuu EI, Nieminen T, Peltomaki P. LINE-1 hypomethylation in familial and sporadic cancer. J Mol Med (Berl). 2012;90:827–35.
36. Brueckner B, Stresemann C, Kuner R, Mund C, Musch T, Meister M, et al. The human let-7a-3 locus contains an epigenetically regulated microRNA gene with oncogenic function. Cancer Res. 2007;67:1419–23.
37. Baer C, Claus R, Frenzel LP, Zucknick M, Park YJ, Gu L, et al. Extensive promoter DNA hypermethylation and hypomethylation is associated with aberrant microRNA expression in chronic lymphocytic leukemia. Cancer Res. 2012;72:3775–85.
38. He XX, Kuang SZ, Liao JZ, Xu CR, Chang Y, Wu YL, et al. The regulation of microRNA expression by DNA methylation in hepatocellular carcinoma. Mol Biosyst. 2015;11:532–9.
39. Lao VV, Grady WM. Epigenetics and colocrectal cancer. Nat Rev Gastroenterol Hepatol. 2011;12:686–700.
40. Anton R, Chatterjee SS, Simunda J, Cowin P, Dasgupta R. A systematic screen for micro-RNAs regulating the canonical Wnt pathway. PLoS One. 2011;10:e26257.

41. Lujambio A, Esteller M. CpG island hypermethylation of tumor suppressor microRNAs in human cancer. Cell Cycle. 2007;6:1455–9.
42. Agirre X, Vilas-Zornoza A, Jimenez-Velasco A, Martin-Subero JI, Cordeu L, Garate L, et al. Epigenetic silencing of the tumor suppressor microRNA Hsa-miR-124a regulates CDK6 expression and confers a poor prognosis in acute lymphoblastic leukemia. Cancer Res. 2009;69:4443–53.
43. Wong KY, So CC, Loong F, Chung LP, Lam WW, Liang R, et al. Epigenetic inactivation of the miR-124-1 in haematological malignancies. PLoS One. 2011;6:e19027.
44. Wilting SM, van Boerdonk RA, Henken FE, Meijer CJ, Diosdado B, Meijer GA, et al. Methylation-mediated silencing and tumour suppressive function of hsa-miR-124 in cervical cancer. Mol Cancer. 2010;9:1–14. doi:10.1186/1476-4598-9-167.
45. Zheng F, Liao YJ, Cai MY, Liu YH, Liu TH, Chen SP, et al. The putative tumour suppressor microRNA-124 modulates hepatocellular carcinoma cell aggressiveness by repressing ROCK2 and EZH2. Gut. 2012;2:278–89.
46. Lehmann U, Hasemeier B, Christgen M, Muller M, Romermann D, Langer F, et al. Epigenetic inactivation of microRNA gene hsa-mir-9-1 in human breast cancer. J Pathol. 2008;214:17–24.
47. Omura N, Li CP, Li A, Hong SM, Walter K, Jimeno A, et al. Genome-wide profiling of methylated promoters in pancreatic adenocarcinoma. Cancer Biol Ther. 2008;7:1146–56.
48. Bandres E, Agirre X, Bitarte N, Ramirez N, Zarate R, Roman-Gomez J, et al. Epigenetic regulation of microRNA expression in colorectal cancer. Int J Cancer. 2009;125:2737–43.
49. Hildebrandt MA, Gu J, Lin J, Ye Y, Tan W, Tamboli P, et al. Hsa-miR-9 methylation status is associated with cancer development and metastatic recurrence in patients with clear cell renal cell carcinoma. Oncogene. 2010;29:5724–8.
50. Tsai KW, Liao YL, Wu CW, Hu LY, Li SC, Chan WC, et al. Aberrant hypermethylation of miR-9 genes in gastric cancer. Epigenetics. 2011;6:1189–97.
51. Rodriguez-Otero P, Roman-Gomez J, Vilas-Zornoza A, Jose Eneriz ES, Martin-Palanco V, Rifon J, et al. Deregulation of FGFR1 and CDK6 oncogenic pathways in acute lymphoblastic leukaemia harbouring epigenetic modifications of the MIR9 family. Br J Haematol. 2010;1:73–83.
52. Rotkrua P, Akiyama Y, Hashimoto Y, Otsubo T, Yuasa Y. Mir-9 downregulation CDX2 expression in gastric cancer cells. Int J Cancer. 2011;11:2611–20.
53. Ma L, Young J, Prabhala MV, Pan E, Mestdagh P, Muth D, et al. miR-9, a MYC/MYCN-activated microRNA, regulates E-cadherin and cancer metastasis. Nat Cell Biol. 2010;3:247–56.
54. Chen X, Hu H, Guan X, Xiong G, Wang Y, Wang K, Li J, et al. CpG island methylation status of miRNAs in esophageal squamous cell carcinoma. Int J Cancer. 2012;130:1607–13.
55. Kozaki K, Imoto I, Mogi S, Omura K, Inazawa J. Exploration of tumor-suppressive microRNAs silenced by DNA hypermethylation in oral cancer. Cancer Res. 2008;68:2094–105.
56. Lujambio A, Calin GA, Villanueva A, Ropero S, Sanchez-Cespedes M, Blanco D, Lujambio A, et al. A microRNA DNA methylation signature for human cancer metastasis. Proc Natl Acad Sci U S A. 2008;105:13556–61.
57. Toyota M, Suzuki H, Sasaki Y, Maruyama R, Imai K, Shinomura Y, et al. Epigenetic silencing of microRNA-34b/c and B-cell translocation gene 4 is associated with CpG island methylation in colorectal cancer. Cancer Res. 2008;68:4123–32.
58. Park SM, Gaur AB, Lengyel E, Peter ME. The miR-200 family determines the epithelial phenotype of cancer cells by targeting the E-cadherin repressors ZEB1 and ZEB2. Genes Dev. 2008;22:894–907.
59. Davalos V, Moutinho C, Villanueva A, Boque R, Silva P, Carneiro F, et al. Dynamic epigenetic regulation of the microRNA-200 family mediates epithelial and mesenchymal transitions in human tumorigenesis. Oncogene. 2012;31:2062–74.
60. Althoff K, Beckers A, Odersky A, Mestdagh P, Köster J, Bray IM, et al. MiR-137 functions as a tumor suppressor in neuroblastoma by downregulating KDM1A. Int J Cancer. 2013;5:1064–673.
61. Denis H, Van Grenbergen O, Delatte B, Dedeurwaerder S, Putmans S, Calonne E, et al. MicroRNAs regulate KDM5 histone demethylases in breast cancer cells. Mol Biosyst. 2015;12:404–13.
62. Ren X, Bai X, Zhang X, Li Z, Tang L, Zhao X, et al. Quantitative nuclear proteomics identifies that miR-137-mediated EZH2 reduction regulates resveratrol-induced apoptosis of neuroblastoma cells. Mol Cell Proteomics. 2015;2:316–28.
63. Balaguer F, Link A, Lozano JJ, Cuatrecasas M, Nagasaka T, Boland RC, et al. Epigenetic silencing of miR-137 is an early event in colorectal carcinogenesis. Cancer Res. 2010;70:6609–18.
64. Datta J, Kutay H, Nasser MW, Nuovo GJ, Wang B, Majumder S, et al. Methylation mediated silencing of MicroRNA-1 gene and its role in hepatocellular carcinogenesis. Cancer Res. 2008;68:5049–58.
65. Rao PK, Missiaglia E, Shields L, Hyde G, Yuan B, Shepherd CJ, et al. Distinct roles for miR-1 and miR-133a in the proliferation and differentiation of rhabdomyosarcoma cells. FASEB J. 2010;24:3427–37.
66. Lanza G, Ferracin M, Gafa R, Veronese A, Spizzo R, Pichiorri F, et al. mRNA/microRNA gene expression profile in microsatellite unstable colorectal cancer. Mol Cancer. 2007;6:54.
67. Earle JS, Luthra R, Romans A, Abraham R, Ensor J, Yao H, et al. Association of microRNA expression with microsatellite instability status in colorectal adenocarcinoma. J Mol Diagn. 2010;12:433–40.

68. Kaur S, Lotsari JE, Al-Sohaily S, Warusavitarne J, Kohonen-Corish MR, Peltomaki P. Identification of subgroup-specific miRNA patterns by epigenetic profiling of sporadic and Lynch syndrome-associated colorectal and endometrial carcinoma. Clin Epigenetics. 2015;7:20. doi:10.1186/s13148-015-0059-3.
69. Taft RJ, Pang KC, Mercer TR, Dinger M, Mattick JS. Non-coding RNAs: regulators of disease. J Pathol. 2010;2:126–39.
70. Di Ruscio A, Ebralidze AK, Benoukraf T, Amabile G, Goff LA, Terragni J, et al. DNMT1-interacting RNAs block gene-specific DNA methylation. Nature. 2013;503:371–6.
71. Merry CR, Forrest ME, Sabers JN, Beard L, Gao XH, Hatzoglou M, et al. DNMT1-associated long non-coding RNAs regulate global gene expression and DNA methylation in colon cancer. Hum Mol Genet. 2015;24:6240–53.
72. Khalil AM, Guttman M, Huarte M, Garber M, Raj A, Rivea Morales D, et al. Many human large intergenic noncoding RNAs associate with chromatin-modifying complexes and affect gene expression. Proc Natl Acad Sci U S A. 2009;106:11667–72.
73. Saito Y, Liang G, Egger G, Friedman JM, Chuang JC, Coetzee GA, et al. Specific activation of microRNA-127 with downregulation of the proto-oncogene BCL6 by chromatin-modifying drugs in human cancer cells. Cancer Cell. 2006;9:435–43.
74. Tanaka T, Arai M, Wu S, Kanda T, Miyauchi H, Imazeki F, et al. Epigenetic silencing of microRNA-373 plays an important role in regulating cell proliferation in colon cancer. Oncol Rep. 2011;26:1329–35.
75. Tang JT, Wang JL, Du W, Hong J, Zhao SL, Wang YC, et al. MicroRNA 345, a methylation-sensitive microRNA is involved in cell proliferation and invasion in human colorectal cancer. Carcinogenesis. 2011;32:1207–15.
76. Lv LV, Zhou J, Lin C, Hu G, Yi LU, DU J, et al. DNA methylation is involved in the aberrant expression of miR-133b in colorectal cancer cells. Oncol Lett. 2015;10:907–12.
77. Hashimoto Y, Akiyama Y, Otsubo T, Shimada S, Yuasa Y. Involvement of epigenetically silenced microRNA-181c in gastric carcinogenesis. Carcinogenesis. 2010;31:777–84.
78. Saito Y, Suzuki H, Tsugawa H, Nakagawa I, Matsuzaki J, Kanai Y, et al. Chromatin remodeling at Alu repeats by epigenetic treatment activates silenced microRNA-512-5p with downregulation of Mcl-1 in human gastric cancer cells. Oncogene. 2009;28:2738–44.
79. Yan H, Choi AJ, Lee BH, Ting AH. Identification and functional analysis of epigenetically silenced microRNAs in colorectal cancer cells. PLoS One. 2011;6:e20628.
80. Shen J, Wang S, Siegel AB, Remotti H, Wang Q, Sirosh I, et al. Genome-Wide Expression of MicroRNAs Is Regulated by DNA Methylation in Hepatocarcinogenesis. Gastroenterol Res Pract. 2015;2015:1–12.
81. Vrba L, Munoz-Rodriguez JL, Stampfer MR, Futscher BW, et al. miRNA gene promoters are frequent targets of aberrant DNA methylation in human breast cancer. PLoS One. 2013;8:e54398.
82. Nygren AO, Ameziane N, Duarte HMB, Vijzelaar RNCP, Waisfisz Q, Hess CJ, et al. Methylation-specific MLPA (MS-MLPA): simultaneous detection of CpG methylation and copy number changes of up to 40 sequences. Nucleic Acids Res. 2005;33:e128.
83. Nieminen TT, Gylling A, Abdel-Rahman WM, Nuorva K, Aarnio M, Renkonen-Sinisalo L, et al. Molecular analysis of endometrial tumorigenesis: importance of complex hyperplasia regardless of atypia. Clin Cancer Res. 2009;15:5772–83.
84. Suijkerbuijk KP, Pan X, van der Wall E, van Diest PJ, Vooijs M. Comparison of different promoter methylation assays in breast cancer. Anal Cell Pathol. 2010;33:274–6.
85. Homig-Holzel C, Savola S. Multiplex ligation-dependent amplification (MLPA) in tumor diagnostics and prognostics. Diagn Mol Pathol. 2012;4:189–206.
86. Tournier B, Chapusot C, Courcet E, Martin L, Lepage C, Faivre J, et al. Why do results conflict regarding the prognostic value of the methylation status on colon cancer? The role of the preservation method. BMC Cancer. 2012;12:1–12.
87. Xiong Z, Laird PW. COBRA: a sensitive and quantitative DNA methylation assay. Nucleic Acids Res. 1997;12:2532–4.
88. Morita S, Takahashi RU, Yamashita R, Toyoda A, Horii T, Kimura M, et al. Genome-wide analysis of DNA methylation and expression of microRNAs in breast cancer cells. Int J Mol Sci. 2012;13:8259–72.
89. Zhao Y, Sun J, Zhang H, Guo S, Gu J, Wang W, et al. High-frequency aberrantly methylated targets in pancreatic adenocarcinoma identified via global DNA methylation analysis using methylCap-seq. Clin Epigenetics. 2014;6:18.
90. Tsuruta T, Kozaki K, Uesugi A, Furuta M, Hirasawa A, Imoto I, et al. miR-152 is a tumor suppressor microRNA that is silenced by DNA hypermethylation in endometrial cancer. Cancer Res. 2011;71:6450–62.
91. Uesugi A, Kozaki K, Tsuruta T, Furuta M, Morita K, Imoto I, et al. The tumor suppressive microRNA miR-218 targets the mTOR component Rictor and inhibits AKT phosphorylation in oral cancer. Cancer Res. 2011;71:5765–78.
92. Kong YW, Ferland-McCollough D, Jackson TJ, Bushell M. microRNAs in cancer management. Lancet Oncol. 2012;6:e249–58.
93. Iorio MV, Croce CM. MicroRNA dysregulation in cancer, diagnostics, monitoring and therapeutics. A comprehensive review. Mol Med. 2012;3:143–59.

94. Wang Z, Chen Z, Gao Y, Li N, Li B, Tan F, et al. DNA hypermethylation of microRNA-34b/c has prognostic value for stage non-small cell lung cancer. Cancer Biol Ther. 2011;11:490–6.
95. Roman-Gomez J, Agirre X, Jimenez-Velasco A, Arqueros V, Vilas-Zornoza A, Rodriguez-Otero P, et al. Epigenetic regulation of microRNAs in acute lymphoblastic leukemia. J Clin Oncol. 2009;27:1316–22.
96. Kitano K, Watanabe K, Emoto N, Kage H, Hamano E, Nagase T, et al. CpG island methylation of microRNAs is associated with tumor size and recurrence of non-small-cell lung cancer. Cancer Sci. 2011;12:2126–31.
97. Grady WM, Parkin RK, Mitchell PS, Lee JH, Kim YH, Tsuchiya KD, et al. Epigenetic silencing of the intronic microRNA hsa-miR-342 and its host gene EVL in colorectal cancer. Oncogene. 2008;27:3880–8.
98. Deng G, Kakar S, Kim YS. MicroRNA-124a and microRNA-34b/c are frequently methylated in all histological types of colorectal cancer and polyps, and in the adjacent normal mucosa. Oncol Lett. 2011;2:175–80.
99. Kalimutho M, Di Cecilia S, Del Vecchio BG, Roviello F, Sileri P, Cretella M, et al. Epigenetically silenced miR-34b/c as a novel faecal-based screening marker for colorectal cancer. Br J Cancer. 2011;104:1770–8.
100. Yin Y, Zhang B, Wang W, Fei B, Quan C, Zhang J, et al. miR-204-5p inhibits proliferation and invasion and enhances chemotherapeutic sensitivity of colorectal cancer cells by downreagulating RAB22A. Clin Cancer Res. 2014;23:6187–99.
101. Cheasley D, Jorissen RN, Liu S, et al. Genomic approach to translational studies in colorectal cancer. Transl Cancer Res. 2015;4:235–55.
102. Ji Q, Hao X, Zhang M, Tang W, Yang M, Li L, et al. MicroRNA miR-34 inhibits human pancreatic cancer tumor-initiating cells. PLoS One. 2009;4:e6816.
103. Okugawa Y, Grady WM, Goel A. Epigenetic alterations in colorectal cancer: emerging biomarkers. Gastroenterology. 2015;5:1204–25.
104. Duursma AM, Kedde M, Schrier M, le Sage C, Agami R. miR-148 targets human DNMT3b protein coding region. RNA. 2008;5:872–7. doi:10.1261/rna.972008.
105. Braconi C, Huang N, Patel T. MicroRNA-dependent regulation of DNA methyltransferase-1 and tumor suppressor gene expression by interleukin-6 in human malignant cholangiocytes. Hepatology. 2010;3:881–90. doi:10.1002/hep.23381.
106. Varambally S, Cao Q, Mani RS, Shankar S, Wang X, Ateeq B, et al. Genomic loss of microRNA-101 leads to overexpression of histone methyltransferase EZH2 in cancer. Science. 2008;322:1695–9. doi:10.1126/science.1165395.
107. Chen JF, Mandel EM, Thomson JM, Wu Q, Callis TE, Hammond SM, et al. The role of microRNA-1 and microRNA-133 in skeletal muscle proliferation and differentiation. Nat Genet. 2006;38:228–33.
108. Noonan EJ, Place RF, Pookot D, Basak S, Whitson JM, Hirata H, et al. miR-449a targets HDAC-1 and induces growth arrest in prostate cancer. Oncogene. 2009;14:1714–24. doi:10.1038/onc.2009.19.

Polymorphisms in Non-coding RNA Genes and Their Targets Sites as Risk Factors of Sporadic Colorectal Cancer

7

Pavel Vodicka, Barbara Pardini, Veronika Vymetalkova, and Alessio Naccarati

Abstract

Colorectal cancer (CRC) is a complex disease that develops as a consequence of both genetic and environmental risk factors in interplay with epigenetic mechanisms, such as microRNAs (miRNAs). CRC cases are predominantly sporadic in which the disease develops with no apparent hereditary syndrome. The last decade has seen the progress of genome-wide association studies (GWAS) that allowed the discovery of several genetic regions and variants associated with weak effects on sporadic CRC. Collectively these variants may enable a more accurate prediction of an individual's risk to the disease and its prognosis. However, the number of variants contributing to CRC is still not fully explored.

SNPs in genes encoding the miRNA sequence or in 3′UTR regions of the corresponding binding sites may affect miRNA transcription, miRNA processing, and/or the fidelity of the miRNA–mRNA interaction. These variants could plausibly impact miRNA expression and target mRNA translation into proteins critical for cellular integrity, differentiation, and proliferation.

P. Vodicka (✉)
Institute of Experimental Medicine, Academy of Sciences of the Czech Republic,
Videnska 1083, 142 00 Prague, Czech Republic

Institute of Biology and Medical Genetics, 1st Medical Faculty, Charles University,
Albertov 4, 128 00 Prague, Czech Republic

Biomedical Centre, Faculty of Medicine in Pilsen, Charles University in Prague,
323 00 Pilsen, Czech Republic
e-mail: pvodicka@biomed.cas.cz

B. Pardini • A. Naccarati
Institute of Experimental Medicine, Academy of Sciences of the Czech Republic,
Videnska 1083, 142 00 Prague, Czech Republic

Human Genetics Foundation - Torino (HuGeF),
via Nizza 52, 10126 Turin, Italy

V. Vymetalkova
Institute of Experimental Medicine, Academy of Sciences of the Czech Republic,
Videnska 1083, 142 00 Prague, Czech Republic

Institute of Biology and Medical Genetics, 1st Medical Faculty, Charles University,
Albertov 4, 128 00 Prague, Czech Republic

O. Slaby, G.A. Calin (eds.), *Non-coding RNAs in Colorectal Cancer*, Advances in Experimental Medicine and Biology 937, DOI 10.1007/978-3-319-42059-2_7

In the present chapter, we describe the different aspects of variations related to miRNAs and other non-coding RNAs (ncRNAs) and evidence from studies investigating these candidate genetic alterations in support to their role in CRC development and progression.

Keywords
Colorectal cancer • Risk factors • Polymorphism • SNP • miRSNP • miRNA target site

7.1 Introduction

Human genomic variation consists of changes in the sequence and structure of DNA that include single-nucleotide or multi-nucleotide variants, short insertions/deletions, copy number variants, and copy neutral inversions and translocations. These changes underlie a phenotypic variability either in the sense of adaptive traits and selective advantage or inherited susceptibility towards various diseases, including malignancy [1]. Single-nucleotide polymorphisms (SNPs), the most frequent form of genetic variation, are single-base-pair changes without any net gain or loss of genetic material and occur with a frequency >1 % in individuals. Recently, the 1000 Genomes Project has summed up more than 38 million SNPs (nearly a half was previously unknown) in 14 different populations worldwide [2]. The advent of new technologies for massive mapping of the entire spectrum of genetic variants at the genome-wide level has led to a better understanding of human genomic variation from the functional point of view (i.e. the effect on gene expression) and its contribution to disease development. However, genetic differences do not successfully explain the entire phenotypic variability. Epigenetic changes (such as DNA methylation and chromatin modification) are also important players in modulating gene expression [3, 4].

Colorectal cancer (CRC) is a complex disease that develops as a consequence of both genetic and environmental risk factors. Diet, smoking and drinking habit are among environmental factors frequently associated with CRC risk [5]. The majority of CRC cases is "sporadic" where the disease develops with no obvious hereditary syndrome (only 2–5 % of cases are in fact hereditary) [6]. Epidemiological studies have suggested interactions between genetic variations, environmental factors and sporadic CRC risk [4, 7, 8]. However, still little is known about the genetic mechanisms that contribute to the risk modulation, especially by common variants [9]. Polymorphisms in oncogenes, tumor suppressor genes, or genes modifying the colon microenvironment and those involved in the DNA repair, signaling as well as in genes implicated in other important pathways have been for long considered as possible candidates for modulating CRC susceptibility. Interestingly, those candidate SNPs have been rarely associated with CRC [10–12]. GWAS have enabled the discovery of over 40 genetic regions and variants associated with weak effects on sporadic CRC [6]. Some of the identified variants are in genes involved in important biological pathways (such as *SMAD7*, *MLH1*, *APC*) while others reside in the so-called gene desert regions. For the latter the mechanism by which they contribute to CRC remains unclear. Experimental validations have identified the variant(s) related to the disease in a limited number of these genetic regions. Thus, it remains still unknown whether the observed variants are causal or just surrogates that are in linkage disequilibrium with the real functional loci that could be represented by other coding/noncoding regions of the genome, with still undisclosed role [3].

Currently, over 2000 human microRNA (miRNA) sequences are reported in repositories (http://www.mirbase.org/) and additional thousands constitute other non-coding RNAs (ncRNAs). The functional association of noncoding RNAs with cancer, their small gene size and their potential to simultaneously affect a multitude of genes make them good candidate loci for conferring cancer susceptibility. SNPs in miRNA genes or miRNA machinery and in different categories of non-coding RNAs may have effects on CRC risk, prognosis and treatment response. In addition, miRNAs alone are thought to regulate expression of more than 1/3 of human protein-coding genes. Thus, in turn, each miRNA may potentially regulate hundreds of potential targets in the human genome. The identification of allele-specific miRNA:mRNA interactions may help us to understand the role of many SNPs for which functionality is still unknown.

7.2 Variations in miRNA Genes and Binding Sites: Modulation of miRNA-Gene Regulation

Nearly 98 % of human RNA is not translated into proteins and constitutes the so-called ncRNA. Small ncRNAs (miRNAs, small nucleolar RNAs (snoRNAs), transfer RNAs (tRNAs), endogenous small interfering RNAs (siRNAs) and PiWi-interacting RNAs (piRNAs)) were initially discarded as RNA turnover artefacts. However, the potentialities of small ncRNAs are enormous, since accumulating evidence revealed that they may have functional activities in normal and malignant cells [13]. miRNAs in particular, are a family of endogenous, short ncRNAs (~22 nucleotides in length) that modulate posttranscriptional gene regulation by binding to either full or partial complementary sequences primarily in the 3′ untranslated region (UTR). miRNAs constitutes an attractive biomarker source for cancer research since their altered expression have been observed in several cancer tissues [14]. The role of miRNAs in human cancer pathogenesis has been further confirmed by the identification of genetic alterations in miRNA loci, miRNA expression signatures that define different neoplastic phenotypes, and numerous oncogenes and tumor suppressor genes as miRNA targets [15].

Deregulation of miRNAs in cancer can occur through epigenetic changes (for example, promoter CpG island hypermethylation in the case of the miR-200 family [16]) and genetic alterations, which can affect the production of the primary miRNA transcripts, their processing to mature miRNAs and/or interactions with mRNA targets. The changes in miRNA/mRNA expression levels observed in CRC may be finely influenced or modulated by genetic variations that are residing in one or more of the elements involved in the regulation of gene expression.

The 3′UTR heterogeneity can derive from the presence of polymorphisms or alternative polyadenylation (APA) of mRNAs and strongly impacts miRNA-mediated posttranscriptional regulation. This aspect has to be taken into account for the prediction of miRNA target sites within mammalian cells. In humans, 3′UTR associated SNPs have been linked to malignant disease susceptibility [17], but the subsets of polymorphisms with a functional role in regulating gene expression are yet to be defined. Polymorphisms within miRNA binding sites may potentially disrupt miRNA binding or even to introduce novel binding sites in 3′UTRs, and the biological relevance of these polymorphisms is currently being examined in large case–control studies [18].

Considering the prevalence of miRNA-mediated gene regulation, sequence variations in miRNAs can significantly contribute to changes in critical cellular pathways and thus impact pathological processes. Genetic variations in miRNA sequences are unique as they can influence both the expression levels and functionality of these molecules (Fig. 7.1a, b). A SNP located in the crucial "seed" sequence of a miRNA affects its complementarity with target genes and leads to deregulation of multiple cellular pathways. Moreover, since the expression of miRNAs is highly tissue- and disease-specific, changes within the miRNA sequence can indeed specifically predispose particular organs to cancers and mediate different molecular changes in various tissues [19].

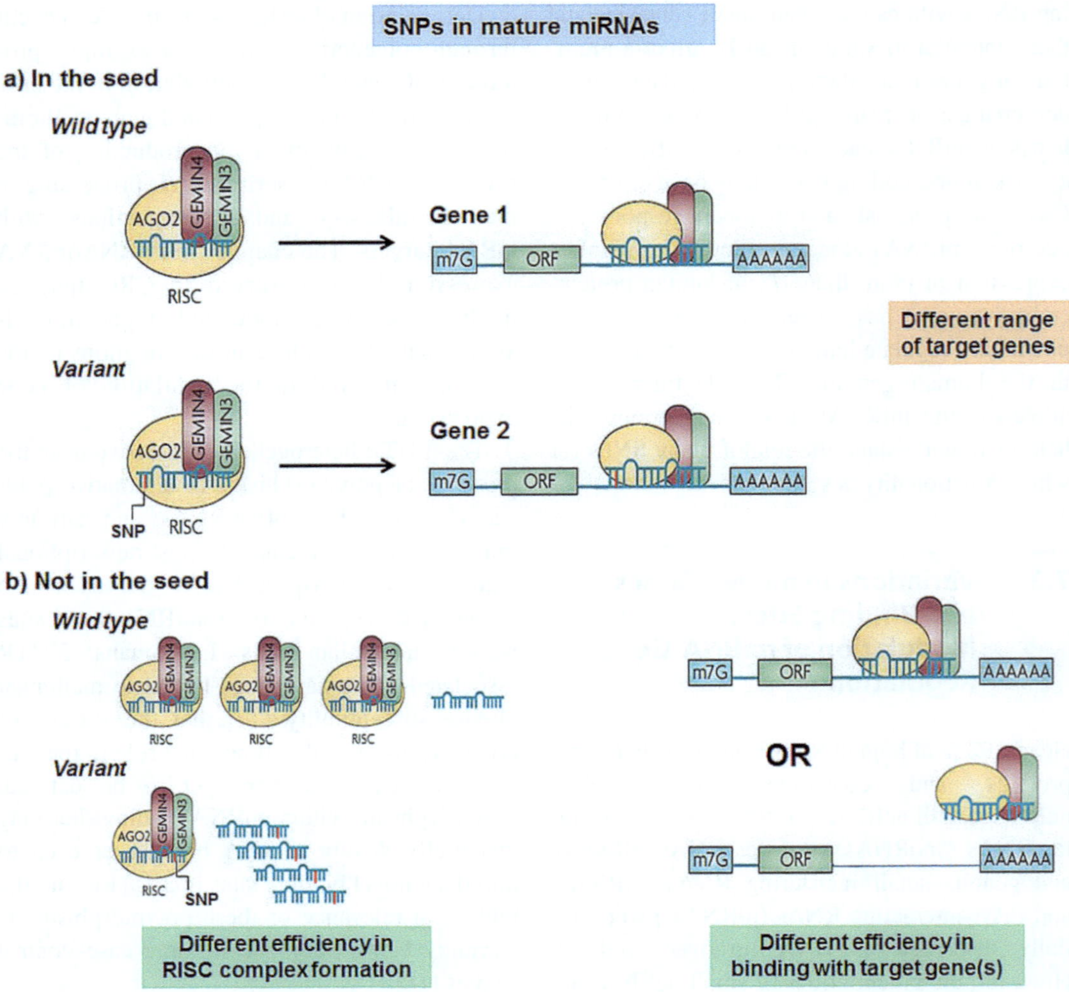

Fig. 7.1 Effect of genetic variations in mature miRNA sequences: (**a**) a SNP located in the crucial seed sequence may affect its complementarity with target genes and lead to deregulation of multiple cellular pathways; (**b**) variations outside the seed region may modify the efficiency of binding with RISC complex element (therefore changing the efficacy of post-transcriptional regulation) or alter the complementarity of binding with the target region

SNPs in miRNA target genes can influence miRNA binding by two distinct mechanisms. SNPs can directly disrupt or create a miRNA binding site altering the range of miRNAs targeting one gene (Fig. 7.2a). Alternatively, the presence of a SNP in target region may modify the efficacy of miRNA binding, with a consequent increased/decreased miRNA:mRNA binding efficiency (Fig. 7.2b). On the other hand, SNPs can interrupt or create novel polyadenylation (APA) signals that lead to transcript isoforms indirectly altering a miRNA binding site availability [20]. A screening of human SNPs indicates that a substantial fraction of them may potentially create or disrupt APA signals. Therefore, these last polymorphisms have been defined as APA-SNPs [21]. The polymorphisms altering miRNA seed region or modifying a miRNA binding site on mRNA are called miRSNPs [22, 23].

A single miRSNP may interfere with the function of a miRNA and consequently affect the expression of multiple genes involved in pathways regulating drug absorption, metabolism, disposition, stem cell function, DNA damage repair and the cell cycle, and impact the overall sensitivity/resistance towards the chemothera-

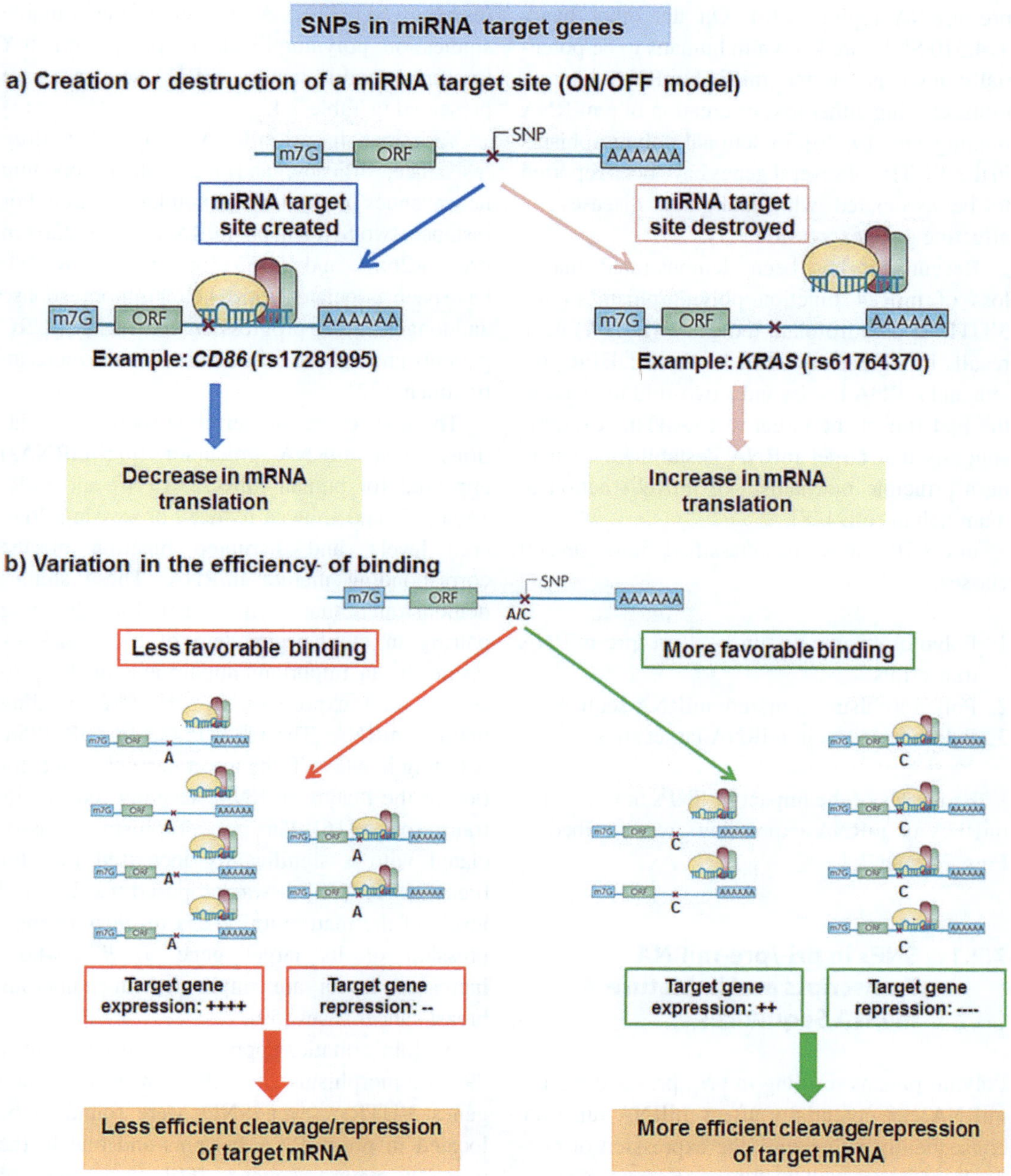

Fig. 7.2 Effect of genetic variations in miRNA binding site: (**a**) SNPs can disrupt or create a novel miRNA binding site generating selective mRNA translation; (**b**) SNPs can variate efficiency of binding creating an alteration/modulation of repression of target mRNA

peutic drug. It has been described that conserved miRNA binding sites may contribute to natural negative selection more than other conserved sequence motifs in 3′UTRs, demonstrating the contribution of miRNAs to Darwinian fitness. [24]. Analysis of the publicly available SNP database revealed the presence of a relatively high level of variations in the 3′UTR regions of miRNA target genes [25]. Conversely, there is relatively low level of variation in the miRNA seed region of a functional miRNA [25]. To date, there are only 1940 SNPs included in the pre-miRNA regions [26]. On the other hand, 414,510 SNPs are known in humans to be potentially involved in the miRNA:mRNA interactions, causing either loss or creation of a miRNA binding site [19, 26]. Functional polymorphisms in the 3′UTRs of several genes have been reported to be associated with malignant diseases by affecting gene expression [27].

Recently it has been demonstrated that a loss of miR24 function polymorphism in the 3′UTR of dihydrofolate reductase (*DHFR*) gene results in high steady-state levels of DHFR protein and mRNA levels, and a two-fold increase in the half-life of the target mRNA. This example suggests that target mRNA destabilization may be a principle mechanism of miRNA action in mammalian cells [28].

miRSNPs may be classified into several classes:

1. Polymorphisms in pri-/pre-miRNA transcripts
2. Polymorphisms in mature miRNA sequences
3. Polymorphisms in miRNA target sites

Examples of the impact of SNPs in relation to miRNA or mRNA expression are described in Figs. 7.1 and 7.2.

7.2.1 SNPs in pri-/pre-miRNA Transcripts and in Mature miRNA Sequences

Polymorphisms residing in pri-, pre- and mature miRNA can potentially affect miRNA function, consequently influencing the expression of hundreds of genes and pathways. Sequence variations in miRNA genes may influence the processing and/or target selection of miRNAs (Fig. 7.1a, b) [16]. To date, several studies have suggested that the functional consequences of SNPs in pri-miRNAs are related to molecule processing and affect levels of the mature form. For example, SNPs in the pri- regions of let-7e and miR-16 lead to decreased levels of their mature miRNA. Some of these polymorphisms have also been associated with cancer, but not with CRC (as reviewed in [29]). A summary of the available studies on polymorphisms in pri-/pre-miRNA transcripts and in mature miRNA sequences are presented in Table 7.1.

Variations in pri-miRNAs may affect drug resistance, efficacy, and metabolism, opening new avenues of pharmacogenomics research. For instance, two SNPs in pri-miRNAs (rs7372209 in pri-miR26a-1 and rs1834306 in pri-miR-100) have been associated with a better tumor response and longer time to progression in metastatic CRC patients undergoing 5-fluorouracil and irinotecan treatment [32].

The first reports on germline sequence variations in a miRNA precursor (pre-miRNAs) appeared for human miR-125a [16] and miR-146a [57] and showed reduced processing, lowered levels, and disrupted function of the corresponding mature miRNA. These studies demonstrated that a SNP disrupting the base pairing in the hairpin stem of a pre-miRNA results in an important impairment in the processing and expression of the corresponding mature miRNA. The rs12975333 in miR-125a, resulting in a G → T change at the eighth nucleotide of the mature miRNA, severely reduces its transcription [16]. This polymorphism was associated with a significantly increased risk for breast cancer [58]. It was proposed that lowered levels of the mature miR-125a result in overexpression of its target gene, *HER2*, whose increased levels are implicated in numerous breast cancer cases [59].

A bioinformatics approach was used to study 79 polymorphisms in 129 cancer-associated genes 3′UTRs: seven SNPs were found to be located in pre-miRNA hairpins and one in the miR-608 mature sequence [60]. In a screen of 227 known human miRNAs, a total of 323 SNPs were identified, of which 12 were found to be located within the pre-miRNA sequences [16]. A C → T germline alteration in the primary transcript of miR-15a/miR-16 was found in some patients with familial chronic lymphocytic leukemia (CLL) [61, 62]. This specific polymorphism resulted associated with miR-15 and miR-16 reduced expression. It is known that approximately 70 % of CLL cases express low levels of

Table 7.1 Overview of the studies investigating SNPs in miRNA encoding genes in relation with CRC risk and prognosis

Study	miRNA	SNP Allele change	Ethnicity	Cases	Controls	Outcome with CRC	Notes
Chen et al. [30]	miR-196a2	rs11614913 C>T	AS	126	407	No association	Frequency of CT or CC carriers ↓ in CRC
Zhan et al. [31]	miR-196a2	rs11614913 C>T	AS	252	543	CC or C allele carriers with ↑ risk	CC or C allele with ↑ level of mature miR-196a
Boni et al. [32]	pri-miR26a-1	rs7372209 T>C	CAU	61	–	CC and CT carriers with favorable tumor response and ↑ PFS	
	pri-miR-100	rs1834306 T>C				CC and CT carriers with ↑ PFS	
Xing et al. [33]	pre-miR-423	rs6505162 A>C	AS	408	–	CC carriers with ↓ OS and PFS	The effects were evident only in patients receiving chemotherapy
	pre-miR-608	rs4919510 C>G				CC carriers with ↑ OS and PFS	The effects were evident only in patients receiving chemotherapy
Zhu et al. [34]	miR-196a2	rs11614913 C>T	AS	573	588	CT or CC carriers with ↑ risk	
Ryan et al. [35]	pre-miR-608	rs4919510 C>G	CAU and AA	245	446	No association	GG carriers with ↑ risk of death in CAU and with ↓ risk of death in AA
Lin et al. [36]	pre-miR-608	rs4919510 C>G	CAU and AA	1097	–	↑ risk for both PFS and death	In patients with stage III disease
	mir219-1	rs213210 A>G				↑ risk for death	In patients with stage III disease
Hu et al. [37]	miR-146a	rs2910164 G>C		276	373	CG carriers with ↓ risk	Association with histological differentiation
Oh et al. [38]	miR-34b/c	rs4938723 T>C	AS	545	428	No association	All genotypes together with diabetes mellitus associated with ↑ CRC risk
Parlayan et al. [39]	miR-146a	rs2910164 G>C	AS	116	524	No association	
	miR-196a2	rs11614913 C>T				No association	

(continued)

Table 7.1 (continued)

Study	miRNA	SNP Allele change	Ethnicity	Cases	Controls	Outcome with CRC	Notes
Wang et al. [40]	pre-miR-27a	rs895819 A>G	AS	205	455	GG carriers ↑ risk	GG and G allele carriers with ↑ risk of metastasis
Pardini et al. [41]	pre-miR-608	rs4919510 C>G	CAU	1083		G carriers with ↑ PFS	In stage III patients and receiving 5-FU chemotherapy
	mir219-1	rs213210 A>G				CT+TT carriers with ↓ OS and PFS	Receiving 5-FU chemotherapy
Cao et al. [42]	miR-27a	rs895819 A>G	AS	254	238	AG+GG with ↑ risk	In older (≥60 years) and male subjects in GG or G allele carriers expression of miR-27a ↑ in tumor tissues
Gao et al. [43]	miR-34b/c	rs4938723 T>C	AS	347	488	CC or C with ↓ risk	
Tang et al. [44]	pre-miR-1307	rs7911488 T>C	AS	1026	1026	C carriers with ↑ CRC risk	
Vinci et al. [45]	miR-146a	rs2910164 G>C	not available	160	178	No association	
	miR-149	rs2292832 T>C				No association	
	miR-196a2	rs11614913 C>T				No association	
	miR-499	rs3746444 A>G				GG carriers with ↑ risk	Reduction of miRNA expression in CRC related to a specific genotype
Dikaiakos et al. [46]	miR-146a	rs2910164 G>C	CAU	157	299	CC or C allele carriers with ↑ risk	
	miR-196a2	rs11614913 C>T				No association	
	miR-499	rs3746444 A>G				No association	
Kupcinskas et al. [47]	miR-146a	rs2910164 G>C	CAU	193	428	CC carriers with ↓ risk	
	miR-492	rs2289030 G>C				No association	
	miR-196a2	rs11614913 C>T				No association	
	miR-608	rs4919510 C>G				No association	

(continued)

Table 7.1 (continued)

Study	miRNA	SNP Allele change	Ethnicity	Cases	Controls	Outcome with CRC	Notes
	miR-27a	rs895819 T>C				No association	
Min et al. [48]	miR-196a2	rs11614913 C>T	AS	446	502	CC carriers with ↑ risk	
Lv et al. [49]	miR-196a2	rs11614913 C>T	AS	353	540	T carriers with ↑ risk	
Hezova et al. [50]	miR-196a2	rs11614913 C>T	CAU	212	197	No association	
	miR-27a	rs895819 T>C				No association	
	miR-146a	rs2910164 G>C				No association	
Wang et al. [51]	miR-603	rs11014002 C>T	AS	102	204	CT or TT carriers with ↑ risk	
Wu et al. [52]	miR-196a2	rs11614913 C>T	meta-analysis	2209	2803	C carriers with ↑ risk	in Asians
	miR-146a	rs2910164 G>C		2349	2663	No association	
	miR-149	rs2292832 T>C		1409	1115	No association	
Liu et al. [53]	miR-146a	rs2910164 G>C	meta-analysis	5486	7184	No association	GC+GG carriers ↑ risk in hospital-based studies
	miR-149	rs2292832 T>C				No association	CT+TT carriers ↑ risk in population-based studied
	miR-196a2	rs11614913 C>T				No association	CC carriers with ↓ risk, after excluding studies with HWE inconsistence
	miR-499	rs3746444 A>G				No association	AA carriers with ↓ risk (in Caucasians)
Xie et al. [54]	miR-146a	rs2910164 G>C	meta-analysis	2978	3576	No association	
Yi et al. [55]	miR-34b/c	rs4938723 T>C	meta-analysis	6036 (all)	7490	CC carriers with ↓ risk	
Li et al. [56]	miR-34b/c	rs4938723 T>C	meta-analysis	7753 (all)	8014	CC carriers with ↓ risk	

CAU Caucasians, *AA* Afro-Americans; *AS* Asians, *EA* European ancestry, *OS* overall survival, *PFS* progression-free survival

these two miRNAs, suggesting a possible role of this genetic polymorphism to leukemogenesis [61, 62]. With only a few exceptions, variations in pre-miRNAs seem not to be crucial in modulating CRC risk. For instance, the GG genotype of rs895819 in pre-miR27a or C-allele carriers of rs7911488 in pre-miR-1307 were more prevalent in CRC patients than in healthy controls in two Chinese populations [40, 44]. However, SNPs in pre-miRNAs are of particular interest for clinical outcomes. As an example, rs4919510, a polymorphism in pre-miR-608, has been repeatedly associated with survival and risk of recurrence in large CRC patient cohorts [29, 32, 33, 41].

SNPs residing in the mature miRNA sequence seem to have the most pronounced effect in modulating the post-transcriptional regulation of its targets (Fig. 7.1). Mature miRNAs consist of two main regions: the seed and the so-called 3′-mismatch tolerant region (3′MTR). The seed, from nucleotides 2–7 in the 5′region of the mature molecule, confers the highest target recognition specificity. The 3′MTR is able to tolerate mismatches with certain flexibility. The seed region is very important for miRNA binding, but it is not a unique predictor of the actual miRNA target [63]. miRNA seed sequences are short and highly conserved, and the probability of the presence of a SNP occurring in a miRNA seed region is less than 1 % [25]. A recent study identified a polymorphism present in the seed region of miR-125a that significantly inhibited the processing of pri-miRNA to pre-miRNA, resulting in reduced miRNA-mediated translational repression [64].

Particularly interesting is the example of rs2910164, a G→C variation in miR-146a. This SNP impairs processing and downregulation of the mature miRNA levels with a consequent decrease of the impact on its target gene regulation. Mature miRNAs may derive from both the 5p and 3p arms of the pre-mir-146a. The presence of the variant rs2910164, located in the seed region of miR-146a-3p, generates two isoforms that regulate distinct sets of target genes. Notably, carriers of the homozygous genotypes with GG or CC alleles produce two mature molecules (miR-146a-5p from the leading strand, and miR-146a-3p(G) or 146a-3p(C), respectively, from the passenger strand). Conversely, GC heterozygotes produce 3 mature miRNAs: miR-146a-5p and both miR-146a-3p(G) and miR-146a-3p(C) [57, 65]. The SNP has been associated with predisposition to various cancers, such as papillary thyroid carcinoma [57], prostate cancer [66], bladder cancer [67], and CRC [68].

Several SNPs in miRNA-related genes have been previously associated with the risk of CRC [69] and a polymorphism in miR-26a-1 was linked to differential response to irinotecan-based chemotherapy in CRC patients [32]. More recently, Lin et al. found that a polymorphism (rs4919510) in miR-608 was associated with increased risk for both recurrence and death and a polymorphism (rs213210) in miR-219-1 was related to increased risk for death in a mixed population of stage III CRC patients from USA undergoing 5-FU-based chemotherapy [36]. These last results were also independently validated on a European population [41] implicating that variations in miRNA-encoding genes may modulate CRC prognosis and predict therapy response.

Although studies have started to disclose the nature of the association between miRSNPs and cancer risk, several points remain to be elucidated: a) most of the studies used a candidate gene approach; b) the few of them using a systematic approach had outdated lists owing to enhanced screening techniques that have identified new miRNA genes and updated builds of genome-wide SNP repositories [29]. Moreover, the minor allele frequencies of many of the miRSNPs already identified have not been determined. Therefore, population studies should be conducted with the aim to ascertain whether these variants are truly polymorphic and what would be their distribution in various populations (one example is rs4919510 in miR-608 that shows significant differences in ethnicities as observed by [41]). This is an important consideration, as data are emerging to suggest that some miRSNPs have evolved to a high level of variance in distinct populations [29].

7.2.2 Polymorphisms in miRNA Target Sites

In contrast to the polymorphisms in miRNA genes, miRSNPs located at the 3′UTR of a target (coding) gene are more abundant in the human genome and have a more defined and limited range of effects (Fig. 7.2). MiRSNPs in miRNA target sites will impact only their encoded target-mRNA and their downstream effectors, meaning that they are more specific. A GWAS suggested that a gene presenting more than two miRNA target sites can show an increased variability in its expression when compared with a gene not regulated by a miRNA. This observed variability may be furtherly increased by the presence of SNPs in the miRNA target sites [70]. Considering the large number of less conserved 3′UTR target sequences, SNPs in these regions may potentially harbor a higher frequency of target miRSNPs, being therefore more important from an epidemiological standpoint (reviewed in [18, 71]). Interestingly, it should be considered that polymorphisms can be within the target site or nearby and in both cases they could be efficient in the modulation of the binding.

The miRNA seed sequence plays an important role in target recognition and binding, a polymorphism in this region or in the 3′-MTR binding region may therefore have a high probability of affecting a miRNA function. However, also nearby polymorphisms outside the miRNA target site can affect accessibility of the miRNA. Unlike DNA–protein interactions, mRNA–protein interactions are based on the presence/absence of secondary structure motifs in mRNAs. Most of the miRNAs binding sites in the 3′UTRs of a target mRNA lack a complex secondary structure, thereby facilitating access for a miRNA [72]. Variations that can create or abolish a secondary structure near to miRNA binding site may potentially influence miRNA-mediated translational repression of a target gene by affecting the accessibility of a miRNA to its binding site [73].

It has been demonstrated that under certain cellular conditions (oxidative stress, pH variation, etc.) a stable secondary structure could be unfolded to provide access to a miRNA target site [74]. This regulation mediated by miRNA can be then exploited by a cell during stress response or in case of specific tissues. In addition, two miRNAs may bind to the same target mRNA working in coordination. In fact, the binding of a miRNA to its target site may induce a remodeling of secondary structures in the neighboring regions, facilitating binding of other miRNAs that otherwise did not have access to the sequence due to the mRNA spatial conformation [75].

Hence, polymorphisms near to miRNA target site can potentially influence the accessibility of a miRNA–RISC complex by affecting the RNA structural motifs necessary for RNA–protein interaction. Further analyses on the interactions between miRNA and other regulatory elements present in 3′UTRs will shed more light on the function of miRNA polymorphisms and will eventually establish whether 3′UTRs could be considered as hotspots for pathology [23].

7.2.3 Alternative Polyadenylation (APA)

A polymorphism near to a miRNA target site could disrupt the association of a miRNA with other regulatory elements present in the 3′UTR of the target transcript. The length of the 3′UTR of a miRNA target provides significant potential for miRNA-mediated, transcript-specific gene regulation, where a target gene can be regulated by more than one miRNA.

In addition, more than 70 % of human genes encode primary transcripts that contain multiple polyadenylation sites (PA sites). APA is a common regulatory mechanism of gene expression that generates mRNAs with distinct 3′UTRs as well as coding sequences [76, 77]. A systematic screening of 3′UTRs produced by APA in murine cells revealed that approximately half of all miRNA target sites are located downstream of the first poly(A) site [78]. Several studies emphasized the implications of APA in human diseases. For instance, the deregulated expression of the gene encoding brain-derived neurotrophic factor (BDNF) contributes to several neurodegenerative diseases [79]. The 3′UTR of the *BDNF* gene harbors two PA sites: one distal (~3 Kb from 3′UTR start, *BDNF-L*) and one proximal (~350 bps from 3′UTR start, *BDNF-S*) site [77]. The long isoform (*BDNF-L*) presents ten predicted potential miRNA binding sites, whereas the short isoform (*BDNF-S*) includes only six. In a study on human embryonic kidney 293 cells (HEK-293) by using the luciferase reporter assays (see ***paragraph 7.5***) a direct interaction of *BDNF-L* with miR-1, miR-10b, miR-155 and miR-191 was confirmed while

BDNF-S interacted only with miR-1 and miR-10b. This is consistent with the observation that the short *BDNF* transcript isoform neither carries a miR-155 nor a miR-191 binding site. Furthermore, after transfection with miR-1 precursor, the luciferase activity was significantly lower for the *BDNF-L* isoform that carries three predicted binding sites for this miRNA, compared to BDNF-S which carries only a single miR-1 binding site [80]. The protein level of *BDNF*, thus, largely depends on PA site usage associated with an altered post-transcriptional regulation of the encoding mRNA by miRNAs.

The 3′UTR shortening by APA is associated with carcinogenesis and may represent a mechanism for genes to escape miRNA-mediated post-transcriptional repression in cancer [81]. Most cancer cell lines express significantly shorter APA isoforms with an associated increase of protein-expression levels that could be attributed to a loss of post-transcriptional regulation via miRNAs due to 3′UTR shortening (25–70% range) [81].

The number of proteins contributing to or directly interacting with the pre-mRNA 3′ end for APA, is quite large (about 90 proteins) [82]. Transcription of mRNA isoforms with distinct 3′UTRs may also modulate the post-transcriptional fate of these mRNAs through inclusion or exclusion of miRNA binding sites [20]. Thus, a miRNA-mediated post-transcriptional regulation may be affected by the altered accessibility of miRNA binding sites in shorter/longer 3′UTRs due to secondary structure variations, and by the varying proximity to the translation machinery [83].

miRNAs are shown to promote polyadenylation by interacting with cytoplasmic PA elements and other proteins or protein complexes within the 3′UTR [22]. MiRSNPs may potentially affect these interactions as well.

7.3 miRSNPs Relevant in CRC Diagnosis, Progression, and Prognosis

In the last years, a growing number of studies on miRNA-related polymorphisms in association with CRC risk and clinical outcome have been performed. However, the number of investigated SNPs is still limited. A list of studies investigating SNPs in miRNA encoding genes and CRC risk and prognosis is reported in Table 7.1.

A recent meta-analysis retrieved all the available studies investigating SNPs in miRNA genes in association with cancer risk, progression, and prognosis [84]. Interestingly, a SNP in miR-219-1 (rs213210) resulted associated with an increased risk of CRC recurrence and death [36, 41]. On the other hand, rs4919510 in miR-608 was linked with risk of death in Caucasian (increased risk) and African American (decreased risk) CRC patients [29] and with an increased risk of CRC recurrence and death in a mixed American population [36]. In contrast, the same SNP, investigated in another Caucasian population, was associated with a decreased risk of recurrence [41] and with favorable overall and recurrence-free survival in Chinese patients [33]. Rs6505162 in pre-miR-423 was associated with unfavorable overall and recurrence-free survival in Chinese CRC patients [33]. Other SNPs (rs11614913, rs895819, and rs2910164) respectively in miR-196a2, miR-27a and miR-146a were not associated with CRC risk [50].

Rs4938723 in the promoter region of pri-miR-34b/c was not associated with CRC alone in a case-control study performed in Korea. However, when combined with another SNP in *TP53* (rs1042522), a well-known tumor suppressor gene involved in CRC, a reduced risk of this cancer was observed for a particular combination of genotypes. Individuals carrying the combined rs4938723 CC and rs1042522 GG genotypes showed a lower risk of CRC. miR-34 family is a direct transcriptional target of *TP53* and loss of miR-34 function may impair TP53–mediated functions [38].

Several studies have also highlighted the impact of SNPs in miRNA binding sites on CRC. A list of studies investigating SNPs in miRNA binding sites and CRC risk and prognosis is reported in Table 7.2. The first demonstration of an implication of polymorphisms within miRNA binding sites in CRC susceptibility was presented in a study on sporadic CRC cases and controls from the Czech Republic [85, 86]. More

Table 7.2 A brief overview of studies investigating SNPs in 3′UTR predicted miRNA binding sites and CRC risk and prognosis

Study	Ethnicity	Gene	SNP	Predicted miRNA binding	Cases	Controls	Outcomes with CRC
Landi et al. [86]	CAU	*CD86*	rs17281995 G>C	miR-337, miR-582, miR-200a*, miR-184 and miR-212	697	624	CC carriers with ↑ risk
		INSR	rs1051690 G>A	miR-618, miR-612			AA carriers with ↑ risk
Landi et al. [90]	CAU	*KIAA0182*	rs709805 G>A	miR-324-3p	717	739	AA carriers with ↑risk
		NUP210	rs354476 T>C	miR-152a, miR-125b			CC carriers with ↑ risk
Naccarati et al. [85]	CAU	*RPA2*	rs7356 A>G	miR-3149, miR-1183	1098	1469	GG carriers with ↑ risk of CRC and RC
		GTF2H1	rs4596 G>C	miR-518a-5p, miR-527, miR-1205			G carriers with ↓ risk of CRC or RC
Pardini et al. [27]	CAU	*NEIL2*	rs1534862 C>T	miR-218-3p, miR-1224-5p, miR-3605-3p, miR-1294, miR-4464	1098	1469	T carriers with ↓ risk of relapse
		NEIL2	rs6997097 T>C	miR-6124, miR-541-5p, miR-548an, miR-4484, miR-1185-5p, miR-5191			TC carriers with ↓ OS
		SMUG1	rs2233921 G>T	miR-770-5p, miR-665, miR-455-3p			TT carriers with ↑ OS
		SMUG1	rs971 G>A	miR-548ag, miR-548ai, miR-548 m, miR-570-5p, miR-1208, miR-610, miR-3179, miR-342-5p			GA carriers with ↓ OS
Vymetalkova et al. [88]	CAU	*MLH3*	rs108621 T>C	miR-3190-5p, miR-3194-5p, miR-5689, miR-588, miR-429, miR-4701-5p, miR-200b-3p, miR-200c-3p, miR-4693-5p	1098	1469	CC carriers with ↑ OS

(continued)

Table 7.2 (continued)

Study	Ethnicity	Gene	SNP	Predicted miRNA binding	Cases	Controls	Outcomes with CRC
Naccarati et al. [89]	CAU	*MRE11A*	rs2155209 T>C	miR-27a-5p, miR-3158-3p, miR-3655, miR-4446-3p, miR-5683	1111	1469	CC carriers with ↓ risk and shorter OS
		RAD52	rs1051669 G>A	miR-277, miR-450b-5p, miR-106a-3p, miR-367-3p, miR-363-3p, miR-451b, miR-25-3p			AA carriers with ↑ risk
		RAD52	rs11571475 T>C	miR-3162-3p, miR-3180-5p, miR-24-1-5p, miR-150-5p, miR-629-3p, miR-708-3p			C carriers with ↓ risk
		RAD52	rs11226 C>T	miR-635, miR-3614-5p, miR-6724-5p, miR-3163-3p, miR-3147, miR-4640-5p, miR-496, miR-296-5p, miR-6075, miR-890, miR-1587, miR-4260			TT carriers with ↑ OS
		NBN	rs14448 T>C	miR-640, miR-4319, miR-587, miR-4725-5p, miR-345-3p			TC carriers with ↓ risk
		RAD51	rs12593359 T>G	miR-548ay-3p, miR-548o-3p, miR-548at-3p, miR-548a-3p, miR-424-3p			TG carriers with ↓ risk
		XRCC5	rs1051685 A>G	miR-4731-3p, miR-505-3p, miR-125b-2-3p, miR-382-5p, miR-758-3p, miR-21-3p, miR-329			GG carriers with ↓ OS
Winder et al. [91]	CAU	*KRAS*	rs61764370 T>G	let-7	130 186		TT carriers with mt KRAS and treated with irinotecan/cetuximab in the EPIC trial with ↑ PFS
Graziano et al. [92]	CAU	*KRAS*	rs61764370 T>G	let-7	134		G carriers with ↓ OS and PFS
Smits et al. [93]	CAU	*KRAS*	rs61764370 T>G	let-7	660	1497	GG carriers with early-stage of CRC with ↓ risk and ↑ OS

Sebio et al. [94]	CAU	*KRAS*	rs61764370 T>G	let-7	100		31.9 % of patients with the TT genotype presented a complete or a partial response to anti-EGFR-based treatment
Ryan et al. [95]	AA	*KRAS*	rs61764370 T>G	let-7	237	441	G carriers with ↑ OS in late stages of CRC
	EA						
Zhang et al. [96]	CAU	*KRAS*	rs61764370 T>G	let-7	130		TT carriers with ↓ object response rate and ↓ OS and PFS
	AA						
	AS						
Kjersem et al. [97]	CAU	*KRAS*	rs61764370 T>G	let-7	197	358	No association
					1060		
					180		
					355		
Ruzzo et al. [98]	CAU	*KRAS*	rs61764370 T>G	let-7	172		TT carriers with high level of *let-7* with ↑ OS and PFS
Saridaki et al. [99]		*KRAS*	rs61764370 T>G	let-7	512		G carriers with anti-EGFR mAb monotherapy treatment with ↑ PFS
Pan et al. [100]	AS	*KRAS*	rs712 G>T	let-7, miR-3125, miR-374c-3p	339	313	T carriers with ↑ risk
Song et al. [101]	AS	*NFkBIA*	rs696 A>G	miR-449, miR-34	1001	1005	GG carriers with ↑ risk
Zanetti et al. [102]	AA	*MBL2*	rs10082466 T>C	mir-27a	103	127	CC carriers with ↑ risk
			rs2120132 A>G	miR-625-3p, miR-422a, miR-6507-5p			G carriers with ↑ risk
			rs2099902 A>G	miR-3133, miR-539-5p, miR-4263, miR-6080, miR-628-3p, miR-5571-5p, miR-4760-3p, miR-576-5p			G carriers with ↑ risk
			rs10450310 C>T				T carriers with ↑ risk
Schmit et al. [103]	AJ	*UHRF1BP1*	rs2985 T>C	miR-4529-5p, miR-885-5p	596	429	C carriers with ↓ risk
		AKR1C1	rs1139139 C>T	miR-451b, miR-556-5p			T carriers with ↑ risk
		RAPGEF2	rs6827968 C>A				A carriers with ↑ risk
		KIAA0090	rs12130051 C>T	miR-222, miR-1244, miR-3129			T carriers with ↑ risk

(continued)

Table 7.2 (continued)

Study	Ethnicity	Gene	SNP	Predicted miRNA binding	Cases	Controls	Outcomes with CRC
		RALGPS2	rs80350662 T>A	miR-32-3p, miR-1277-5p, miR-4775, miR-889			A carriers with ↑ risk
		IL18	rs1834481 G>C	miR-637, miR-5009-5p, miR-541-3p			G carriers with ↓ risk
		RNF217	rs1044724 A>C	miR-3978			C carriers with ↓ risk
			rs7746892 C>G	miR-545, miR-1252, miR-4476			G carriers with ↓ risk
			rs7746860 C>G	miR-2681, miR-1295			G carriers with ↓ risk
			rs471429 C>G	miR-3126-5p, miR-4270			G carriers with ↓ risk
		PTPN11	rs4766991 C>T	miR-1302-1			T carriers with ↑ risk
		LOC100129055	rs2489495 C>T	miR-635			T carriers with ↓ risk
		ARHGAP26	rs853158 T>C	miR-3926, miR-4480			C carriers with ↓ risk
		CEP57L1	rs9374072 A>G	miR-605			G carriers with ↓ risk
		CCDC7	rs12268559 A>C	miR-578			C carriers with ↓ risk
			rs56391924 A>C	miR-4273			C carriers with ↓ risk
			rs12247495 G>C	miR-539-5p			C carriers with ↓ risk
		C1GALT1C1	rs142004998 T>C	miR-1284, miR-337-3p, miR-520d-5p			C carriers with ↓ risk
		HS6ST1	rs3180466 T>G	miR-4758-5p, miR-4669, miR-574-5p, miR-3659, miR-615-3p, miR1238-5p, miR-4745-5p, miR-3677-5p			G carriers with ↓ risk
		C1orf220	rs79029362 T>G	miR-455-3p			G carriers with ↑ risk
		PTPN11	rs4766992 G>A	miR-1302			A carriers with ↑ risk
		FLJ30403	rs117299563 T>C	miR-3684			C carriers with ↓ risk
		TOP1	rs6072275 G>A	GWAS			A carriers with ↑ risk
		FLJ41941	rs107321 T>C	miR-1284, miR-337-3p, miR-374a-5p			T carriers with ↓ risk
		ERBB4	rs1972820 G>A	miR-4633-5p, miR-3144-3p, miR-532-5p, miR-875-5p			G carriers with ↑ risk

CAU Caucasians, *AA* Afro-Americans, *AS* Asians, *AJ* Ashkenazi Jewish, *EA* European ancestry, *OS* overall survival, *PFS* progression-free survival precisely in the 3′UTR of *CD86* gene, a C→G polymorphism (rs17281995), predicted to affect the binding with miR-337, miR-582, miR-200a, miR-184, and miR-212, was significantly associated with increased CRC. The study also identified rs1051690 in *INSR* predicted to affect miR-618 and miR-612, previously found to be associated with breast cancer [87]. More recently, several studies conducted on the same population investigated the role of a large number of variants in the 3′UTRs of genes involved in different pathways of the DNA repair [27, 85, 88, 89].

Interestingly, several variants were associated with cancer risk or clinical outcome and the majority of them were never previously observed in association with CRC. In particular, a variant in the *MRE11A* gene, involved in the double-strand break repair pathway, resulted associated with decreased CRC risk [89], while a variant in *SMUG1* (a gene of the base-excision repair) was associated with increased overall survival of CRC patients [27]. For both SNPs, the effect of each different allele was assayed by functional tests (see ***paragraph 7.5***) and significantly different expression levels of the reporter gene were observed.

Regulation of *KRAS* gene is critical in colorectal carcinogenesis. Genetic variations may also contribute to the regulation of this oncogene and potentially affect response to therapy among patients with wild-type *KRAS*. For instance, miRNA such as has-let-7 may prevent its translation, and it has been found that a common polymorphism (rs61764370) in the *KRAS* 3′UTR stemming from a T→G transversion affects the binding of let-7 family miRNA and results in lower levels of KRAS expression [104]. A recent review found no clear association between the rs6174370 and overall or progression-free survival among CRC patients. Due to the conflicting body of literature surrounding clinical utility of testing for this SNP, rs6174370 by itself is an insufficient CRC predictor of outcome.

Most probably, more complex sets of molecular markers may enable the optimization of therapeutic regimens within guidelines of precision medicine.

7.4 Identification of miRSNPs

In 2008, when the first studies on CRC and miRSNPs started to emerge [86] a certain number of specialized algorithms were available (e.g. miRBase (http://www.mirbase.org/), miRanda (http://www.microrna.org/microrna/getDownloads.do), PicTar (http://pictar.mdc-berlin.de/), MicroInspector [105], Diana-MicroT [106], and TargetScanS [107] (http://genes.mit.edu/tscan/targetscanS2005.html). These tools allowed to find putative miRNA binding sites within the 3′UTR of a gene of interest relatively rapidly. On the contrary, the search of polymorphisms in the target sequences or any other variation related to miRNAs or ncRNAs was performed "manually" by retrieving the particular 3′UTR or the sequence of interest and by mining all possible SNPs present in the region by dsSNPs (http://www.ncbi.nlm.nih.gov/projects/SNP), using BLAST (http://www.ncbi.nlm.nih.gov/BLAST/Blast) and BLAST-SNP algorithms (http://www.ncbi.nlm.nih.gov/SNP/snpblast), with the aim to identify those within a miRNA predicted to bind in the area. Over the years, the search for polymorphisms related to miRNAs has taken advantage of the growing number of publicly available bioinformatics tools [26, 108–114]. Currently, several different tools provide more rapid and comprehensive information about existing miRSNPs inside gene of interests: they include for instance features such as the prediction of miRNA binding and different approaches to evaluate the effect of a particular allele in the locus of interest. We report in Table 7.3 a list and a short description of the most common tools available. These databases/tools use algorithms similar to those utilized by miRNA prediction tools previously listed in order to detect the effects of the SNPs on miRNA binding. The algorithms run on the whole genome for all SNPs present in a genomic database like dbSNP. Users can query the results using SNP, gene or miRNA IDs. The applications interrogate mainly the 3′UTRs and predict if a SNP within the target site will disrupt/eliminate or enhance/create a miRNA binding site. Tools compute these sites and examine the

Table 7.3 List of some bioinformatics tools investigating variations in relation to noncoding RNAs available online

Tool	SNPs in	Website	Ref
miRNAS			
MicroSNiPer	3′UTR miRNA predicted binding sites	http://epicenter.ie-freiburg.mpg.de/services/microsniper/	Barenboim et al. [114]
miRdSNP	3′UTRs of human genes from available publications in on 204 disease types	http://mirdsnp.ccr.buffalo.edu/	Bruno et al. [116]
miRNASNIPER	miRNA genes in vertebrates	http://www.integratomics-time.com/miRNA-SNiPer/	
miRNASNP	(1) in human pre-miRNAs;	http://www.bioguo.org/miRNASNP/search.php	Gong et al. [117]
	(2) in human pre-miRNA flanking regions;		
	(3) in pre-miRNAs of other eight species;		
	(4) targets gain/loss by SNPs in miRNA seeds		
	(5) targets gain/loss by SNPs in target 3UTRs		
MiRSNP	Predicted miRNA-mRNA binding sites	http://202.38.126.151/hmdd/mirsnp/search/	Liu et al. [26]
mirSNPscore	miRNA target sites and uses linkage disequilibrium to map candidate mirSNPs to disease data from GWAS	http://www.bigr.medisin.ntnu.no/mirsnpscore/	Thomas et al. [118]
mrSNP	SNP in 3′UTR on miRNA binding and predicting their impact	http://mrsnp.osu.edu/	Deveci et al. [119]
PolymiRTS	microRNA (miRNA) seed regions and miRNA target sites	http://compbio.uthsc.edu/miRSNP/	Bhattacharya et al. [108]
Long non coding RNAs			
lncRNASNP	lncRNAs and their potential functions in human and mouse	http://bioinfo.life.hust.edu.cn/lncRNASNP/	Gong et al. [110]
SNP@lincTFBS	human lincRNA transcription factor binding sites	http://bioinfo.hrbmu.edu.cn/SNP_lincTFBS	Ning et al. [109, 111]

effects of SNPs in real time. The main advantages of these tools include an incredible ease of use, flexibility, and straightforward graphical representation of the results. Among the main limitations are the possibility to investigate only SNPs that already exist in databases and the preclusion to work with novel or unreported SNPs. The newly discovered polymorphisms can be investigated manually; however, in the case of large list of SNPs the web interface of tools may require an infeasible amount of manual labor. Additionally, databases are not always updated to the last release of either dbSNP or miRBase and it is not unusual to find discrepancies in the outcomes of search for both SNPs identified and miRNAs predicted to bind. Thus, in the end, the recommended way to perform a SNP selection in miRNA target sites is to compare results from different databases, waiting for a sort of implementation of tools like miRWalk [115], which compare and integrate data from multiple algorithms.

After SNPs identification, additional steps are required before starting an association study on cancer risk or clinical outcome. The frequency of a particular SNP in a given population should be considered for instance in order to reach the appropriate statistical power of a study. The polymorphism could also be within a site of binding of different regulatory elements, in which case it may be more complicated to understand the impact of the different alleles on the mechanism modulating gene expression.

With the advent of next-generation sequencing technologies such as RNA-Seq, exome and whole genome sequencing, thousands of novel SNPs in 3′UTRs are being identified. RNA-Seq, which sequences all expressed genes in a sample, provides concordant gene expression and SNP data. Since a substantial number of the detected SNPs is sometimes undocumented, the use of algorithms that require a SNP to be present in dbSNP may not meet the needs of researchers using RNA-Seq or other next-generation sequencing methods. Currently, when a novel SNP is encountered, a user can compare the location of the SNP against the predicted and validated miRNA target sites using the current prediction tools. However, this approach is fairly labor intensive.

The probability of the SNP disturbing a binding site can be considered to be proportional to the distance of the SNP to the seed of the target site. However, a SNP may not affect binding even when it is very close to the miRNA target seed region. Moreover, a SNP may introduce a totally new binding with a new miRNA, which is impossible to capture with the current databases.

Thus, next-generation sequencing data require new computational tools to relate the identified SNPs and gene expression data. In this respect a new web-based tool, named mrSNP, has been recently introduced to overcome the shortcomings of existing tools [119].

For variation in miRNA encoding genes, polymorphisms are easier to be identified, being the number of miRNAs and the regions to scan much smaller. Catalogues of polymorphisms in miRNA genes are also currently available (see Table 7.3).

7.5 Functional Analyses to Test miRSNPs Effect

To understand miRNA functions and to functionally evaluate a miRSNP effect, it is important to experimentally test the interaction between the miRNA and its mRNA targeting site(s). The luciferase reporter gene assay has been recently adapted also to this scope. One of the most common application of luciferase reporter gene assay is to test the regulation of transcriptional activities by promoters and transcription factors. To evaluate the effect of miRNA-mediated, post-transcriptional regulation on target genes a luciferase gene construct containing the predicted miRNA targeting sequence (often the 3′UTR of a target gene) has been engineered. For many human genes, luciferase constructs containing the entire 3′UTR can be obtained from a number of commercial sources (e.g., OriGene Technologies, GeneCopoeia, SwitchGear Genomics) [120]. This *in vitro* assay permits to measure whether a 3′UTRs with different allele variants could affect, to some extent, the levels of the proteins. Briefly, the assay implicates the creation of chimeric constructs of the 3′UTRs of the gene of interest in which the common or variant allele of the SNP of interest is present. These constructs are placed as 3′UTRs of the reporter gene firefly luciferase and co-transfected into cells together with Renilla luciferase used as reference. The quantification of each luciferase luminescence in relation to the luminescence from the Renilla and, then, the calculation of the ratio between the construct carrying the common allele and that with the variant allele are the measurement of the effect of the SNP in the host cell taking into account its miRNome. A complete description of the methodology has been reviewed by Jin and colleagues [120]. There are examples in CRC research in which the luciferase reporter gene assay was used to demonstrate the effect of various miRSNPs in genes important for colorectal tumorigenesis [27, 89, 90, 121].

Reporter assays enable the identification of direct interactions between a given miRNA and specific mRNA targets. These assays pose some limitations due to the fact that are usually restricted to a fixed set of interactions. Therefore, promising candidate interactions have to be defined in advance based on either additional experimental data or by *in silico* prediction of potential miRNA target genes. The identification of miRNA-mRNA interactions requires sophisticated bioinformatics analyses and/or previous knowledge of promising candidate interactions. As a consequence, computational methods need to be constantly improved and modulated depending on the available experimental data. Current computational approaches for *in silico* prediction of miRNA targets are based on several algorithms included in the following tools: TargetScan [107], miRanda [122], PITA [73] and PicTar [123]. All of them rely on mRNA 3′UTR sequence complementarity with the seed region of a given miRNA. They also consider the secondary structure of the miRNA and/or its target site (for example, the free energy of the miRNA–mRNA binding: ΔG or costs to unfold the secondary structure of the target site $\Delta\Delta G$), the number of potential binding sites within one transcript, and if applicable, conservation of the miRNA and/or the target site across mammals (assuming that conservation increases the likelihood of a functional site). An overview of the most commonly used databases for miRNA-mRNA interactions has been provided by [20] and [124]. However, all these databases are not taking into account the heterogeneity of 3′UTR length among different transcripts of the same genes due to APA or the variability conferred by the presence of polymorphisms both in the miRNA and mRNA sequences.

This heterogeneity in 3′UTR represents one of the major reasons for the relatively high false positive rate of available predictions in databases, which is estimated to reach up to 70 % [125], and consequently complicates reliable predictions for interactions in a given tissue.

It has been calculated that the overlaps between prediction results from different tools/databases can vary (from 5 to 70 %). The launch of databases such as miRo [126] or miRWalk [115], that highlight only predicted interactions commonly identified by more tools, has helped to increase the probability of identifying true interactions. More recent tools have attempted to implement the prediction algorithms with co-expression information in the effort to consider also specific biological characteristics of a given sample. For instance, MirCox [127] and miR-Connect [128] use miRNA and mRNA sequencing data publicly available to calculate negative correlations between mRNA and miRNA expression levels. Although these correlations are not completely correct from the biological point of view (miRNA-mRNA interaction/regulation is post-transcriptional) this is a first attempt to move from a "simple" mathematical/physical calculation to the biological environment.

7.6 Variations in Other ncRNAs

Recently it has become clear that only a small percentage (7 %) of disease-associated SNPs are located in protein-coding regions, while the remaining 93 % are located in gene regulatory regions or in intergenic regions. Thus, the understanding of how genetic variations control the expression of ncRNAs (in a tissue-dependent manner) has far-reaching implications. The association of SNPs with expression levels (eQTLs) of large intergenic non-coding RNAs (lincRNAs) was tested, using genome-wide gene expression and genotype data from five different tissues. One hundred and twelve cis-regulated lincRNAs were identified, and 45 % of them could be replicated in an independent dataset. Of the total number of SNPs affecting lincRNA expression (lincRNA cis-eQTLs), 75 % were specific to lincRNA alone and did not affect the expression of neighboring protein-coding genes. This specific genotype-lincRNA expression correlation was tissue-dependent and many of these lincRNA cis-eQTL SNPs were also associated with complex traits and diseases [129].

Five SNPs in the novel lncRNA PRNCR1, located in the 8q24 region, were genotyped in 908 subjects (313 CRC cases and 595 controls). Rs13252298 and rs1456315 were associated with significantly decreased risk of CRC. Additionally, patients with the rs7007694 C allele and rs16901946 G allele had decreased risk to develop poorly differentiated CRC, whereas the G allele carriers of rs1456315 in the same group of patients showed an increased risk of recurrence/progression [130].

Another lncRNA, CCAT2, encompassing the rs6983267 SNP, is highly overexpressed in microsatellite-stable CRC and promotes tumor growth, metastasis, and chromosomal instability. Ling et al. demonstrated that CCAT2 may induce up-regulation of MYC, miR-17-5p, and miR-20a through TCF7L2-mediated transcriptional regulation. The physical interaction between CCAT2 and TCF7L2 results in an enhancement of WNT signaling activity. The presence of one of the 2 alleles of rs6983267 affects CCAT2 expression and the risk G allele produces more CCAT2 transcript. These findings support a new mechanism of MYC and WNT regulation by the novel lncRNA CCAT2 in CRC pathogenesis, and provide an alternative explanation of the SNP-conferred cancer risk [131].

piRNAs, another type of identified small ncRNA, also play a crucial role in germline development and carcinogenesis. Seven common SNPs were found in 9 piRNAs by Chu and colleagues in a systematical screening of all known piRNAs. The role of these polymorphisms in CRC susceptibility was tested in 1147 cancer patients and 1203 controls from China. Rs11776042 in piR-015551 was significantly associated with a decreased risk of CRC; however, this protective effect was not significant after correction for multiple comparisons. A marginal protective effect was observed in individuals who never drank alcoholic beverages and in CRC patients who had tumors with low differentiation or Dukes stage A and B. Interestingly, authors also noted that piR-015551 expression was positively correlated with expression levels of LNC00964-3, suggesting that piR-015551 may be generated from this lncRNA [132].

The potentiality of investigating variations in all classes of ncRNAs is rather huge, but at present still not exhaustively explored. In Table 7.3 we show some databases/tool available to mine these SNPs [109–113].

7.7 Conclusions and Perspectives

miRNAs are involved in fine-tuning of fundamental cellular processes such as proliferation, cell death and cell cycle control and are believed to confer robustness to biological responses. Genetic variations related to miRNAs serve an additional way to modulate gene expression of either single target gene or multiple genes. However, subtle regulation interplay comprising genetic and epigenetic mechanisms in affecting fundamental tumor suppressor/oncogenic pathways have to be addressed in order to understand the nature of tumorigenesis. miRSNPS might also substantially affect therapy response or may lead to therapy failure, which makes the understanding of the underlying processes a necessity. Characterization of miRNA polymorphisms and identification of their functional impact may provide a good basis for miRNA-based therapeutic approaches in the future.

Several elements must converge for a miRNA binding site variation to be considered functional:

(1) a SNP should be associated with cancer (risk or prognosis); (2) both miRNA(s) and predicted binding site(s) should be expressed in a particular tissue under investigation and at the same developmental stage; (3) the allelic changes must result in a differential binding of one or more miRNAs, and affect the expression of the target gene; (4) a mechanistic verification and proof of principle have to be substantial preconditions of each study design.

In respect to CRC, we have seen several studies exploring the potentiality of ncRNAs variations in relation to cancer onset and clinical outcomes, though mainly miRNA-related. First meta-analyses are providing evidence for the relevance of these variants. The analysis of large study populations of different ethnic groups in multicentric design is necessary to verify the associations and answer questions regarding the possible impact of ncRNA variation on this cancer and the importance of the results in clinical practice.

Acknowledgement This work was supported by Internal Grant Agency of the Ministry of Health of the Czech Republic (AZV MZ 15-26535A) and Czech Science Foundation (GA15-08239S).

References

1. Haraksingh RR, Snyder MP. Impacts of variation in the human genome on gene regulation. J Mol Biol. 2013;425(21):3970–7. doi:10.1016/j.jmb.2013.07.015.
2. Abecasis GR, Auton A, Brooks LD, DePristo MA, Durbin RM, Handsaker RE, et al. An integrated map of genetic variation from 1,092 human genomes. Nature. 2012;491(7422):56–65. doi:10.1038/nature11632.
3. Khurana E, Fu Y, Chakravarty D, Demichelis F, Rubin MA, Gerstein M. Role of non-coding sequence variants in cancer. Nat Rev Genet. 2016;17(2):93–108. doi:10.1038/nrg.2015.17.
4. Zhang K, Civan J, Mukherjee S, Patel F, Yang H. Genetic variations in colorectal cancer risk and clinical outcome. World J Gastroenterol. 2014;20(15):4167–77. doi:10.3748/wjg.v20.i15.4167.
5. Haggar FA, Boushey RP. Colorectal cancer epidemiology: incidence, mortality, survival, and risk factors. Clin Colon Rectal Surg. 2009;22(4):191–7. doi:10.1055/s-0029-1242458.
6. Peters U, Bien S, Zubair N. Genetic architecture of colorectal cancer. Gut. 2015;64(10):1623–36. doi:10.1136/gutjnl-2013-306705.
7. de la Chapelle A. Genetic predisposition to human disease: allele-specific expression and lowpenetrance regulatory loci. Oncogene. 2009;28(38):3345–8. doi:10.1038/onc.2009.194.
8. Srinivasan S, Clements JA, Batra J. Single nucleotide polymorphisms in clinics: fantasy or reality for cancer? Crit Rev Clin Lab Sci. 2015;53(1):29–39. doi:10.3109/10408363.2015.1075469.
9. Tenesa A, Dunlop MG. New insights into the aetiology of colorectal cancer from genome-wide association studies. Nat Rev Genet. 2009;10(6):353–8. doi:10.1038/nrg2574.
10. Bhalla A, Zulfiqar M, Weindel M, Shidham VB. Molecular diagnostics in colorectal carcinoma. Clin Lab Med. 2013;33(4):835–59. doi:10.1016/j.cll.2013.10.001.
11. Diaz Jr LA, Williams RT, Wu J, Kinde I, Hecht JR, Berlin J, et al. The molecular evolution of acquired resistance to targeted EGFR blockade in colorectal cancers. Nature. 2012;486(7404):537–40. doi:10.1038/nature11219.
12. Tomlinson IP, Houlston RS, Montgomery GW, Sieber OM, Dunlop MG. Investigation of the effects of DNA repair gene polymorphisms on the risk of colorectal cancer. Mutagenesis. 2012;27(2):219–23. doi:10.1093/mutage/ger070.
13. Martens-Uzunova ES, Olvedy M, Jenster G. Beyond microRNA–novel RNAs derived from small non-coding RNA and their implication in cancer. Cancer Lett. 2013;340(2):201–11. doi:10.1016/j.canlet.2012.11.058.
14. Ambros V. The functions of animal microRNAs. Nature. 2004;431(7006):350–5. doi:10.1038/nature02871.
15. Spizzo R, Nicoloso MS, Croce CM, Calin GA. SnapShot: microRNAs in cancer. Cell. 2009;137(3):586–e1. doi:10.1016/j.cell.2009.04.040.
16. Duan R, Pak C, Jin P. Single nucleotide polymorphism associated with mature miR-125a alters the processing of pri-miRNA. Hum Mol Genet. 2007;16(9):1124–31. doi:10.1093/hmg/ddm062.
17. Nicoloso MS, Sun H, Spizzo R, Kim H, Wickramasinghe P, Shimizu M, et al. Single-nucleotide polymorphisms inside microRNA target sites influence tumor susceptibility. Cancer Res. 2010;70(7):2789–98. doi:10.1158/0008-5472.CAN-09-3541.
18. Chen K, Song F, Calin GA, Wei Q, Hao X, Zhang W. Polymorphisms in microRNA targets: a gold mine for molecular epidemiology. Carcinogenesis. 2008;29(7):1306–11. doi:10.1093/carcin/bgn116.
19. Wojcicka A, de la Chapelle A, Jazdzewski K. MicroRNA-related sequence variations in human cancers. Hum Genet. 2014;133(4):463–9. doi:10.1007/s00439-013-1397-x.

20. Afonso-Grunz F, Muller S. Principles of miRNA-mRNA interactions: beyond sequence complementarity. Cell Mol Life Sci. 2015. doi:10.1007/s00018-015-1922-2.
21. Thomas LF, Saetrom P. Single nucleotide polymorphisms can create alternative polyadenylation signals and affect gene expression through loss of microRNA-regulation. PLoS Comput Biol. 2012;8(8):e1002621. doi:10.1371/journal.pcbi.1002621.
22. Mishra PJ, Banerjee D, Bertino JR. MiRSNPs or MiR-polymorphisms, new players in microRNA mediated regulation of the cell: Introducing microRNA pharmacogenomics. Cell Cycle. 2008;7(7):853–8.
23. Mishra PJ, Bertino JR. MicroRNA polymorphisms: the future of pharmacogenomics, molecular epidemiology and individualized medicine. Pharmacogenomics. 2009;10(3):399–416. doi:10.2217/14622416.10.3.399.
24. Chen K, Rajewsky N. Natural selection on human microRNA binding sites inferred from SNP data. Nat Genet. 2006;38(12):1452–6. doi:10.1038/ng1910.
25. Saunders MA, Liang H, Li WH. Human polymorphism at microRNAs and microRNA target sites. Proc Natl Acad Sci U S A. 2007;104(9):3300–5. doi:10.1073/pnas.0611347104.
26. Liu C, Zhang F, Li T, Lu M, Wang L, Yue W, et al. MirSNP, a database of polymorphisms altering miRNA target sites, identifies miRNA-related SNPs in GWAS SNPs and eQTLs. BMC Genomics. 2012;13:661. doi:10.1186/1471-2164-13-661.
27. Pardini B, Rosa F, Barone E, Di Gaetano C, Slyskova J, Novotny J, et al. Variation within 3′-UTRs of base excision repair genes and response to therapy in colorectal cancer patients: a potential modulation of microRNAs binding. Clin Cancer Res. 2013;19(21):6044–56. doi:10.1158/1078-0432.CCR-13-0314.
28. Mishra PJ, Humeniuk R, Longo-Sorbello GS, Banerjee D, Bertino JR. A miR-24 microRNA bindingsite polymorphism in dihydrofolate reductase gene leads to methotrexate resistance. Proc Natl Acad Sci U S A. 2007;104(33):13513–8. doi:10.1073/pnas.0706217104.
29. Ryan BM, Robles AI, Harris CC. Genetic variation in microRNA networks: the implications for cancer research. Nat Rev Cancer. 2010;10(6):389–402. doi:10.1038/nrc2867.
30. Chen H, Sun LY, Chen LL, Zheng HQ, Zhang QF. A variant in microRNA-196a2 is not associated with susceptibility to and progression of colorectal cancer in Chinese. Intern Med J. 2012;42(6):e115–9. doi:10.1111/j.1445-5994.2011.02434.x.
31. Zhan JF, Chen LH, Chen ZX, Yuan YW, Xie GZ, Sun AM, et al. A functional variant in microRNA- 196a2 is associated with susceptibility of colorectal cancer in a Chinese population. Arch Med Res. 2011;42(2):144–8. doi:10.1016/j.arcmed.2011.04.001.
32. Boni V, Zarate R, Villa JC, Bandres E, Gomez MA, Maiello E, et al. Role of primary miRNA polymorphic variants in metastatic colon cancer patients treated with 5-fluorouracil and irinotecan. Pharmacogenomics J. 2011;11(6):429–36. doi:10.1038/tpj.2010.58.
33. Xing J, Wan S, Zhou F, Qu F, Li B, Myers RE, et al. Genetic polymorphisms in pre-microRNA genes as prognostic markers of colorectal cancer. Cancer Epidemiol Biomark Prev. 2012;21(1):217–27. doi:10.1158/1055-9965.EPI-11-0624.
34. Zhu L, Chu H, Gu D, Ma L, Shi D, Zhong D, et al. A functional polymorphism in miRNA-196a2 is associated with colorectal cancer risk in a Chinese population. DNA Cell Biol. 2012;31(3):350–4. doi:10.1089/dna.2011.1348.
35. Ryan BM, McClary AC, Valeri N, Robinson D, Paone A, Bowman ED, et al. rs4919510 in hsa-mir-608 is associated with outcome but not risk of colorectal cancer. PLoS One. 2012;7(5):e36306. doi:10.1371/journal.pone.0036306.
36. Lin M, Gu J, Eng C, Ellis LM, Hildebrandt MA, Lin J, et al. Genetic polymorphisms in microRNA related genes as predictors of clinical outcomes in colorectal adenocarcinoma patients. Clin Cancer Res. 2012;18(14):3982–91. doi:10.1158/1078-0432.CCR-11-2951.
37. Hu X, Li L, Shang M, Zhou J, Song X, Lu X, et al. Association between microRNA genetic variants and susceptibility to colorectal cancer in Chinese population. Tumour Biol. 2014;35(3):2151–6. doi:10.1007/s13277-013-1285-y.
38. Oh J, Kim JW, Lee BE, Jang MJ, Chong SY, Park PW, et al. Polymorphisms of the pri-miR-34b/c promoter and TP53 codon 72 are associated with risk of colorectal cancer. Oncol Rep. 2014;31(2):995–1002. doi:10.3892/or.2013.2926.
39. Parlayan C, Ikeda S, Sato N, Sawabe M, Muramatsu M, Arai T. Association analysis of single nucleotide polymorphisms in miR-146a and miR-196a2 on the prevalence of cancer in elderly Japanese: a case-control study. Asian Pac J Cancer Prev. 2014;15(5):2101–7.
40. Wang Z, Sun X, Wang Y, Liu X, Xuan Y, Hu S. Association between miR-27a genetic variants and susceptibility to colorectal cancer. Diagn Pathol. 2014;9:146. doi:10.1186/1746-1596-9-146.
41. Pardini B, Rosa F, Naccarati A, Vymetalkova V, Ye Y, Wu X, et al. Polymorphisms in microRNA genes as predictors of clinical outcomes in colorectal cancer patients. Carcinogenesis. 2015;36(1):82–6. doi:10.1093/carcin/bgu224.
42. Cao Y, Hu J, Fang Y, Chen Q, Li H. Association between a functional variant in microRNA-27a and susceptibility to colorectal cancer in a Chinese Han population. Genet Mol Res. 2014;13(3):7420–7. doi:10.4238/2014.September.12.8.
43. Gao LB, Li LJ, Pan XM, Li ZH, Liang WB, Bai P, et al. A genetic variant in the promoter region of miR-34b/c is associated with a reduced risk of colorectal

cancer. Biol Chem. 2013;394(3):415–20. doi:10.1515/hsz-2012-0297.
44. Tang R, Qi Q, Wu R, Zhou X, Wu D, Zhou H, et al. The polymorphic terminal-loop of pre-miR-1307 binding with MBNL1 contributes to colorectal carcinogenesis via interference with Dicer1 recruitment. Carcinogenesis. 2015;36(8):867–75. doi:10.1093/carcin/bgv066.
45. Vinci S, Gelmini S, Mancini I, Malentacchi F, Pazzagli M, Beltrami C, et al. Genetic and epigenetic factors in regulation of microRNA in colorectal cancers. Methods. 2013;59(1):138–46. doi:10.1016/j.ymeth.2012.09.002.
46. Dikaiakos P, Gazouli M, Rizos S, Zografos G, Theodoropoulos GE. Evaluation of genetic variants in miRNAs in patients with colorectal cancer. Cancer Biomark. 2015;15(2):157–62. doi:10.3233/CBM-140449.
47. Kupcinskas J, Bruzaite I, Juzenas S, Gyvyte U, Jonaitis L, Kiudelis G, et al. Lack of association between miR-27a, miR-146a, miR-196a-2, miR-492 and miR-608 gene polymorphisms and colorectal cancer. Sci Rep. 2014;4:5993. doi:10.1038/srep05993.
48. Min KT, Kim JW, Jeon YJ, Jang MJ, Chong SY, Oh D, et al. Association of the miR-146aC>G, 149C>T, 196a2C>T, and 499A>G polymorphisms with colorectal cancer in the Korean population. Mol Carcinog. 2012;51 Suppl 1:E65–73. doi:10.1002/mc.21849.
49. Lv M, Dong W, Li L, Zhang L, Su X, Wang L, et al. Association between genetic variants in pre-miRNA and colorectal cancer risk in a Chinese population. J Cancer Res Clin Oncol. 2013;139(8):1405–10. doi:10.1007/s00432-013-1456-7.
50. Hezova R, Kovarikova A, Bienertova-Vasku J, Sachlova M, Redova M, Vasku A, et al. Evaluation of SNPs in miR-196-a2, miR-27a and miR-146a as risk factors of colorectal cancer. World J Gastroenterol. 2012;18(22):2827–31. doi:10.3748/wjg.v18.i22.2827.
51. Wang FJ, Ding Y, Mao YY, Jing FY, Zhang ZY, Jiang LF, et al. Associations between hsa-miR-603 polymorphism, lifestyle-related factors and colorectal cancer risk. Cancer Biomark. 2014;14(4):225–31. doi:10.3233/CBM-140395.
52. Wu Y, Hao X, Feng Z, Liu Y. Genetic polymorphisms in miRNAs and susceptibility to colorectal cancer. Cell Biochem Biophys. 2015;71(1):271–8. doi:10.1007/s12013-014-0195-y.
53. Liu XX, Wang M, Xu D, Yang JH, Kang HF, Wang XJ, et al. Quantitative assessment of the association between genetic variants in MicroRNAs and colorectal cancer risk. BioMed Res Int. 2015;2015:276410. doi:10.1155/2015/276410.
54. Xie WQ, Tan SY, Wang XF. Effect of a common genetic variant microRNA-146a rs2910164 on colorectal cancer: a meta-analysis. J Dig Dis. 2014;15(12):647–53. doi:10.1111/1751-2980.12201.
55. Yi DH, Wang BG, Zhong XP, Liu H, Liu YF. Pri-miR-34b/c rs4938723 TC heterozygote is associated with increased cancer risks: evidence from published data. Tumour Biol. 2014;35(12):11967–75. doi:10.1007/s13277-014-2493-9.
56. Li X, Wang L, Yu J, Xu J, Du J. The genetic association between pri-miR-34b/c polymorphism (rs4938723 T > C) and susceptibility to cancers: evidence from published studies. Tumour Biol. 2014;35(12):12525–34. doi:10.1007/s13277-014-2572-y.
57. Jazdzewski K, Murray EL, Franssila K, Jarzab B, Schoenberg DR, de la Chapelle A. Common SNP in pre-miR-146a decreases mature miR expression and predisposes to papillary thyroid carcinoma. Proc Natl Acad Sci U S A. 2008;105(20):7269–74. doi:10.1073/pnas.0802682105.
58. Li W, Duan R, Kooy F, Sherman SL, Zhou W, Jin P. Germline mutation of microRNA-125a is associated with breast cancer. J Med Genet. 2009;46(5):358–60. doi:10.1136/jmg.2008.063123.
59. Lehmann TP, Korski K, Ibbs M, Zawierucha P, Grodecka-Gazdecka S, Jagodzinski PP. rs12976445 variant in the pri-miR-125a correlates with a lower level of hsa-miR-125a and ERBB2 overexpression in breast cancer patients. Oncol Lett. 2013;5(2):569–73. doi:10.3892/ol.2012.1040.
60. Landi D, Gemignani F, Barale R, Landi S. A catalog of polymorphisms falling in microRNA-binding regions of cancer genes. DNA Cell Biol. 2008;27(1):35–43. doi:10.1089/dna.2007.0650.
61. Raveche ES, Salerno E, Scaglione BJ, Manohar V, Abbasi F, Lin YC, et al. Abnormal microRNA-16 locus with synteny to human 13q14 linked to CLL in NZB mice. Blood. 2007;109(12):5079–86. doi:10.1182/blood-2007-02-071225.
62. Calin GA, Ferracin M, Cimmino A, Di Leva G, Shimizu M, Wojcik SE, et al. A MicroRNA signature associated with prognosis and progression in chronic lymphocytic leukemia. N Engl J Med. 2005;353(17):1793–801. doi:10.1056/NEJMoa050995.
63. Didiano D, Hobert O. Perfect seed pairing is not a generally reliable predictor for miRNA-target interactions. Nat Struct Mol Biol. 2006;13(9):849–51. doi:10.1038/nsmb1138.
64. Hausser J, Zavolan M. Identification and consequences of miRNA-target interactions–beyond repression of gene expression. Nat Rev Genet. 2014;15(9):599–612. doi:10.1038/nrg3765.
65. Jazdzewski K, de la Chapelle A. Genomic sequence matters: a SNP in microRNA-146a can turn anti-apoptotic. Cell Cycle. 2009;8(11):1642–3.
66. Xu B, Feng NH, Li PC, Tao J, Wu D, Zhang ZD, et al. A functional polymorphism in Pre-miR-146a gene is associated with prostate cancer risk and mature miR-146a expression in vivo. Prostate. 2010;70(5):467–72. doi:10.1002/pros.21080.

67. Wang M, Chu H, Li P, Yuan L, Fu G, Ma L, et al. Genetic variants in miRNAs predict bladder cancer risk and recurrence. Cancer Res. 2012;72(23):6173–82. doi:10.1158/0008-5472.CAN-12-0688.
68. Ma L, Zhu L, Gu D, Chu H, Tong N, Chen J, et al. A genetic variant in miR-146a modifies colorectal cancer susceptibility in a Chinese population. Arch Toxicol. 2013;87(5):825–33. doi:10.1007/s00204-012-1004-2.
69. Goel A, Boland CR. Recent insights into the pathogenesis of colorectal cancer. Curr Opin Gastroenterol. 2010;26(1):47–52. doi:10.1097/MOG.0b013e328332b850.
70. Zhang R, Su B. MicroRNA regulation and the variability of human cortical gene expression. Nucleic Acids Res. 2008;36(14):4621–8. doi:10.1093/nar/gkn431.
71. Sethupathy P, Collins FS. MicroRNA target site polymorphisms and human disease. Trends Genet. 2008;24(10):489–97. doi:10.1016/j.tig.2008.07.004.
72. Zhao Y, Samal E, Srivastava D. Serum response factor regulates a muscle-specific microRNA that targets Hand2 during cardiogenesis. Nature. 2005;436(7048):214–20. doi:10.1038/nature03817.
73. Kertesz M, Iovino N, Unnerstall U, Gaul U, Segal E. The role of site accessibility in microRNA target recognition. Nat Genet. 2007;39(10):1278–84. doi:10.1038/ng2135.
74. Wexler Y, Zilberstein C, Ziv-Ukelson M. A study of accessible motifs and RNA folding complexity. J Comput Biol. 2007;14(6):856–72. doi:10.1089/cmb.2007.R020.
75. Hon LS, Zhang Z. The roles of binding site arrangement and combinatorial targeting in microRNA repression of gene expression. Genome Biol. 2007;8(8):R166. doi:10.1186/gb-2007-8-8-r166.
76. Elkon R, Ugalde AP, Agami R. Alternative cleavage and polyadenylation: extent, regulation and function. Nat Rev Genet. 2013;14(7):496–506. doi:10.1038/nrg3482.
77. Tian B, Manley JL. Alternative cleavage and polyadenylation: the long and short of it. Trends Biochem Sci. 2013;38(6):312–20. doi:10.1016/j.tibs.2013.03.005.
78. Sandberg R, Neilson JR, Sarma A, Sharp PA, Burge CB. Proliferating cells express mRNAs with shortened 3′ untranslated regions and fewer microRNA target sites. Science. 2008;320(5883):1643–7. doi:10.1126/science.1155390.
79. Muller S. In silico analysis of regulatory networks underlines the role of miR-10b-5p and its target BDNF in huntington's disease. Transl Neurodegener. 2014;3:17. doi:10.1186/2047-9158-3-17.
80. Varendi K, Kumar A, Harma MA, Andressoo JO. miR-1, miR-10b, miR-155, and miR-191 are novel regulators of BDNF. Cell Mol Life Sci. 2014;71(22):4443–56. doi:10.1007/s00018-014-1628-x.
81. Mayr C, Bartel DP. Widespread shortening of 3′UTRs by alternative cleavage and polyadenylation activates oncogenes in cancer cells. Cell. 2009;138(4):673–84. doi:10.1016/j.cell.2009.06.016 S0092-8674(09)00716-8 [pii].
82. Shi Y, Di Giammartino DC, Taylor D, Sarkeshik A, Rice WJ, Yates 3rd JR, et al. Molecular architecture of the human pre-mRNA 3′ processing complex. Mol Cell. 2009;33(3):365–76. doi:10.1016/j.molcel.2008.12.028.
83. Grimson A, Farh KK, Johnston WK, Garrett-Engele P, Lim LP, Bartel DP. MicroRNA targeting specificity in mammals: determinants beyond seed pairing. Mol Cell. 2007;27(1):91–105. doi:10.1016/j.molcel.2007.06.017.
84. Pipan V, Zorc M, Kunej T. MicroRNA polymorphisms in cancer: a literature analysis. Cancers (Basel). 2015;7(3):1806–14. doi:10.3390/cancers7030863.
85. Naccarati A, Pardini B, Stefano L, Landi D, Slyskova J, Novotny J, et al. Polymorphisms in miRNAbinding sites of nucleotide excision repair genes and colorectal cancer risk. Carcinogenesis. 2012;33(7):1346–51. doi:10.1093/carcin/bgs172.
86. Landi D, Gemignani F, Naccarati A, Pardini B, Vodicka P, Vodickova L, et al. Polymorphisms within micro-RNA-binding sites and risk of sporadic colorectal cancer. Carcinogenesis. 2008;29(3):579–84. doi:10.1093/carcin/bgm304.
87. Iorio MV, Ferracin M, Liu CG, Veronese A, Spizzo R, Sabbioni S, et al. MicroRNA gene expression deregulation in human breast cancer. Cancer Res. 2005;65(16):7065–70. doi:10.1158/0008-5472.CAN-05-1783.
88. Vymetalkova V, Pardini B, Rosa F, Di Gaetano C, Novotny J, Levy M, et al. Variations in mismatch repair genes and colorectal cancer risk and clinical outcome. Mutagenesis. 2014;29(4):259–65. doi:10.1093/mutage/geu014.
89. Naccarati A, Rosa F, Vymetalkova V, Barone E, Jiraskova K, Di Gaetano C, et al. Double-strand break repair and colorectal cancer: gene variants within 3′ UTRs and microRNAs binding as modulators of cancer risk and clinical outcome. Oncotarget. 2015. doi:10.18632/oncotarget.6804.
90. Landi D, Gemignani F, Pardini B, Naccarati A, Garritano S, Vodicka P, et al. Identification of candidate genes carrying polymorphisms associated with the risk of colorectal cancer by analyzing the colorectal mutome and microRNAome. Cancer. 2012;118(19):4670–80. doi:10.1002/cncr.27435.
91. Winder T, Scheithauer W, Lang A. K-ras mutations and cetuximab in colorectal cancer. N Engl J Med. 2009;360(8):834–5; author reply 5–6.
92. Graziano F, Canestrari E, Loupakis F, Ruzzo A, Galluccio N, Santini D, et al. Genetic modulation of the Let-7 microRNA binding to KRAS 3′-untranslated region and survival of metastatic colorectal cancer patients treated with salvage cetuximab-

irinotecan. Pharmacogenomics J. 2010;10(5):458–64. doi:10.1038/tpj.2010.9.
93. Smits KM, Paranjape T, Nallur S, Wouters KA, Weijenberg MP, Schouten LJ, et al. A let-7 microRNA SNP in the KRAS 3′UTR is prognostic in early-stage colorectal cancer. Clin Cancer Res. 2011;17(24):7723–31. doi:10.1158/1078-0432.CCR-11-0990.
94. Sebio A, Pare L, Paez D, Salazar J, Gonzalez A, Sala N, et al. The LCS6 polymorphism in the binding site of let-7 microRNA to the KRAS 3′-untranslated region: its role in the efficacy of anti-EGFR-based therapy in metastatic colorectal cancer patients. Pharmacogenet Genomics. 2013;23(3):142–7. doi:10.1097/FPC.0b013e32835d9b0b.
95. Ryan BM, Robles AI, Harris CC. KRAS-LCS6 genotype as a prognostic marker in early-stage CRC–letter. Clin Cancer Res. 2012;18(12):3487–8. doi:10.1158/1078-0432.CCR-12-0250.
96. Zhang W, Labonte MJ, Lenz HJ. KRAS let-7 LCS6 SNP predicts cetuximab efficacy in KRASwt metastatic colorectal cancer patients: Does treatment combination partner matter? Ann Oncol. 2011;22(2):484–5. doi:10.1093/annonc/mdq704.
97. Kjersem JB, Ikdahl T, Guren T, Skovlund E, Sorbye H, Hamfjord J, et al. Let-7 miRNA-binding site polymorphism in the KRAS 3′UTR; colorectal cancer screening population prevalence and influence on clinical outcome in patients with metastatic colorectal cancer treated with 5-fluorouracil and oxaliplatin +/– cetuximab. BMC Cancer. 2012;12:534. doi:10.1186/1471-2407-12-534.
98. Ruzzo A, Graziano F, Vincenzi B, Canestrari E, Perrone G, Galluccio N, et al. High let-7a microRNA levels in KRAS-mutated colorectal carcinomas may rescue anti-EGFR therapy effects in patients with chemotherapy-refractory metastatic disease. Oncologist. 2012;17(6):823–9. doi:10.1634/theoncologist.2012-0081.
99. Saridaki Z, Weidhaas JB, Lenz HJ, Laurent-Puig P, Jacobs B, De Schutter J, et al. A let-7 microRNAbinding site polymorphism in KRAS predicts improved outcome in patients with metastatic colorectal cancer treated with salvage cetuximab/panitumumab monotherapy. Clin Cancer Res. 2014;20(17):4499–510. doi:10.1158/1078-0432.CCR-14-0348.
100. Pan XM, Sun RF, Li ZH, Guo XM, Zhang Z, Qin HJ, et al. A let-7 KRAS rs712 polymorphism increases colorectal cancer risk. Tumour Biol. 2014;35(1):831–5. doi:10.1007/s13277-013-1114-3.
101. Song FJ, Chen KX. Single-nucleotide polymorphisms among microRNA: big effects on cancer. Chin J Cancer. 2011;30(6):381–91.
102. Zanetti KA, Haznadar M, Welsh JA, Robles AI, Ryan BM, McClary AC, et al. 3′-UTR and functional secretor haplotypes in mannose-binding lectin 2 are associated with increased colon cancer risk in African Americans. Cancer Res. 2012;72(6):1467–77. doi:10.1158/0008-5472.CAN-11-3073.
103. Schmit SL, Gollub J, Shapero MH, Huang SC, Rennert HS, Finn A, et al. MicroRNA polymorphisms and risk of colorectal cancer. Cancer Epidemiol Biomark Prev. 2015;24(1):65–72. doi:10.1158/1055-9965.EPI-14-0219.
104. Langevin SM, Christensen BC. Let-7 microRNA-binding-site polymorphism in the 3′UTR of KRAS and colorectal cancer outcome: a systematic review and meta-analysis. Cancer Med. 2014;3(5):1385–95. doi:10.1002/cam4.279.
105. Rusinov V, Baev V, Minkov IN, Tabler M. MicroInspector: a web tool for detection of miRNA binding sites in an RNA sequence. Nucleic Acids Res. 2005;33(Web Server issue):W696–700. doi:10.1093/nar/gki364.
106. Maragkakis M, Reczko M, Simossis VA, Alexiou P, Papadopoulos GL, Dalamagas T, et al. DIANAmicroT web server: elucidating microRNA functions through target prediction. Nucleic Acids Res. 2009;37(Web Server issue):W273–6. doi:10.1093/nar/gkp292.
107. Lewis BP, Burge CB, Bartel DP. Conserved seed pairing, often flanked by adenosines, indicates that thousands of human genes are microRNA targets. Cell. 2005;120(1):15–20. doi:10.1016/j.cell.2004.12.035.
108. Bhattacharya A, Ziebarth JD, Cui Y. PolymiRTS Database 3.0: linking polymorphisms in microRNAs and their target sites with human diseases and biological pathways. Nucleic Acids Res. 2014;42(Database issue):D86–91. doi:10.1093/nar/gkt1028.
109. Ning S, Zhao Z, Ye J, Wang P, Zhi H, Li R, et al. SNP@lincTFBS: an integrated database of polymorphisms in human LincRNA transcription factor binding sites. PLoS One. 2014;9(7):e103851. doi:10.1371/journal.pone.0103851.
110. Gong J, Liu W, Zhang J, Miao X, Guo AY. lncRNASNP: a database of SNPs in lncRNAs and their potential functions in human and mouse. Nucleic Acids Res. 2015;43(Database issue):D181–6. doi:10.1093/nar/gku1000.
111. Ning S, Zhao Z, Ye J, Wang P, Zhi H, Li R, et al. LincSNP: a database of linking disease-associated SNPs to human large intergenic non-coding RNAs. BMC Bioinform. 2014;15:152. doi:10.1186/1471-2105-15-152.
112. Bhartiya D, Jalali S, Ghosh S, Scaria V. Distinct patterns of genetic variations in potential functional elements in long noncoding RNAs. Hum Mutat. 2014;35(2):192–201. doi:10.1002/humu.22472.
113. Ning S, Wang P, Ye J, Li X, Li R, Zhao Z, et al. A global map for dissecting phenotypic variants in human lincRNAs. Eur J Hum Genet. 2013;21(10):1128–33. doi:10.1038/ejhg.2013.7.
114. Barenboim M, Zoltick BJ, Guo Y, Weinberger DR. MicroSNiPer: a web tool for prediction of SNP effects on putative microRNA targets. Hum Mutat. 2010;31(11):1223–32. doi:10.1002/humu.21349.

115. Dweep H, Sticht C, Pandey P, Gretz N. miRWalk–database: prediction of possible miRNA binding sites by "walking" the genes of three genomes. J Biomed Inform. 2011;44(5):839–47. doi:10.1016/j.jbi.2011.05.002.
116. Bruno AE, Li L, Kalabus JL, Pan Y, Yu A, Hu Z. miRdSNP: a database of disease-associated SNPs and microRNA target sites on 3′UTRs of human genes. BMC Genomics. 2012;13:44. doi:10.1186/1471-2164-13-44.
117. Gong J, Tong Y, Zhang HM, Wang K, Hu T, Shan G, et al. Genome-wide identification of SNPs in microRNA genes and the SNP effects on microRNA target binding and biogenesis. Hum Mutat. 2012;33(1):254–63. doi:10.1002/humu.21641.
118. Thomas LF, Saito T, Saetrom P. Inferring causative variants in microRNA target sites. Nucleic Acids Res. 2011;39(16):e109. doi:10.1093/nar/gkr414.
119. Deveci M, Catalyurek M, Toland AE. mrSNP: software to detect SNP effects on microRNA binding. BMC Bioinform. 2014;15:73. doi:10.1186/1471-2105-15-73.
120. Jin Y, Chen Z, Liu X, Zhou X. Evaluating the microRNA targeting sites by luciferase reporter gene assay. Methods Mol Biol. 2013;936:117–27. doi:10.1007/978-1-62703-083-0_10.
121. Landi D, Moreno V, Guino E, Vodicka P, Pardini B, Naccarati A, et al. Polymorphisms affecting microRNA regulation and associated with the risk of dietary-related cancers: a review from the literature and new evidence for a functional role of rs17281995 (CD86) and rs1051690 (INSR), previously associated with colorectal cancer. Mutat Res. 2011;717(1–2):109–15. doi:10.1016/j.mrfmmm.2010.10.002.
122. John B, Enright AJ, Aravin A, Tuschl T, Sander C, Marks DS. Human microRNA targets. PLoS Biol. 2004;2(11):e363. doi:10.1371/journal.pbio.0020363.
123. Krek A, Grun D, Poy MN, Wolf R, Rosenberg L, Epstein EJ, et al. Combinatorial microRNA target predictions. Nat Genet. 2005;37(5):495–500. doi:10.1038/ng1536.
124. Cipollini M, Landi S, Gemignani F. MicroRNA binding site polymorphisms as biomarkers in cancer management and research. Pharmgenomics Pers Med. 2014;7:173–91. doi:10.2147/PGPM.S61693.
125. Muniategui A, Pey J, Planes FJ, Rubio A. Joint analysis of miRNA and mRNA expression data. Brief Bioinform. 2013;14(3):263–78. doi:10.1093/bib/bbs028.
126. Lagana A, Forte S, Giudice A, Arena MR, Puglisi PL, Giugno R et al. miRo: a miRNA knowledge base. Database (Oxford). 2009;2009:bap008. doi:10.1093/database/bap008.
127. Giles CB, Girija-Devi R, Dozmorov MG, Wren JD. mirCoX: a database of miRNA-mRNA expression correlations derived from RNA-seq meta-analysis. BMC Bioinform. 2013;14 Suppl 14:S17. doi:10.1186/1471-2105-14-S14-S17.
128. Hua Y, Duan S, Murmann AE, Larsen N, Kjems J, Lund AH, et al. miRConnect: identifying effector genes of miRNAs and miRNA families in cancer cells. PLoS One. 2011;6(10):e26521. doi:10.1371/journal.pone.0026521.
129. Kumar V, Westra HJ, Karjalainen J, Zhernakova DV, Esko T, Hrdlickova B, et al. Human diseaseassociated genetic variation impacts large intergenic noncoding RNA expression. PLoS Genet. 2013;9(1):e1003201. doi:10.1371/journal.pgen.1003201.
130. Li L, Sun R, Liang Y, Pan X, Li Z, Bai P, et al. Association between polymorphisms in long noncoding RNA PRNCR1 in 8q24 and risk of colorectal cancer. J Exp Clin Cancer Res. 2013;32:104. doi:10.1186/1756-9966-32-104.
131. Ling H, Spizzo R, Atlasi Y, Nicoloso M, Shimizu M, Redis RS, et al. CCAT2, a novel noncoding RNA mapping to 8q24, underlies metastatic progression and chromosomal instability in colon cancer. Genome Res. 2013;23(9):1446–61. doi:10.1101/gr.152942.112.
132. Chu H, Xia L, Qiu X, Gu D, Zhu L, Jin J, et al. Genetic variants in noncoding PIWI-interacting RNA and colorectal cancer risk. Cancer. 2015;121(12):2044–52. doi:10.1002/cncr.29314.

Part II

Non-coding RNAs: New Class of Biomarkers in Colorectal Cancer

8

Non-coding RNAs as Biomarkers for Colorectal Cancer Screening and Early Detection

Ondrej Slaby

Abstract

Early detection of colorectal cancer (CRC) is the key for prevention and the ability to impact long-term survival of CRC patients. Current CRC screening modalities are inadequate for global application because of low sensitivity and specificity in case of conventional stool-based screening tests, and high costs and a low participation compliance in colonoscopy. An accurate stool- or blood-based screening test with use of innovative biomarkers is an appealing alternative as it is non-invasive and poses minimal risk to patients. It is easy to perform, can be repeated at shorter intervals, and therefore would likely lead to a much higher compliance rates. Non-coding RNAs (ncRNAs) have recently gained attention because of their involvement in different biological processes, such as proliferation, differentiation, migration, angiogenesis and apoptosis. An increasing number of studies have demonstrated that mutations or abnormal expression of ncRNAs are closely associated with various cancers, including CRC. The discovery that ncRNAs (mainly microRNAs) are stable in stool and in blood plasma and serum presents the opportunity to develop novel strategies taking advantage of circulating ncRNAs as early diagnostic biomarkers of CRC. This chapter is a comprehensive examination of aberrant ncRNAs expression levels in tumor tissue, stool and blood of CRC patients and a summary of the current findings on ncRNAs, including microRNAs, small nucleolar RNAs, small nuclear RNAs, Piwi-interacting RNAs, circular RNAs and long ncRNAs in regards to their potential usage for screening or early detection of CRC.

O. Slaby (✉)
Central European Institute of Technology, Masaryk University, University Campus Bohunice, Building A35, Kamenice 5, 625 00 Brno, Czech Republic
e-mail: on.slaby@gmail.com

O. Slaby, G.A. Calin (eds.), *Non-coding RNAs in Colorectal Cancer*, Advances in Experimental Medicine and Biology 937, DOI 10.1007/978-3-319-42059-2_8

Keywords

Colorectal cancer • Non-coding RNA • microRNA • lncRNA • piRNAs • snoRNAs • snRNAs • Screening • Early diagnosis

8.1 Introduction

Colorectal cancer (CRC) is the second most commonly diagnosed cancer and the second leading cause of cancer death in Europe, with an incidence of 43 600 new cases between 2007 and 2008 [1]. In the United States, CRC- related deaths account for approximately 9 % of all cancer mortality. The lifetime incidence of CRC in the average-risk population living in North America and Western Europe is 5 %. The majority of cases (90 %) occur after the age of 50 [2]. CRC actually fulfills the World Health Organization conditions required for mass screening, since it is a very common disease, with major morbidity and mortality rates and is almost always preceded by a slow progressive premalignant lesion (the adenomatous polyp) which can readily be removed leading to true cancer prevention [3, 4].

Screening strategies for CRC involve separation of the population into two main categories: average risk and high-risk populations. Each of these categories is targeted using a different screening program. In the first group, adults over 50 without a personal or family history of CRC, polyps or inflammatory bowel diseases (IBD), are screened. The high- risk population includes subjects with a family history of CRC, personal history of CRC or polyps or are index cases affected by IBD. There is, however, a third category, more specifically characterized by hereditary or familial risk and represented by hereditary cancer syndromes, such as familial adenomatous polyposis (FAP) or its variants, MYH-associated polyposis, Lynch syndrome (hereditary non-polyposis colon cancer – HNPCC), BRCA2, juvenile polyposis, or any personal or family history of sporadic CRC or adenomatous polyps [3, 4]. Such cases should be screened directly with total colonoscopy (TC). The average risk population reflects vast majority of the population and needs to be screened by less-invasive, low-cost techniques with acceptable patient compliance.

The dramatically high social and economic impact of CRC on human health makes identification of a reliable screening tool of great importance. As an ideal approach for CRC screening, the method must possess a very high degree of sensitivity and specificity for the early detection of cancer. In recent years, extended efforts were made by researchers to look for more reliable and effective screening tests based on a systems biology approach, using easily available biological samples, such as urine, breath, serum and stool [3].

There is a growing evidence suggesting that detection of non-coding RNAs, mainly microRNAs (miRNAs), in stool or circulating in blood provides a novel and promising early diagnostic option for CRC screening [5]. Significance of non-coding RNAs in CRC screening and early diagnosis is summarized and discussed in this chapter.

8.2 Traditional Approaches in Colorectal Cancer Screening

Clinically validated screening strategies currently available in practice include fecal occult blood testing (FOBT), total colonoscopy (TC), flexible sigmoidoscopy (FS) and radiographic imaging, such as double contrast barium enema and virtual TC.

Guidelines from the European Union recommend annual or biennial screening with high-sensitivity FOBTs or the newer fecal immunochemical tests (FITs) [1, 2], while guidelines from North America and eastern Asia recommend any of several screening tests, including sigmoidoscopy, colonoscopy, double-contrast barium enema, CT colonography, and fecal DNA in addition to FOBTs and FITs [2]. German guidelines recommend FOBT annually and colonoscopy every 10 years. In Germany, Poland, and the USA, colonoscopy every 10 years is currently the most prevalent screening strategy. Although colonoscopy reduces CRC incidence [6] and mortality [7], and is cost-effective [8], it requires considerable provider and financial resources [9]. Further, adherence to screening remains suboptimal [10], at least in part to a perceived lack of effective options to colonoscopy [2].

FOBT is the most commonly used method for CRC screening. Since FOBT is safe and acceptable to patients, it has been widely used as noninvasive screening tool for CRC. However, FOBT has relatively poor sensitivity and specificity for CRC diagnosis; moreover, FOBT screen reduces relative risk associated with CRC-related mortality only by 16–25 % [11].

Two types of FOBTs are used in clinical practice. One type is guaiac FOBT (gFOBT) (available as Hemoccult II and Hemoccult SENSA), which works by detecting peroxidase-like activity of the heme molecule. The test is not automated, nor is its interpretation objective [2]. The more recent type is the FITs (available as HemeSelect, InSure, Flexsure OBT, etc.), which uses antibodies to globin or albumin to detect human blood from the lower gastrointestinal tract [12]. In contrast to gFOBTs, FITs processing and interpretation are automated and objective. There are several cross sectional studies that provide test characteristics of FITs with or without a comparative gFOBT group and with colonoscopy as the reference standard [13]. These studies show that FITs are associated with higher participation rates, higher positivity rates and greater sensitivity for CRC and advanced adenoma, but slightly lower specificity.

Conventional cancer biomarkers, such as carcino-embryonic antigen (CEA) or CA19-9, were developed by quantifying a small amount of circulating proteins. These markers are specific for certain types of cancer, permitting early detection of cancers and monitoring cancer relapse and cancer prognosis. However, this approach suffers from well-documented limited sensitivity and specificity [14].

8.3 Emerging Approaches in Colorectal Cancer Screening

Current screening methods, such as the FOBT and colonoscopy, do not adequately meet the ideal requisites of a screening test because, even if they are effective, they are limited first by too low specificity and sensitivity, or second by high invasiveness, costs and risk. Extended effort has been dedicated by researchers at looking for more reliable and effective screening tests using genomic (genetic and epigenetic alteration), transcriptomic (mRNA), proteomic (cancer related antigens, new antibodies against tumor-associated antigens, mutated proteins) and metabolomic (volatile organic metabolites) techniques [3].

Following the Vogelstein sequential model of colorectal cancer pathogenesis, mutations in the APC, K-Ras and p53 genes were initially investigated in stool samples of CRC patients [15]. Several other hyper-methylated genes isolated from stool samples have been identified as potential biomarkers for the detection of CRC or colorectal adenoma, including p16, hMLH1, MGMT, SFRP1 and VIM [16]; however, sensitivity and specificity reported for these DNA methylation markers was highly variable. One fecal DNA mutation panel included 21 point mutations, BAT-26 (a marker of microsatellite instability), and a DNA integrity assay, which is a marker for apoptotic DNA. Other FDA approved screening tests consist of molecular assays for aberrantly methylated BMP3 and NDRG4 promotor regions, mutant KRAS, B-actin (as a reference gene for human DNA quantity) [17]. A recent interesting approach involves the use of fluorescent long DNA (FL-DNA) measurement, designed to identify cancer DNA fragments greater than 150–200 bp [18]. Changes are noted since cancer cells do not undergo apoptosis, which in normal epithelial cells typically initiate DNA cleavage and degradation producing small measurable fragments [for review 14].

Another molecular screening strategy is to screen for fecal RNA and amplify cancer specific molecular markers using RT-PCR. Fecal mRNA frequently investigated as a potential CRC marker in stool are guanylyl cyclase C, PYPAF5 or prostaglandin synthase 2 [19]. Several studies have investigated also mRNA markers in blood. Marshall et al. developed a blood-based test using a seven-gene biomarker panel (ANXA3, CLEC4D, LMNB1, PRRG4, TNFAIP6, VNN1 and IL2RB) testing RNA extracted from peripheral blood cells [20]. However, most studies involve performance of mRNA molecules encoding for CEA, CK19, and CK20 for detection of CRC [21]. Literature in this area is heterogeneous with respect to same sizes, sample collection and sample preparation, making it difficult to compare results.

A further approach for early detection and screening of CRC is to study modified "proteome" as a direct effect of mutated gene expression or as occurrence of new antibodies against tumor-associated antigens (TAAs) identified in CRC. Other studies have focused on the use of autoantibodies antibodies against TAAs as serological markers for cancer diagnosis because they are absent in healthy subjects and other non-cancer conditions [3]. Antigens considered for screening purposes included the sialylated Lewis antigen X, CO 29.11, urokinase-type plasminogen activator and small intestinal mucin antigen, but none of these serological antigens have so far demonstrated an acceptable reliability in clinical testing. Many autoantibodies against known or unknown TAAs have been found in sera of patients with a range of malignancies [for review 22], but any particular autoantibody was always detectable only in a limited proportion of patients (<40 %). Mutated or abnormal proteins have been detected also in the feces as potential biomarkers for screening, including tumor pyruvate kinase type M2, S100 calcium binding protein A12 and metallopeptidase inhibitor 1 [for review 22].

More recently, the study of specific metabolomic biomarkers for cancers has developed as a new frontier in cancer screening. Metabolomics are the endpoint of the "omics" cascade and incorporate the comprehensive study of low-molecular-weight metabolites, using high-throughput technologies, such as gas chromatography mass spectrometry, or other analytical platforms. Urine and serum are ideal tools for metabolomic analyses. Some studies using high-throughput techniques and artificial neural network statistics have identified some volatile organic metabolites as potential biomarkers for CRC in urine [23]. Very recently, a Japanese group has developed a CRC-prediction model based on serum metabolomics analysis which demonstrated high sensitivity as a novel potential screening test for CRC [24]. A similar metabolomic approach was carried out by Altomare et al., looking at the volatile organic compounds (VOCs) contained in breath [25].

As with other types of markers, majority of these studies based on emerging technological approaches are case-control studies that include subjects with advanced CRC cases and colonoscopy-negative controls. Since investigation of these markers is in the discovery phase, with exception of one FDA-approved DNA mutation panel, none of the findings has been independently validated nor investigated in the screening setting.

8.4 Deregulation of Non-coding RNAs in Colorectal Adenoma and Carcinoma: Rationale for Their Application in Screening

Nowadays, there is overwhelming evidence indicating that transcriptional, post-transcriptional and translational controls, mediated by various non-coding RNAs, exert critical pleiotropic actions on different features of CRC biology. This has opened space for discovery and characterization of non-coding RNAs as biomarkers in CRC and led to hundreds of studies published in this field in the last 10 years. The list of aberrantly expressed non-coding RNAs in colorectal adenoma or carcinoma tissues is long and continuously growing [5, 26, 27] creating a biological rationale for determination of non-coding RNAs also in the non-invasively accessible body fluids or stool and their clinical application in CRC screening.

8.4.1 MicroRNAs

The most frequently studied subclass of non-coding RNAs are microRNAs (miRNAs). These single-stranded RNA molecules, 18–25 nucleotides in length, evolutionary conserved, are involved in post-transcriptional regulation of gene expression [28]. It is predicted that miRNA genes account for 1–2 % of the human genome and control the activity of more than 50 % of all protein-coding genes. These days, the miRNAs biology is well known and was repeatedly reviewed [29]. Here, we will focus only on deregulation of miRNA expression levels associated with CRC tumor tissue [30, 31].

Deregulation of miRNAs can influence CRC pathogenesis if their mRNA targets are encoded by tumor suppressor genes or oncogenes. As shown in many studies, miRNAs might perfectly fit and integrate the model initially postulated by Vogelstein by controlling several pathways involved in CRC development. Both overexpression and silencing or switching off of specific miRNAs have been described in the pathogenesis of CRC. Up-regulation of mature miRNA may occur owing to transcriptional activation or amplification of the miRNA encoding gene, whereas silencing or reduced expression may result from deletion of a particular chromosomal region, epigenetic silencing or defects in their biogenesis. Altered miRNA expression plays an etiological role in the initiation and progression of CRC: global miRNA expression patterns can discriminate between normal colonic tissues and CRC tissues more efficiently than mRNA expression patterns. Furthermore, several investigations have shown the ability of miRNA expression patterns to distinguish microsatellite stable (MSS) and microsatellite instable (MSI) CRC tumors [32] and to improve diagnosis of poorly differentiated tumors [33].

In the comprehensive review of Mazeh et al. [31], 46 different studies focused on miRNA expression analysis in CRC tumor tissue were reviewed and discussed [31]. These studies differed in the number of evaluated miRNAs (range 1–723 miRNAs), number of samples used (4–197 samples), and number of deregulated miRNAs found (1–71 miRNAs). Table 8.1 summarizes miRNAs that were found to be deregulated at least in two independent studies. Overall, 170 different miRNAs were found to be up-regulated in CRC tumor tissue in comparison to paired normal colonic tissue. Up-regulation of miR-21 was demonstrated in 15 studies followed by miR-31, miR-135b, and miR-183 (up-regulated in 11, 9, and 8 studies, respectively). Another 110 different miRNAs were found to be up-regulated in only one study each (for review [31]). A total of 127 different miRNAs were found to be down-regulated in CRC. Down-regulation of miR-145 was demonstrated in 15 studies, followed by miR-143, down-regulated in 9 studies, followed by miR-1, miR-195, and miR-378 down-regulated in six studies each [31]. As the objective here is not to summarize hundreds of studies focused on miRNA deregulation in tumor tissue, we do not provide references to these studies and quote only one highly comprehensive review [30].

Accumulating evidence indicates that a deregulation of miRNAs is associated also with colorectal adenomas, particularly advanced colorectal adenomas (colorectal polyps greater than 1 cm in diameter and/or villous component and/or severe dysplasia), which are recognized as critical premalignant lesions for CRC development and are the primary target lesions for CRC screening [30]. MiR-21 is one of the most well-established oncogenic miRNAs in CRC, and it is frequently overexpressed also in colorectal adenoma tissues compared with normal colonic mucosa [34]. In more comprehensive analysis using 41 adenoma and 55 normal tissue specimens, miR-31 and miR-135b were found to be overexpressed, while miR-1, miR-9, miR-99b and miR-137 were down-regulated in the adenoma tissues [35]. In accordance with this study, later studies confirmed that expression of miR-135a and miR-135b was up-regulated in adenoma [36]. Also, increased expression of miR-92a in adenoma specimens has been demonstrated in other studies [37].

Table 8.1 MicroRNAs deregulated in colorectal cancer tumor tissue

	Number of studies	miRNAs deregulated in colorectal cancer
Up-regulated	15	miR-21
	11	miR-31
	9	miR-135b
	8	miR-183, miR-20a
	7	miR-19a, miR-203, miR-96
	5	miR-18a, miR-92, miR-181b
	4	miR-15b, miR-17, miR-17-5p, miR-19b, miR-20, miR-25, miR-93, miR-106a, miR-182, miR-200c, miR-224
	3	miR-15a, miR-29a, miR-95, miR-103, miR-106b, miR-130b, miR-142-3p, miR-148a, miR-221, miR-191
	2	let-7f, let-7 g, miR-10a, miR-17-3p, miR-27a, miR-29b, miR-32, miR-34a, miR-92a, miR-98, miR-105, miR-107, miR-133b, miR-135a, miR-182*, miR-188, miR-200a*, miR-210, miR-213, miR-223, miR-301b, miR-320, miR-324-5p, miR-424, miR-493, miR-513a-5p, miR-552, miR-584
Down-regulated	15	miR-145
	9	miR-143
	7	miR-1, miR-195, miR-378
	5	miR-133a, miR-133b, miR-139-5p, miR-192, miR-215
	4	miR-30a-3p, miR-375, miR-422a
	3	miR-10b, miR-26b, miR-30b, miR-30c, miR-138, miR-139, miR-194, miR-363, miR-378*, miR-490-3p, miR-497, miR-551b
	2	miR-9, miR-9*, miR-16, miR-28-3p, miR-30a*, miR-30a-5p, miR-30e, miR-101, miR-125b, miR-137, miR-149, miR-150, miR-192*, miR-204, miR-320a, miR-328, miR-365, miR-486-5p, miR-598, miR-642

Adapted from Ref. [30]

8.4.2 PIWI-Interacting RNAs

Recently, a new class of non-coding small RNAs (26–31 nt in length), which interact with a subset of Argonaute proteins related to Piwi (the P element-induced wimpy testis), was described [26, 38]. These Piwi-interacting RNAs (piRNAs) can control and silence the transposable elements (TEs) to protect the genome, whereas the uncontrolled expression of TEs can lead to the loss of genome integrity. The abnormal piRNA pathway increases the repeats of retrotransposons, the component parts of the telomeres. Piwi is highly conserved during evolution and plays essential roles in stem cell self-renewal and RNA silencing in diverse organisms [38]. Recent reports suggested that piRNAs may play a role in the biogenesis of cancer [39], and piR-651 was found to have significantly increased expression levels in colon cancer tissues [40]. In some cancer types, deregulated piRNAs were identified also in body fluids indicating their potential usage for screening purposes [41].

8.4.3 Small Nucleolar RNAs

The snoRNA are well-conserved, abundant, short non-coding RNA molecules, 60–300 nucleotides in length, which localize to a specific compartment of the eukaryotic cell nucleus-the nucleolus, and are involved in the chemical modification of ribosomal RNAs (rRNAs).

In vertebrates, the majority of snoRNAs are encoded in introns of protein-coding or non-coding genes and are transcribed simultaneously by RNA Polymerase II [26]. SnoRNAs are classified as C/D box snoRNAs and H/ACA box snoRNAs. C/D box snoRNAs are responsible for 2′-O-ribose methylation of rRNAs, whereas H/ACA box snoRNAs are responsible for pseudouridylation of rRNAs [42].

SnoRNAs are well-known from miRNA expression studies, where they are frequently used as reference genes for normalization of the miRNA expression [43]. In the last years, several snoRNAs were shown to have significantly changed expression levels in many human diseases, including cancer. We have already summarized these studies together with snoRNAs biological behaviors and functioning in cancer in a review paper [42].

When hypermethylation-associated inactivation of snoRNAs was studied in CRC, the host gene-associated 5′-CpG islands of the snoRNAs, such as SNORD123, U70C and ACA59B were hypermethylated in the cancer cells but not in the corresponding normal colon tissue. CpG island hypermethylation was associated with the transcriptional silencing of the respective snoRNAs. However, the hypermethylation of snoRNAs was not limited to CRC, and it was a common phenomenon in other cancers, especially in leukemias [44].

Very recently, next generation sequencing of small RNA was performed in the paired samples of tumor and non-tumor tissue of CRC patients to identify deregulated snoRNAs. Among 32 snoRNAs differentially expressed in tumor tissue, the up-regulation of SNORD12B expression in tumors bore the greatest statistically significant difference from normal samples ($P<0.0000$) [45]. SNORD12B was also reported as up-regulated in rectal cancer in another study [46]. Consistent with up-regulation of SNORD12B, the host gene of this snoRNA, ZNFX1 antisense RNA 1 (ZFAS1), a long non-coding RNA, was up-regulated in all examined colorectal carcinomas. ZFAS1 was observed to be strongly over-expressed in CRC tumor tissue also in our recent study [47].

8.4.4 Circular RNAs

Circular RNAs (circRNAs) belong to an odd, but extremely interesting class of lncRNA molecules, which has been recently described. Animal genomes can express thousands of circRNAs from different genomic locations, and approximately 2000 human, 1900 mice, and 700 nematode circRNAs were identified using sequencing; however, the true number of circRNAs is thought to be higher [48]. CircRNAs play important roles in the regulation of gene expression at the transcriptional or post-transcriptional level. CircRNAs act as natural miRNA sponges and so compete with other RNAs for binding to miRNAs (such as CiRS-7 and SRY for miR-7 and miR-138, respectively) and RNA binding proteins (RBPs): they may have a role in modulating local concentration of RBPs and RNAs, as part of the competing endogenous RNA network [49]. Accordingly, they perform a critical role modulating the connection between genotype and molecular phenotype. Moreover, in contrast with classical competing endogenous RNAs (ceRNAs), circRNAs have no accessible termini, which makes them resistant to miRNA-mediated RNA degradation or other exonucleolytic activities. Bachmayr-Heyda et al. found a global reduction of circRNA abundance in CRC cell lines and CRC tissues compared with normal tissues [50]. Among 39 circRNAs differentially expressed in the tumor tissue, 11 were are up-regulated and 28 were down-regulated. The ratio of circRNA isoforms was lower in tumors than in normal samples, and it was even lower in CRC cell lines. Furthermore, this ratio was shown to be negatively correlated with the proliferation index [50]. Very recently, circ_001569 was described to be negatively correlated with miR-145 in CRC tumors and functionally linked to cell proliferation and invasion in CRC [51].

8.4.5 Long Non-coding RNAs

Long non-coding RNAs (lncRNAs) are commonly defined as RNA transcripts of more than 200 nucleotides, usually transcribed by RNA polymerase II, which have no open reading frames and map to intronic and intergenic regions of the genome [26]. It has been estimated that approximately 15,000 lncRNAs are present in the human genome, but the GENCODE v19 catalog of human lncRNAs contains 13,870 lncRNA genes that produce 23,898 lncRNAs [52]. In contrast to small non-coding RNAs, such as

Table 8.2 Long non-coding RNAs deregulated in colorectal cancer tumor tissue

	Number of studies	Long non-coding RNA	Size [bp]	Locus
Up-regulated	4	H19	2322	11p15.5
	3	MALAT1	8708	11q13.1
	3	CCAT1	2407	8q24.21
	2	HOTAIR	2158	12q13.13
	2	PVT1	>300 kb	8q21.21
	2	PRNCR1	13 kb	8q24
	2	CRNDE	1017	16q12.2
	2	uc. 73A	201	2q22.3
	2	uc. 388	590	12q13.13
	2	UCA1/CUDR	2314	19q12.12
Down-regulated	2	TUSC7	2105	3q13.31
	2	MEG3	1595	14q32.2

miRNAs, which have been extensively studied for their roles in cancer, lncRNAs are relatively less well described. However, the inherent biology of lncRNAs, often referred to as the dark matter of the genome, is gradually being elucidated and a growing body of literature suggests that they have a wide variety of roles in controlling the gene and miRNA expression in cancer. LncRNAs are involved in a variety of regulatory activities, including chromatin remodeling, transcriptional activation, decoy (transcriptional repressor) and RNA degradation. They can also act as miRNA sponges and affect translational efficacy [52].

These days, there is no doubt that the deregulation of lncRNAs affects various cancer-related signaling pathways and has a significant role in tumor development. A variety of large-scale genomic studies, such as TCGA, are being used to investigate the abnormal expression profiles of lncRNAs in human tumors [26]. Several recent reports and reviews have described the role of lncRNAs in CRC [52]. Due to their tissue specificity, lncRNAs could potentially be more sensitive for diagnosis than the current DNA, protein-coding RNA or protein biomarkers.

Based on its unique expression characteristics in CRC, the lncRNA CCAT1 has emerged as a potential biomarker for screening precancerous lesions. CCAT1 is markedly overexpressed in CRC and is also up-regulated in precancerous tissues, including benign inflammatory colonic tissues and adenomatous polyps [53]. By use of high-throughput microarray technology, 762 significantly deregulated lncRNAs, including well known oncogenic lncRNA- HOTAIR, were identified in CRC tumor tissue [54]. Another study focused on screening lymph node metastasis-associated lncRNAs in CRC patients, resulting in identification of 1133 lncRNAs differentially expressed in metastatic lymph nodes compared with normal lymph nodes, of which 260 were up-regulated and 873 were down-regulated [55]. Table 8.2 summarizes lncRNAs that were found to be deregulated at least in two independent studies. LncRNAs CCAT2 [56] and ZFAS1 [47] are not listed in Table 8.2 as they were observed only once so far, but were described to have significantly increased expression levels in tumors of large cohorts of CRC patients.

8.5 Colorectal Cancer Screening Based on Non-coding RNAs in Stool

Stool has been widely used as a potential substrate for developing non-invasive molecular screening tests for CRC patients. As mentioned earlier, considering the limitations of conventional stool-based screening tests including low sensitivity and specificity for detection of CRC and advanced adenoma, fecal DNA-based testing for CRC has been an area of active investigation since the 1990s [14]. Although several studies have reported stool mRNA-based assays, to date, no optimal method that offers superior detection accuracy compared to conventional screening tests (FOBT/FIT) has been established [3].

There is a strong underlying biological and analytical rationale for determination of non-coding RNAs expression levels in stool. This rationale includes the following observations. First, colonocytes are continuously shed into the fecal stream, with a periodicity of replacement roughly every 3–4 days, and neoplastic cells exfoliate at even a higher rate. In addition, tumor-secreted non-coding RNAs (mainly miRNAs) are directly and continuously released from the tumors into intestinal lumen. Second, deregulations in the expression of oncogenic or tumor suppressive non-coding RNAs are very specific for pre-cancer or cancer. Third, small non-coding RNAs, such as miRNAs, piRNAs or snoRNAs are extremely stable, enabling accurate and reproducible detection in the stool without need of special stabilization or logistical requirements [5]. Currently, only miRNAs were studied as stool biomarkers in CRC, but based on above mentioned observations, it is likely that studies focused on the other classes of non-coding RNAs as stool biomarkers are ongoing in CRC and will be published soon.

8.5.1 MicroRNAs in Stool

The first study reporting stool miRNAs as biomarkers in CRC was conducted by Ahmed et al in 2009 [57]. In this study, miRNA expression was determined in colonocytes extracted from stool specimens of CRC and ulcerative colitis patients as well as healthy controls. Authors identified seven up-regulated miRNAs (miR-20a, miR-21, miR-92, miR-96, miR-106a, miR-203, and miR-326), and seven down-regulated (miR-16, miR-125b, miR-126, miR-143, miR-145, miR-320, and miR-484-5p) in the stool of CRC patients [57]. The same laboratory further extended the study to establish a standardized protocol for measuring miRNA levels in stool specimens [58]. In contrast, another group focused on cell-free miRNA in stool and conducted a study analyzing stool miRNA expression using a much simpler approach. They demonstrated the feasibility of a one-step miRNA extraction and amplification method and showed increased expression levels of miR-21 and miR-106a in stool of CRC patients [59].

To measure miRNA expression in exfoliated colonocytes isolated from stool, others extracted total RNA from stool colonocytes isolated by immunomagnetic beads conjugated with epithelial cell adhesion molecule monoclonal antibody and evaluated expression status of 10 miRNAs [60]. This study showed that expression levels of miR-17-92a cluster and miR-135 was significantly higher in CRC patients than in healthy controls and suggested miRNA expression profile could be assessed in stool colonocytes as a potential screening test in patients with CRC [60]. The same group of investigators extended their research to further optimize the approach in pursuit to develop a clinically viable assay and evaluated stool miRNA expression in residual material collected from FOBT kits. They determined that miR-106a expression enhanced sensitivity of FOBT in identifying patients with CRC by approximately 10 % [61].

In another study, 253 of 648 miRNAs were successfully detected in stool samples, and it was demonstrated that miRNAs are stable in the fecal microenvironment. Among detected miRNAs, miR-144* was up-regulated in the stool of CRC patients [62]. A subsequent study that assessed stool miR-21 and miR-92a levels, presented miRNA expression as miRNA copy number per ng of extracted stool RNA, and found that fecal miR-92a expression could differentiate patients with CRC or advanced adenoma from those with lower-risk polyps or healthy subjects [63]. MiR-221 and miR-18a, known to be up-regulated in CRC tumor tissue, showed increased expression levels also in stool samples of stages I-IV CRC patients independently on the location of the tumor, or previous antibiotic intake [64]. In a smaller study, miR-223 and miR-451 were identified to have significantly higher levels in the stool of CRC patients than in healthy controls [65]. Also, stool miR-135b, up-regulated in CRC patients, indicated good analytical performance and could also be used as a potential non-invasive biomarker for CRC screening [66]. Moreover, the expression of miR-135b was significantly down-regulated after tumor removal, and there was no relationship between the levels of miR-135b and localization of colorectal lesions [66].

In a recent large-scale study, miR-20a expression was significantly higher in CRC tumors compared to their respective adjacent normal tissues (P=0.0065), and its expression levels were also significantly higher in stool samples from CRC patients (P<0.0001). No significant difference in the level of miR-20a was found between patients with proximal, distal and rectal cancer;use of antibiotics did not influence stool miR-20a levels [67]. Very recently, the complementary effect of combined analysis of miR-223 and miR-92a expression levels in stool and blood plasma was evaluated. This combined approach yielded the highest sensitivity of 96.8% and specificity of 75% for CRC (AUC=0.907). These results allowed to establish a two-miRNA signature in two types of CRC clinical specimens with a high sensitivity for CRC detection [68].

Only a few studies describe remarkable down-regulation of miRNAs in the stool of CRC patients, e.g. miR-143 and miR-145 [69], let-7a-5p and let-7f-5p [70], or miR-4487 and miR-1295b-3p [71] were described to have decreased expression levels in stool specimens of CRC patients. The abnormal DNA methylation of miR-34a in the stool was also found to be useful for CRC detection, as 63 of 82 samples were methylated, whereas only 2 of 40 healthy samples were methylated. Furthermore, the abnormal methylation of miR-34a was found to be correlated with lymph node metastasis. Regarding miR-34b/c methylation, it was found in 74 of 79 cancer stool samples. These results indicate that methylation of miR-34a and miR-34b/c might play a role in the non-invasive screening and diagnosis of CRC [72].

When Table 8.3 is surveyed, some miRNAs occur there repeatedly, e.g. miR-106a, miR-21, miR-135b, miR-223 or miR-92a, indicating the most promising candidates for future large-scale population-based validation studies.

Table 8.3 MicroRNAs detected in stool of patients with colorectal carcinoma, adenoma and healthy controls

miRNA	Sample size CRC/ adenoma/control	Expression levels	Sensitivity [%]	Specificity [%]	References
miR-16a/miR-21	10/9/10	Up, up	–	–	[59]
miR-17-92 cluster/miR-135b	197/0/119	Up, up	74	79	[60]
miR-144*	35/0/40	Up	74	87	[62]
miR-21	88/57/101	Up	56	73	[63]
miR-92a	88/57/101	Up	72	73	[63]
miR-143	38/0/13	Down	–	–	[69]
miR-145	38/0/13	Down	–	–	[69]
miR-106a	117/0/107	Up	34	97	[61]
miR-221	199/199/198	Up	62	74	[64]
miR-18a	199/199/198	Up	61	69	[64]
miR-223	17/0/28	Up	77	96	[65]
miR-451	17/0/28	Up	88	100	[65]
miR-135b	104/169/109	Up	78	68	[66]
miR-4478	40/0/16	Down	–	–	[71]
miR-1295b-4p	40/0/16	Down	–	–	[71]
let-7a-5p	51/0/26	Down	–	–	[70]
let-7f-5p	51/0/26	Down	AUC=0.71		[70]
miR-20a	199/199/198	Up	55	82	[67]
miR-223/miR-92a	138/0/309	Up, up	97	75	[68]

The main limitations of miRNA analysis in stool is to overcome the complexity of stool density and volume of sample needed for each assay. Since stool conditions are more vulnerable to daily changes compared to blood serum/plasma, standardization of protocols for sample preparation are needed to minimize sample variability. In addition, candidates for stool miRNA test are roughly divided into three types: cell-free miRNAs from stool homogenates, exosomal miRNAs from stool exosomes and stool colonocytes miRNAs [30]. Taking into consideration these differences, further investigation and validation using standardized protocol on a large cohort are necessary before such markers can be seriously considered for adaptation in the clinic for non-invasive CRC screening.

8.6 Colorectal Cancer Screening Based on Non-coding RNAs in Blood Serum and Plasma

Blood serum and plasma belong to the group of easily accessible body fluids, and they are also the most frequently used diagnostic material for the development of surrogate cancer biomarkers [46]. In recent years, tens of studies have demonstrated that colorectal adenoma and carcinoma possess unique expression profiles of various classes of non-coding RNAs in peripheral blood serum and plasma suggesting that determination of circulating non-coding RNAs could provide a novel and promising early diagnostic option for CRC screening.

8.6.1 MicroRNAs in Blood Serum and Plasma

The discovery of miRNAs in blood plasma by Lawrie et al in 2008 [73] triggered a growing number of studies that have not only evaluated their expression in a wide range of diseases, but also focused on the biology and features of circulating miRNAs. Considering circulating miRNAs as a tool employed in the horizontal gene transfer between cells within the tumor or between tumor and host cells, there is a strong biological rationale to use them as a new class of cancer biomarkers. Moreover, in peripheral blood serum and plasma, miRNAs have been shown to be in abundant levels and highly stable even under extreme conditions, such as wide changes of pH, room temperature storage, long-term storage, and multiple freeze-thaw cycles. When compared to their cellular counterparts, circulating miRNAs were also resistant to RNase digestion, and, therefore, suggested to be chemically modified in certain ways to increase their stability. This was proven consequently, and methylation, adenylation and uridylation were described as the main modifications of circulating miRNAs [74]. MiRNAs can leave the cell by passive leakage into circulation, which is minor, and occurs predominantly under pathological circumstances, such as tissue damage and necrosis. But more importantly, there are two major active mechanisms of miRNAs release from the cells, which are also responsible for their stability: (1) secretion of miRNAs containing shedding microvesicles or exosomes, and (2) secretion of miRNAs in the form of ribonucleoprotein complexes [74]. From the biological and also analytical perspective, circulating miRNAs represent a fascinating tool to be used for screening or early detection of CRC or adenomas.

Many studies have already evaluated the feasibility of circulating miRNAs for detection of early CRC or adenomas; however, only a few implemented patient cohorts large enough to provide information of relevant statistical power. Therefore, only studies with sample size higher than 100 CRC cases are discussed and summarized in Table 8.4.

The first study to use blood serum miRNA profiling in CRC by Chen et al. [75] reported 69 serum miRNAs that were differentially expressed in CRC patients. Interestingly, the authors noted that a large number of identified miRNAs were commonly detected in both sera obtained from CRC and lung cancer patients. Only 14 miRNAs were uniquely expressed in CRC patients indicating high degree of CRC-specificity [75].

The first systematic and comprehensive miRNA expression profiling study was con-

Table 8.4 Circulating microRNAs with diagnostic potential in colorectal cancer (only studies with sample size higher than 100 CRC patients were included)

miRNA	Sample	Sample size CRC/ adenoma/control	Expression levels*	Sensitivity/ specificity [%]**	Sensitivity/ specificity [%]***	References
miR-17-3p	Plasma	120/0/75	Up	89/70	–	[76]
miR-92	Plasma	120/0/75	Up	64/70	–	[76]
miR-92a/miR-29a	Plasma	100/37/59	Up, up	83/85	73/80	[77]
miR-221	Plasma	103/0/37	Up	86/41	–	[81]
miR-601/miR-760	Plasma	100/43/68	Down, down	83/69	72/62	[85]
miR-409-3p/miR-7/ miR-93	Plasma	124/0/117	Up, down, down	82/89	–	[86]
miR-21	Serum	186/43/53	Up	83/91	77/81	[82]
miR-21/miR-92a	Serum	200/50/80	Up, up	68/91	70/70	[78]
miR-21/let-7 g/ miR-31/miR-92a/ miR-181b/miR-203	Serum	113/0/89	Up, up, down, down, down, down	96/81	–	[87]
miR-155	Serum	146/0/60	Up	58/95	–	[83]
miR-223/miR-92a	Plasma	215/0/183	Up, up	76/71	–	[68]
miR-145/miR-106a/ miR-17-3p	Serum	195/0/150	Down, up, up	79/83	–	[88]
miR-106a	Plasma	100/0/79	Up	74/44	–	[84]
miR-20a	Plasma	100/0/79	Up	46/73	–	[84]

*all differences in miRNA expression levels were significant ($P<0.05$), **sensitivity and specificity in discrimination of CRC cases in all TNM stages and control, ***sensitivity and specificity in discrimination of adenomas and controls

ducted by Ng and colleagues [76], who evaluated miRNA expression alterations in tissue and plasma samples from CRC, inflammatory bowel diseases and gastric cancer patients as well as healthy subjects. This study revealed that high expression of two miRNAs, miR-92a and miR-17-3p, could discriminate patients with CRC from healthy study subjects. This seminal study further reported that the plasma levels of both miRNAs decreased significantly following surgical resection of the primary tumors, and that plasma miR-92a levels were not elevated in patients with gastric cancer and inflammatory bowel disease [76]. Also, in study of Huang et al. [77], miR-92a independently or in combination with miR-29a successfully discriminated plasma samples of CRC patients, patients with advanced adenomas and healthy controls [77]. Up-regulation of miR-92a in plasma or serum of CRC patients has been observed in other studies [68, 78], while other studies have not confirmed this observations [79, 80].

Further, the first application of the direct amplification of circulating miRNAs from plasma without RNA extraction was established, showing that plasma miR-221 could be used as a potential noninvasive molecular diagnostic biomarker for CRC patients [81].

Another well-characterized oncogenic miRNA is miR-21, which is considered to be one of the promising non-invasive biomarkers for early detection of CRC owing to the following attributes: (i) dysregulation of miR-21 occurs frequently in early stages of the adenoma-carcinoma sequence, (ii) miR-21 is one of the most highly expressed miRNAs in CRC, and (iii) miR-21 is highly secreted by cancer cells and can be measured in exosomes or as free miRNAs in plasma or serum. It was also shown that there is a significant association between lower miR-21 expression in serum and CRC tissues following curative resection of the primary tumor. Several studies confirmed the potential of miR-21 for use as a single miRNA biomarker [82], or in combination with other miRNAs [78], for the early detection of CRC. Among other circulating miRNAs, miR-155 [83], miR-106a [84] and miR-20a [84] were proven as promising individual diagnostic biomarkers in CRC.

The number of miRNAs identified as potential biomarkers for the early diagnosis of CRC is increasing; however, it seems that a single miRNA will be not sufficient to adequately capture the underlying disease heterogeneity in CRC polyps and carcinomas. Accordingly, several studies have proposed combining miRNAs into biomarker panels to improve the detection accuracy of colorectal neoplasms [30].

A panel of 22 miRNAs (miR-10a, miR-19a, miR-22*, miR-24, miR-92a, miR-125a-5p, miR-141, miR-150, miR-188-3p, miR-192, miR-210, miR-221, miR-224*, miR-376a, miR-425*, miR-495, miR-572, miR-601, miR-720, miR-760, let-7a and let-7e) deregulated in CRC plasma samples with fold changes greater than five was described. After validation of this panel on a large cohort of CRC patients, it was noted that miR-601 and miR-760 were significantly down-regulated in CRC plasma samples and could serve as markers accurately differentiating between plasma samples of CRC patients and healthy controls, as well as between plasma of patients with advanced adenomas and plasma of normal controls [85]. Highly comprehensive three-phase biomarker study revealed a plasma panel composed of three miRNAs (miR-409-3p, miR-7, miR-93) showing significant diagnostic value for early non-metastatic CRC detection [86]. Another panel of six diagnostic miRNAs (miR-21, let-7g, miR-31, miR-92a, miR-181b, and miR-203) was identified, showing much higher sensitivity and specificity than currently used biomarkers CEA and CA19-9 in discrimination of the plasma samples of CRC patients and controls [87]. In another study, a panel of miR-145, miR-106a and miR-17-3p was established with significantly different expression between pre- and post-operative CRC patients and between pre-operative CRC patients and normal controls indicating its potential application in non-invasive CRC diagnostics [88]. Very recently, an interesting approach was implemented combining parallel expression analysis of miRNAs in blood plasma and stool. After examining the complementary effect, combined analysis of miR-223 and miR-92a, which were commonly present in stool and plasma samples, yielded high sensitivity of 97 % and specificity of 75 % for CRC (AUC=0.907). These results led to establishment of a 2-miRNA signature in two types of CRC clinical specimens with a high sensitivity for CRC detection [68].

Although increasing number of circulating miRNAs has been identified as potential biomarkers for early diagnosis of CRC, the lack of consistency between biomarker panels in independent studies highlights a major obstacle for the development of robust miRNA biomarkers. It is comprehensible that different experimental designs, procedures and methods, inconsistent normalization approaches, lack of standardization, ethnic and racial differences in patient populations, different instrumentation and lab personnel could contribute to the seemingly contradicting results that have been published so far. It is also important to point out that occurrence of the miRNAs from cellular contaminants, such as platelets or erythrocytes, which is well-established these days, was not considered in the older studies. Finally, most studies published thus far (with very few exceptions) deal with a relatively small sample sizes and no independent cohort(s) validation. In order to verify results obtained in retrospective exploratory cohorts, to achieve true translational relevance and to bring circulating miRNAs into routine diagnostics, multicentric clinical trials have to be performed with experimental design based on a coordinated and synchronized set of experimental procedures (e.g. specimen collection, processing procedures, and storage conditions for the collected specimens) and instrumentation that utilize the same normalization approach. Nevertheless, all the current data underline the enormous potential for circulating miRNAs to serve as new non-invasive biomarkers in early detection and diagnosis of CRC and adenomas.

8.6.2 Small Nuclear RNAs in Blood Serum and Plasma

Small nuclear RNAs (snRNAs) form the core components of the spliceosome and catalyze removal of introns from pre-mRNA. SnRNAs can form complexes with several proteins to form small nuclear ribonucleo-proteins (snRNPs). There are five major classes of snRNAS, including U1, U2, U4, U5 and U6 [26]. The expression of U2 snRNA fragments (RNU2-1f) was shown to be stable both in the serum and plasma of CRC patients. In a cohort of 132 CRC patients and 129 controls, it has shown sensitivity of 97.7 % and specificity of 90.6 % in the discrimination of CRC and control samples, indicating that it could be a potential diagnostic biomarker for CRC. The RNU2-1f assay might correctly identify CRC patients as early as UICC stage II, suggesting that it could function as a potential non-invasive screening method for detecting early CRC with a good prognosis [89].

8.6.3 Long Non-coding RNAs in Blood Serum and Plasma

Until now, research of lncRNAs in CRC has been focused mainly on their roles in carcinogenesis or as the tissue biomarkers. However, several studies have already proven the feasibility of circulating lncRNAs for detection of CRC.

An uncharacterized gene locus (Chr16:hCG_1815491), named colorectal neoplasia differentially expressed (gene symbol CRNDE), is a lncRNA activated early in colorectal neoplasia. CRNDE (splice variant h) was the first lncRNA described to be present in blood plasma of CRC patients, with the expression levels being significantly higher than that of healthy controls, with a sensitivity of 87 % and specificity of 93 % for CRC diagnosis CRC [90].

In another study, the CAHM methylation (Colorectal Adenocarcinoma HyperMethylated, previously LOC100526820) was evaluated in DNA isolated from plasma specimens from 220 colonoscopy-examined patients, and methylated CAHM sequences were detected in plasma of 40/73 (55 %) of CRC patients compared with 3/73 (4 %) from subjects with adenomas and 5/74 (7 %) from subjects without neoplasia. Methylated CAHM DNA shows a promise as a plasma biomarker for use in screening of CRC, but not pre-cancerous lesions [91]. Nuclear-enriched abundant transcript 1 (NEAT1) is the lncRNA proven to be abundant in the whole blood of CRC patients. Expression of NEAT1 variants, NEAT1_v1 and NEAT1_v2 were determined, and the diagnostic value of whole blood NEAT1 expression was evaluated in a test (n = 60) and validation (n = 200) cohorts of CRC patients and controls. NEAT1_v1 and NEAT1_v2 expression were highly accurate in distinguishing CRC patients from controls (area under the curve: 0.787 and 0.871, respectively) [92]. One of the most frequently studied lncRNA in cancer, HOTAIR, was evaluated also in the whole blood of CRC patients. In the group of 84 CRC patients and 40 healthy controls, CRC patients had higher HOTAIR expression in the blood than healthy controls (P=0.0001 at 67 % sensitivity and 92.5 % specificity of tumor detection). Moreover, HOTAIR levels positively correlated between blood and tumor [93].

Although studies of lncRNAs as potential diagnostic or screening biomarkers are quite promising, in comparison to miRNAs they are too preliminary to enable any conclusions in regards to their potential clinical application.

8.7 Conclusions and Future Perspectives

In conclusion, a number of studies have provided compelling evidence that ncRNAs (mainly miRNAs) in stool and blood serum or plasma have a great potential as biomarkers for early detection or CRC screening. However, before ncRNAs are routinely applied to clinical settings, it will be critically important to activate large collaborative efforts to fully validate the clinical potential of this approach. To achieve true translational relevance and bring stool and circulating ncRNAs into routine diagnostics, multicenter clinical trials have to be performed with experimental design based on a coordinated and synchronized set of experimental procedures (e.g. specimen collection, processing procedures, and storage conditions for the collected specimens) and instrumentation that utilize the same normalization approach. As a result, it is possible that for clinical purposes, we may end up using a panel of ncRNAs rather than a single ncRNA, or even combine different types of biomarkers, analysis of ncRNAs in different non-invasively accessible biological specimens (e.g. stool and plasma in parallel) or other available tests, such as CEA, CA19-9 and FOBT, all in an effort to enhance sensitivity and specificity of these analytical approaches and finally develop a novel more accurate CRC screening test.

References

1. Ferlay J, Parkin DM, Steliarova-Foucher E. Estimates of cancer incidence and mortality in Europe in 2008. Eur J Cancer. 2010;46(4):765–81.
2. Imperiale TF. Noninvasive screening tests for colorectal cancer. Dig Dis. 2012;30(2):16–26.
3. Di Lena M, Travaglio E, Altomare DF. New strategies for colorectal cancer screening. World J Gastroenterol. 2013;19(12):1855–60.
4. Ganepola GA, Nizin J, Rutledge JR, Chang DH. Use of blood-based biomarkers for early diagnosis and surveillance of colorectal cancer. World J Gastrointest Oncol. 2014;6(4):83–97.
5. Ragusa M, Barbagallo C, Statello L, Condorelli AG, Battaglia R, Tamburello L, et al. Non-coding landscapes of colorectal cancer. World J Gastroenterol. 2015;21(41):11709–39.
6. Winawer SJ, Zauber AG, Ho MN, O'Brien MJ, Gottlieb LS, Sternberg SS, et al. Prevention of colorectal cancer by colonoscopic polypectomy. The National Polyp Study Workgroup. N Engl J Med. 1993;329(27):1977–81.
7. Zauber AG, Winawer SJ, O'Brien MJ, Lansdorp-Vogelaar I, van Ballegooijen M, Hankey BF, et al. Colonoscopic polypectomy and long-term prevention of colorectal-cancer deaths. N Engl J Med. 2012;366(8):687–96.
8. Lansdorp-Vogelaar I, van Ballegooijen M, Zauber AG, Habbema JD, Kuipers EJ. Effect of rising chemotherapy costs on the cost savings of colorectal cancer screening. J Natl Cancer Inst. 2009;101(20):1412–22.
9. Seeff LC, Richards TB, Shapiro JA, Nadel MR, Manninen DL, Given LS, et al. How many endoscopies are performed for colorectal cancer screening? Results from CDC's survey of endoscopic capacity. Gastroenterology. 2004;127(6):1670–7.
10. Centers for Disease Control and Prevention. Cancer screening – United States, 2010. MMWR Morb Mortal Wkly Rep. 2012;61(3):41–5.
11. Mandel JS, Bond JH, Church TR, Snover DC, Bradley GM, Schuman LM, et al. Reducing mortality from colorectal cancer by screening for fecal occult blood. Minnesota Colon Cancer Control Study. N Engl J Med. 1993;328(19):1365–71.
12. Chiang TH, Lee YC, Tu CH, Chiu HM, Wu MS. Performance of the immunochemical fecal occult blood test in predicting lesions in the lower gastrointestinal tract. CMAJ. 2011;183(13):1474–81.
13. van Rossum LG, van Rijn AF, Laheij RJ, van Oijen MG, Fockens P, van Krieken HH, et al. Random comparison of guaiac and immunochemical fecal occult blood tests for colorectal cancer in a screening population. Gastroenterology. 2008;135(1):82–90.
14. Toiyama Y, Okugawa Y, Goel A. DNA methylation and microRNA biomarkers for noninvasive detection of gastric and colorectal cancer. Biochem Biophys Res Commun. 2014;455(1–2):43–57.
15. Bosch LJ, Carvalho B, Fijneman RJ, Jimenez CR, Pinedo HM, van Engeland M, et al. Molecular tests for colorectal cancer screening. Clin Colorectal Cancer. 2011;10(1):8–23.
16. Yang H, Xia BQ, Jiang B, Wang G, Yang YP, Chen H, et al. Diagnostic value of stool DNA testing for multiple markers of colorectal cancer and advanced adenoma: a meta-analysis. Can J Gastroenterol. 2013;27(8):467–75.
17. Itzkowitz S, Brand R, Jandorf L, Durkee K, Millholland J, Rabeneck L, et al. A simplified, noninvasive stool DNA test for colorectal cancer detection. Am J Gastroenterol. 2008;103(11):2862–70.
18. Calistri D, Rengucci C, Casadei Gardini A, Frassineti GL, Scarpi E, Zoli W, et al. Fecal DNA for noninvasive diagnosis of colorectal cancer in immunochemical fecal occult blood test-positive individuals. Cancer Epidemiol Biomarkers Prev. 2010;19(10):2647–54.
19. Ahmed FE, James SI, Lysle DT, Dobbs Jr LJ, Johnke RM, Flake G, et al. Improved methods for extracting RNA from exfoliated human colonocytes in stool and RT-PCR analysis. Dig Dis Sci. 2004;49(11–12):1889–98.
20. Marshall KW, Mohr S, Khettabi FE, Nossova N, Chao S, Bao W, et al. A blood-based biomarker panel for stratifying current risk for colorectal cancer. Int J Cancer. 2010;126(5):1177–86.
21. Xu D, Li XF, Zheng S, Jiang WZ. Quantitative real-time RT-PCR detection for CEA, CK20 and CK19 mRNA in peripheral blood of colorectal cancer patients. J Zhejiang Univ Sci B. 2006;7(6):445–51.
22. Hundt S, Haug U, Brenner H. Blood markers for early detection of colorectal cancer: a systematic review. Cancer Epidemiol Biomarkers Prev. 2007;16(10):1935–53.
23. Silva CL, Passos M, Camara JS. Investigation of urinary volatile organic metabolites as potential cancer

biomarkers by solid-phase microextraction in combination with gas chromatography-mass spectrometry. Br J Cancer. 2011;105(12):1894–904.
24. Nishiumi S, Kobayashi T, Ikeda A, Yoshie T, Kibi M, Izumi Y, et al. A novel serum metabolomics-based diagnostic approach for colorectal cancer. PLoS One. 2012;7(7):e40459.
25. Altomare DF, Di Lena M, Porcelli F, Trizio L, Travaglio E, Tutino M, et al. Exhaled volatile organic compounds identify patients with colorectal cancer. Br J Surg. 2013;100(1):144–50.
26. Sana J, Faltejskova P, Svoboda M, Slaby O. Novel classes of non-coding RNAs and cancer. J Transl Med. 2012;10:103.
27. Wang J, Song YX, Ma B, Wang JJ, Sun JX, Chen XW, et al. Regulatory roles of Non-coding RNAs in colorectal cancer. Int J Mol Sci. 2015;16(8):19886–919.
28. Slaby O, Svoboda M, Michalek J, Vyzula R. MicroRNAs in colorectal cancer: translation of molecular biology into clinical application. Mol Cancer. 2009;8:102.
29. Ketting RF. MicroRNA biogenesis and function. An overview. Adv Exp Med Biol. 2010;700:1–14.
30. Okugawa Y, Toiyama Y, Goel A. An update on microRNAs as colorectal cancer biomarkers: where are we and what's next? Expert Rev Mol Diagn. 2014;14(8):999–1021.
31. Mazeh H, Mizrahi I, Ilyayev N, Halle D, Brucher B, Bilchik A, et al. The diagnostic and prognostic role of microRNA in colorectal cancer – a comprehensive review. J Cancer. 2013;4(3):281–95.
32. Yamamoto H, Adachi Y, Taniguchi H, Kunimoto H, Nosho K, Suzuki H, et al. Interrelationship between microsatellite instability and microRNA in gastrointestinal cancer. World J Gastroenterol. 2012;18(22):2745–55.
33. Ferracin M, Pedriali M, Veronese A, Zagatti B, Gafa R, Magri E, et al. MicroRNA profiling for the identification of cancers with unknown primary tissue-of-origin. J Pathol. 2011;225(1):43–53.
34. Schetter AJ, Leung SY, Sohn JJ, Zanetti KA, Bowman ED, Yanaihara N, et al. MicroRNA expression profiles associated with prognosis and therapeutic outcome in colon adenocarcinoma. JAMA. 2008;299(4):425–36.
35. Oberg AL, French AJ, Sarver AL, Subramanian S, Morlan BW, Riska SM, et al. miRNA expression in colon polyps provides evidence for a multihit model of colon cancer. PLoS One. 2011;6(6):e20465.
36. Nagel R, le Sage C, Diosdado B, van der Waal M, Oude Vrielink JA, Bolijn A, et al. Regulation of the adenomatous polyposis coli gene by the miR-135 family in colorectal cancer. Cancer Res. 2008;68(14):5795–802.
37. Diosdado B, van de Wiel MA, Terhaar Sive Droste JS, Mongera S, Postma C, Meijerink WJ, et al. MiR-17-92 cluster is associated with 13q gain and c-myc expression during colorectal adenoma to adenocarcinoma progression. Br J Cancer. 2009;101(4):707–14.
38. Iwasaki YW, Siomi MC, Siomi H. PIWI-interacting RNA: its biogenesis and functions. Annu Rev Biochem. 2015;84:405–33.
39. Siddiqi S, Matushansky I. Piwis and piwi-interacting RNAs in the epigenetics of cancer. J Cell Biochem. 2012;113(2):373–80.
40. Cheng J, Guo JM, Xiao BX, Miao Y, Jiang Z, Zhou H, et al. piRNA, the new non-coding RNA, is aberrantly expressed in human cancer cells. Clin Chim Acta. 2011;412(17–18):1621–5.
41. Cui L, Lou Y, Zhang X, Zhou H, Deng H, Song H, et al. Detection of circulating tumor cells in peripheral blood from patients with gastric cancer using piRNAs as markers. Clin Biochem. 2011;44(13):1050–7.
42. Thorenoor N, Slaby O. Small nucleolar RNAs functioning and potential roles in cancer. Tumour Biol. 2015;36(1):41–53.
43. Schwarzenbach H, da Silva AM, Calin G, Pantel K. Data normalization strategies for MicroRNA quantification. Clin Chem. 2015;61(11):1333–42.
44. Ferreira HJ, Heyn H, Moutinho C, Esteller M. CpG island hypermethylation-associated silencing of small nucleolar RNAs in human cancer. RNA Biol. 2012;9(6):881–90.
45. Xu L, Ziegelbauer J, Wang R, Wu WW, Shen RF, Juhl H, et al. Distinct profiles for mitochondrial t-RNAs and small nucleolar RNAs in locally invasive and metastatic colorectal cancer. Clin Cancer Res. 2016;22(3):773–84.
46. Gaedcke J, Grade M, Camps J, Sokilde R, Kaczkowski B, Schetter AJ, et al. The rectal cancer microRNAome–microRNA expression in rectal cancer and matched normal mucosa. Clin Cancer Res. 2012;18(18):4919–30.
47. Thorenoor N, Faltejskova-Vychytilova P, Hombach S, Mlcochova J, Kretz M, Svoboda M, et al. Long non-coding RNA ZFAS1 interacts with CDK1 and is involved in p53-dependent cell cycle control and apoptosis in colorectal cancer. Oncotarget. 2016;7(1):622–37.
48. Ebbesen KK, Kjems J, Hansen TB. Circular RNAs: identification, biogenesis and function. Biochim Biophys Acta. 2016;1859(1):163–8.
49. Li J, Yang J, Zhou P, Le Y, Zhou C, Wang S, et al. Circular RNAs in cancer: novel insights into origins, properties, functions and implications. Am J Cancer Res. 2015;5(2):472–80.
50. Bachmayr-Heyda A, Reiner AT, Auer K, Sukhbaatar N, Aust S, Bachleitner-Hofmann T, et al. Correlation of circular RNA abundance with proliferation–exemplified with colorectal and ovarian cancer, idiopathic lung fibrosis, and normal human tissues. Sci Rep. 2015;5:8057.
51. Xie H, Ren X, Xin S, Lan X, Lu G, Y L, et al. Emerging roles of circRNA_001569 targeting miR-145 in the proliferation and invasion of colorectal cancer. Oncotarget. 2016;7(18):26680–91.
52. Xu MD, Qi P, Du X. Long non-coding RNAs in colorectal cancer: implications for pathogenesis and

clinical application. Mod Pathol. 2014;27(10):1310–20.
53. Alaiyan B, Ilyayev N, Stojadinovic A, Izadjoo M, Roistacher M, Pavlov V, et al. Differential expression of colon cancer associated transcript1 (CCAT1) along the colonic adenoma-carcinoma sequence. BMC Cancer. 2013;13:196.
54. Xue Y, Ma G, Gu D, Zhu L, Hua Q, Du M, et al. Genome-wide analysis of long noncoding RNA signature in human colorectal cancer. Gene. 2015;556(2):227–34.
55. Han J, Rong LF, Shi CB, Dong XG, Wang J, Wang BL, et al. Screening of lymph nodes metastasis associated lncRNAs in colorectal cancer patients. World J Gastroenterol. 2014;20(25):8139–50.
56. Ling H, Spizzo R, Atlasi Y, Nicoloso M, Shimizu M, Redis RS, et al. CCAT2, a novel noncoding RNA mapping to 8q24, underlies metastatic progression and chromosomal instability in colon cancer. Genome Res. 2013;23(9):1446–61.
57. Ahmed FE, Jeffries CD, Vos PW, Flake G, Nuovo GJ, Sinar DR, et al. Diagnostic microRNA markers for screening sporadic human colon cancer and active ulcerative colitis in stool and tissue. Cancer Genomics Proteomics. 2009;6(5):281–95.
58. Ahmed FE, Ahmed NC, Vos PW, Bonnerup C, Atkins JN, Casey M, et al. Diagnostic microRNA markers to screen for sporadic human colon cancer in stool: I. Proof of principle. Cancer Genomics Proteomics. 2013;10(3):93–113.
59. Link A, Balaguer F, Shen Y, Nagasaka T, Lozano JJ, Boland CR, et al. Fecal MicroRNAs as novel biomarkers for colon cancer screening. Cancer Epidemiol Biomarkers Prev. 2010;19(7):1766–74.
60. Koga Y, Yasunaga M, Takahashi A, Kuroda J, Moriya Y, Akasu T, et al. MicroRNA expression profiling of exfoliated colonocytes isolated from feces for colorectal cancer screening. Cancer Prev Res (Phila). 2010;3(11):1435–42.
61. Koga Y, Yamazaki N, Yamamoto Y, Yamamoto S, Saito N, Kakugawa Y, et al. Fecal miR-106a is a useful marker for colorectal cancer patients with false-negative results in immunochemical fecal occult blood test. Cancer Epidemiol Biomarkers Prev. 2013;22(10):1844–52.
62. Kalimutho M, Del Vecchio BG, Di Cecilia S, Sileri P, Cretella M, Pallone F, et al. Differential expression of miR-144* as a novel fecal-based diagnostic marker for colorectal cancer. J Gastroenterol. 2011;46(12):1391–402.
63. Wu CW, Ng SS, Dong YJ, Ng SC, Leung WW, Lee CW, et al. Detection of miR-92a and miR-21 in stool samples as potential screening biomarkers for colorectal cancer and polyps. Gut. 2012;61(5):739–45.
64. Yau TO, Wu CW, Dong Y, Tang CM, Ng SS, Chan FK, et al. microRNA-221 and microRNA-18a identification in stool as potential biomarkers for the non-invasive diagnosis of colorectal carcinoma. Br J Cancer. 2014;111(9):1765–71.
65. Phua LC, Chue XP, Koh PK, Cheah PY, Chan EC, Ho HK. Global fecal microRNA profiling in the identification of biomarkers for colorectal cancer screening among Asians. Oncol Rep. 2014;32(1):97–104.
66. Wu CW, Ng SC, Dong Y, Tian L, Ng SS, Leung WW, et al. Identification of microRNA-135b in stool as a potential noninvasive biomarker for colorectal cancer and adenoma. Clin Cancer Res. 2014;20(11):2994–3002.
67. Yau TO, Wu CW, Tang CM, Chen Y, Fang J, Dong Y, et al. microRNA-20a in human faeces as a non-invasive biomarker for colorectal cancer. Oncotarget. 2016;7(2):1559–68.
68. Chang PY, Chen CC, Chang YS, Tsai WS, You JF, Lin GP, et al. MicroRNA-223 and microRNA-92a in stool and plasma samples act as complementary biomarkers to increase colorectal cancer detection. Oncotarget. 2016;7(9):10663–75.
69. Li JM, Zhao RH, Li ST, Xie CX, Jiang HH, Ding WJ, et al. Down-regulation of fecal miR-143 and miR-145 as potential markers for colorectal cancer. Saudi Med J. 2012;33(1):24–9.
70. Ghanbari R, Mosakhani N, Sarhadi VK, Armengol G, Nouraee N, Mohammadkhani A, et al. Simultaneous underexpression of let-7a-5p and let-7f-5p microRNAs in plasma and stool samples from early stage colorectal carcinoma. Biomark Cancer. 2015;7 Suppl 1:39–48.
71. Ghanbari R, Mosakhani N, Asadi J, Nouraee N, Mowla SJ, Poustchi H, et al. Decreased expression of fecal miR-4478 and miR-1295b-3p in early-stage colorectal cancer. Cancer Biomark. 2015;15(2):189–95.
72. Wu XD, Song YC, Cao PL, Zhang H, Guo Q, Yan R, et al. Detection of miR-34a and miR-34b/c in stool sample as potential screening biomarkers for noninvasive diagnosis of colorectal cancer. Med Oncol. 2014;31(4):894.
73. Lawrie CH, Gal S, Dunlop HM, Pushkaran B, Liggins AP, Pulford K, et al. Detection of elevated levels of tumour-associated microRNAs in serum of patients with diffuse large B-cell lymphoma. Br J Haematol. 2008;141(5):672–5.
74. Redova M, Sana J, Slaby O. Circulating miRNAs as new blood-based biomarkers for solid cancers. Future Oncol. 2013;9(3):387–402.
75. Chen X, Ba Y, Ma L, Cai X, Yin Y, Wang K, et al. Characterization of microRNAs in serum: a novel class of biomarkers for diagnosis of cancer and other diseases. Cell Res. 2008;18(10):997–1006.
76. Ng EK, Chong WW, Jin H, Lam EK, Shin VY, Yu J, et al. Differential expression of microRNAs in plasma of patients with colorectal cancer: a potential marker for colorectal cancer screening. Gut. 2009;58(10):1375–81.
77. Huang Z, Huang D, Ni S, Peng Z, Sheng W, Du X. Plasma microRNAs are promising novel biomarkers for early detection of colorectal cancer. Int J Cancer. 2010;127(1):118–26.

78. Liu GH, Zhou ZG, Chen R, Wang MJ, Zhou B, Li Y, et al. Serum miR-21 and miR-92a as biomarkers in the diagnosis and prognosis of colorectal cancer. Tumour Biol. 2013;34(4):2175–81.
79. Faltejskova P, Bocanek O, Sachlova M, Svoboda M, Kiss I, Vyzula R, et al. Circulating miR-17-3p, miR-29a, miR-92a and miR-135b in serum: evidence against their usage as biomarkers in colorectal cancer. Cancer Biomark. 2012;12(4):199–204.
80. Ramzy I, Hasaballah M, Marzaban R, Shaker O, Soliman ZA. Evaluation of microRNAs-29a, 92a and 145 in colorectal carcinoma as candidate diagnostic markers: an Egyptian pilot study. Clin Res Hepatol Gastroenterol. 2015;39(4):508–15.
81. Pu XX, Huang GL, Guo HQ, Guo CC, Li H, Ye S, et al. Circulating miR-221 directly amplified from plasma is a potential diagnostic and prognostic marker of colorectal cancer and is correlated with p53 expression. J Gastroenterol Hepatol. 2010;25(10):1674–80.
82. Toiyama Y, Takahashi M, Hur K, Nagasaka T, Tanaka K, Inoue Y, et al. Serum miR-21 as a diagnostic and prognostic biomarker in colorectal cancer. J Natl Cancer Inst. 2013;105(12):849–59.
83. Lv ZC, Fan YS, Chen HB, Zhao DW. Investigation of microRNA-155 as a serum diagnostic and prognostic biomarker for colorectal cancer. Tumour Biol. 2015;36(3):1619–25.
84. Chen WY, Zhao XJ, Yu ZF, Hu FL, Liu YP, Cui BB, et al. The potential of plasma miRNAs for diagnosis and risk estimation of colorectal cancer. Int J Clin Exp Pathol. 2015;8(6):7092–101.
85. Wang Q, Huang Z, Ni S, Xiao X, Xu Q, Wang L, et al. Plasma miR-601 and miR-760 are novel biomarkers for the early detection of colorectal cancer. PLoS One. 2012;7(9):e44398.
86. Wang S, Xiang J, Li Z, Lu S, Hu J, Gao X, et al. A plasma microRNA panel for early detection of colorectal cancer. Int J Cancer. 2015;136(1):152–61.
87. Wang J, Huang SK, Zhao M, Yang M, Zhong JL, Gu YY, et al. Identification of a circulating microRNA signature for colorectal cancer detection. PLoS One. 2014;9(4):e87451.
88. Li J, Liu Y, Wang C, Deng T, Liang H, Wang Y, et al. Serum miRNA expression profile as a prognostic biomarker of stage II/III colorectal adenocarcinoma. Sci Rep. 2015;5:12921.
89. Baraniskin A, Nopel-Dunnebacke S, Ahrens M, Jensen SG, Zollner H, Maghnouj A, et al. Circulating U2 small nuclear RNA fragments as a novel diagnostic biomarker for pancreatic and colorectal adenocarcinoma. Int J Cancer. 2013;132(2):E48–57.
90. Graham LD, Pedersen SK, Brown GS, Ho T, Kassir Z, Moynihan AT, et al. Colorectal neoplasia differentially expressed (CRNDE), a novel gene with elevated expression in colorectal adenomas and adenocarcinomas. Genes Cancer. 2011;2(8):829–40.
91. Pedersen SK, Mitchell SM, Graham LD, McEvoy A, Thomas ML, Baker RT, et al. CAHM, a long non-coding RNA gene hypermethylated in colorectal neoplasia. Epigenetics. 2014;9(8):1071–82.
92. Wu Y, Yang L, Zhao J, Li C, Nie J, Liu F, et al. Nuclear-enriched abundant transcript 1 as a diagnostic and prognostic biomarker in colorectal cancer. Mol Cancer. 2015;14(1):191.
93. Svoboda M, Slyskova J, Schneiderova M, Makovicky P, Bielik L, Levy M, et al. HOTAIR long non-coding RNA is a negative prognostic factor not only in primary tumors, but also in the blood of colorectal cancer patients. Carcinogenesis. 2014;35(7):1510–5.

Circulating Non-coding RNA as Biomarkers in Colorectal Cancer

9

Manuela Ferracin, Laura Lupini, Alessandra Mangolini, and Massimo Negrini

Abstract

Recent studies suggested that colorectal cancer influences the types and quantity of nucleic acids - especially microRNAs – detected in the bloodstream. Concentration of circulating (cell-free) microRNAs, and possibly of other non-coding RNAs, could therefore serve as valuable colorectal cancer biomarker and could deliver insight into the disease process. This chapter addresses the recent discoveries on circulating microRNA and long non-coding RNA as diagnostic or prognostic biomarkers in colorectal cancer.

Keywords

MicroRNA • Serum • Plasma • Diagnostics • Colorectal cancer • Biomarkers

9.1 Introduction

Colorectal cancer (CRC) is the third most common cancer in both men and women. It is the second leading cause of cancer-related mortality in the developed countries [1]. Early colorectal cancer diagnosis could increase the chances of early intervention and improve overall survival rate. Colonoscopy, faecal occult blood testing (FOBT) and faecal immunochemical test (FIT) are the most commonly used screening tests worldwide, each one characterized by specific advantages and limits [2]. Colonoscopy displays the higher costs and it is an invasive procedure that has poor compliance among those eligible for colorectal cancer screening. On the other hand, FOBT and FIT have low sensitivity in pre-neoplastic lesions detection and a high rate of false positives detection. Therefore, there is an urgent need for new

M. Ferracin (✉)
Department of Experimental, Diagnostic and Specialty Medicine – DIMES, University of Bologna, Bologna, Italy
e-mail: manuela.ferracin@unibo.it

L. Lupini • A. Mangolini • M. Negrini
Department of Morphology, Surgery and Experimental Medicine, University of Ferrara, Ferrara, Italy
e-mail: laura.lupini@unife.it; alessandra.mangolini@unife.it; ngm@unife.it

O. Slaby, G.A. Calin (eds.), *Non-coding RNAs in Colorectal Cancer*, Advances in Experimental Medicine and Biology 937, DOI 10.1007/978-3-319-42059-2_9

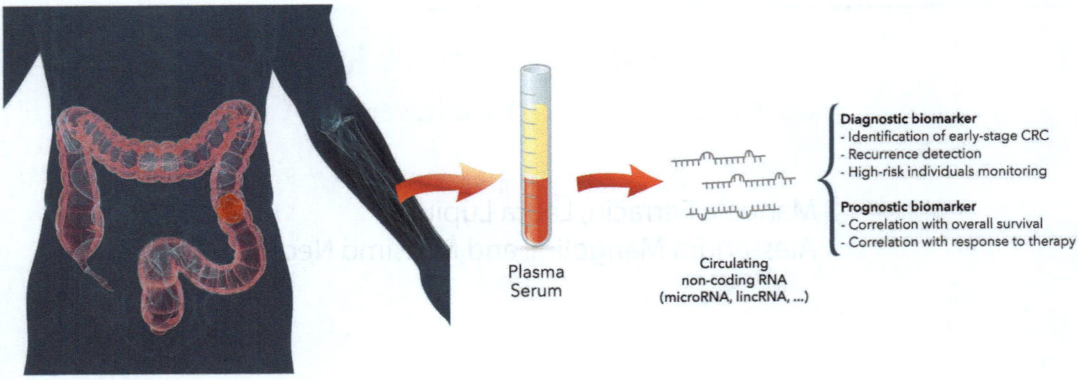

Fig. 9.1 Circulating non-coding RNAs as colorectal cancer biomarkers. Circulating miRNAs and other non-coding RNA are released by cancer cells in the circulation by either active secretory mechanisms (exosomes, microvesicles, apoptotic bodies) or passive secretion (cell necrosis). Non-coding RNA circulating in body fluids (serum, plasma) can be recovered using several existing isolation methods and quantified. Changes in their levels provide useful clinical information about cancer presence, cancer response to therapy and prognosis

non-invasive blood-detectable colorectal cancer biomarkers (liquid biopsy), that could be easily incorporated into routine diagnostic workup, to improve the detection rate or correlate with tumor recurrence and response to therapy (Fig. 9.1).

MicroRNAs (miRNAs) are released into blood circulation by potentially all the cells of the organism, as a consequence of active release or necrotic and apoptotic processes. Cell-free (or circulating) miRNAs have been detected in the bloodstream either as free stable molecules or linked to lipoproteins or enveloped inside exosomes and microvesicles [3–5]. In plasma and serum preparations, circulating miRNAs present an unexpected resistance to degradation. They are believed to function as cell-to-cell communicators [6], and their level changes in the presence of cancer, including colorectal cancer [7].

Their accurate and reproducible quantification is the basis for their use as disease biomarkers. In this context, their low amount and the lack of known endogenous reference genes in body fluids provide a real challenge for every reliable translational application [8, 9]. Indeed, every pre-analytical (tissue preparation, storage condition, extraction method) and analytical (quantification technique, normalization approach) step has previously been proved to affect the final miRNA quantification [8]. For this reason, a generalized lack of concordance between published studies and consensus in cell-free miRNA findings can be observed. However, technological improvements are expected to lead to optimized protocols for circulating miRNA / ncRNA assessment within few years.

9.2 Circulating Non-coding RNAs as Diagnostic Biomarkers

The discovery of non-coding RNAs in biological fluids has attracted considerable attention in oncology. Indeed, the detection of circulating non-coding RNAs such as microRNAs and long non-coding RNA as (lncRNA) is a fast, minimally invasive procedure that could be performed from a blood-based biopsy (liquid biopsy).

Since the early detection is a critical goal in CRC screening programs, circulating non coding RNAs represent a new class of diagnostic biomarkers, which can be used to improve precancerous lesions detection.

9.2.1 Circulating MicroRNAs

MicroRNA is the most investigated class of non-coding RNA and several papers reported the identification of these molecules in serum and/or plasma with diagnostic potential in CRC (Table 9.1). The first

Table 9.1 Circulating diagnostic microRNAs in colorectal cancer

Body fluid	MicroRNA (update name[a])	MicroRNA (previous name)	Reference
Plasma	hsa-miR-15b-5p	hsa-miR-15b	Giráldez et al. [23]
			Kanaan et al. [13]
	hsa-miR-17-5p	hsa-miR-17	Kanaan et al. [13]
	hsa-miR-17-3p	hsa-miR-17-3p	Ng et al. [46]
	hsa-miR-18a-5p	hsa-miR-18a	Giráldez et al. [23]
			Luo et al. [47]
	hsa-miR-19a-3p	hsa-miR-19a	Giráldez et al. [23]
	hsa-miR-19b-3p	hsa-miR-19b	Giráldez et al. [23]
	hsa-miR-20a-5p	hsa-miR-20a	Luo et al. [47]
	hsa-miR-21-5p	hsa-miR-21	Toiyama et al. [41]
			Kanaan et al. [16]
			Luo et al. [47]
	hsa-miR-29a-3p	hsa-miR-29a	Giráldez et al. [23]
			Huang et al. [48]
			Luo et al. [47]
	hsa-miR-92a-3p	hsa-miR-92a	Ng et al. [46]
			Huang et al. [48]
			Luo et al. [47]
	hsa-miR-106b-5p	hsa-miR-106b	Luo et al. [47]
	hsa-miR-133a-3p	hsa-miR-133a	Luo et al. [47]
	hsa-miR-142-3p	hsa-miR-142-3p	Kanaan et al. [13]
	hsa-miR-143-3p	hsa-miR-143	Luo et al. [47]
	hsa-miR-145-5p	hsa-miR-145	Luo et al. [47]
	hsa-miR-195-5p	hsa-miR-195	Kanaan et al. [13]
	hsa-miR-331-3p	hsa-miR-331	Kanaan et al. [13]
	hsa-miR-335-5p	hsa-miR-335	Giráldez et al. [23]
	hsa-miR-378a-3p	hsa-miR-378	Zanutto et al. [18]
	hsa-miR-532-5p	hsa-miR-532	Kanaan et al. [13]
	hsa-miR-532-3p	hsa-miR-532-3p	Kanaan et al. [13]
	hsa-miR-652-3p	hsa-miR-652	Kanaan et al. [13]
	hsa-miR-601	hsa-miR-601	Wang et al. [49]
	hsa-miR-760	hsa-miR-760	Wang et al. [49]
Serum	hsa-let-7 g-5p	hsa-let-7 g	Wang et al. [50]
	hsa-miR-21-5p	hsa-miR-21	Wang et al. [50]
	hsa-miR-23a-3p	hsa-miR-23a	Yong et al. [51]
	hsa-miR-31-5p	hsa-miR-31	Wang et al. [50]
	hsa-miR-92a-3p	hsa-miR-92a	Wang et al. [50]
	hsa-miR-181b-5p	hsa-miR-181b	Wang et al. [50]
	hsa-miR-193a-3p	hsa-miR-193a-3p	Yong et al. [51]
	hsa-miR-203a	hsa-miR-203	Wang et al. [50]
	hsa-miR-338-5p	hsa-miR-338-5p	Yong et al. [51]

[a]Based on Release v.20 of miRBase datadase

study on circulating miRNAs in CRC was performed in 2009. The study involved 90 CRC patients and 50 healthy individuals and demonstrated an increase in miR-17-3p and miR-92a plasma levels. In addition, the authors suggested that the higher levels of miR-92a in CRC patients were not referable to an inflammatory status or other gastrointestinal cancers but were related only to the presence CRC [10]. Increased levels of miR-92a and miR-29a were observed in plasma of CRC patients by Huang and coworkers [11]. They demonstrated that plasma levels of both miRNAs were significantly reduced after surgery in the same patients. Soon afterwards the same authors found that miR-601 and miR-760 were significantly reduced in CRC plasma compared to healthy controls [12]. It is worth noting that the Receive Operation Curve (ROC) analysis obtained by combining the levels of miR-29a, miR-92a and miR-760 improved the overall signature performance to a final 83.3 % sensitivity and 93.1 % specificity in discriminating CRC patients from controls. This was the proof that a combination of several circulating miRNAs could constitute a better diagnostic tool than each individual miRNA.

A panel of 8 plasma miRNAs (miR-15b, miR-17, miR-142-3p, miR-195, miR-331, miR-532-3p, miR-532, and miR-652) was able to distinguish polyps from healthy controls with high accuracy (AUC=0.868) [13]. A different study proposed a nine-miRNAs signature (miR-18a, miR-20a, miR-21, miR-29a, miR-92a, miR-106b, miR-133a, miR-143, miR-145) that was detected with higher levels in plasma of CRC patients [14].

In addition to many other tumor types, circulating miR-21 was investigated in CRC with promising results: several papers demonstrated an increased level of this miRNA both in plasma and in serum. Specifically, miR-21 increased levels in serum robustly distinguished adenoma and CRC from control subjects [15, 16]. Another study demonstrated that the levels of cell-free miRNAs could change in relation to blood drawn site. MiR-21 levels collected nearby a colorectal cancer lesion (mesenteric vein) were higher than in peripheral veins for the same patient, suggesting that the concentration of miR-21 in the blood could be progressively diluted in the circulatory system [17]. Zanutto and co-workers identified plasma miR-378 as higher in CRC patients compared to healthy donors and demonstrated that its levels decreased 4–6 months after surgery in non-relapsing patients. Moreover, the ability of miR-378 to discriminate CRC patients and controls was not influenced by the haemolysis rate of plasma samples [18].

A reduced number of studies were performed using serum as a source of circulating miRNAs (Table 9.1). Wang et al. proposed a panel of six miRNAs that were increased or decreased in CRC patients. This was the first study to describe a serum miRNA-based signature combining both up- and down-regulated miRNAs. The six-miRNA signature (let-7 g, miR-21, miR-31, miR-92a, miR-181b, and miR-203) was able to detect CRC serum samples with a sensitivity and specificity of 93 % and 91 %, respectively [19].

In CRC, a panel of three miRNAs (miR-23a, miR-193a-3p, and miR-338-5p) was proposed to be differentially expressed in both tissues and blood samples and a significant positive correlations were described [20].

A main limit of these studies could be identified in the assumption that miRNAs detectable into blood circulation matched that dysregulated in solid tumors. Indeed, presuming that cell-free miRNAs could be mostly released by neoplastic cells, researchers tried to validate as circulating cancer biomarkers only the same miRNAs that were de-regulated in solid tumors. This hypothesis was not confirmed by several recent studies [8, 21, 22] and may have prevented the identification of more powerful and reliable miRNA biomarkers.

In this regard, Giraldez et al. proposed a panel of 6 upregulated microRNAs (miR-15b,miR-18a, miR-19a, miR-19b, mir-29a and miR-335) as CRC diagnostic biomarkers after performing a genome-wide miRNA profiling on a pilot group of 60 samples and a validation with RT-qPCR on 123 patients with sporadic CRC and 73 controls [23].

From the analysis of all published studies, a generalized lack of concordance and a poor consensus in cell-free miRNA findings can be observed. For this reason, several groups tried to perform meta-analyses of published data, in order to assess the real diagnostic performance of circulating microRNAs. After a careful study selection and data extraction from 19 papers including 1558 CRC patients and 1085 controls, Zeng and co-workers found that the pooled sensitivity and specificity of circulating miRNAs were 0.8 (95 % CI 0.77–0.85) and 0.84 (95 % CI 0.78–

0.88) respectively [24]. In addition, they verified that multiple miRNAs had higher predictive accuracy than single miRNA, probably because a microRNA panel may comprise different aspects of tumorigenesis. Moreover, they didn't find a significant difference between Asian and Caucasian ethnicity and they proposed that different sample specimen could contribute significantly to the heterogeneity between studies. Indeed, they found that serum could be considered a better matrix for miRNA assays in CRC screening compared to plasma. The same results were proposed by another group in a meta-analysis based on 42 articles [19].

Eventually, both meta-analyses suggested that circulating microRNAs could be useful tools for early detection of colorectal cancer.

9.2.2 Other Non-coding RNAs

The diagnostic utility of long non coding RNAs (lncRNAs) in CRC was recently reported in literature. The CAHM (Colorectal Adenocarcinoma HyperMethylated) gene encode a long non-coding RNA (lncRNA) whose methylated sequence was recently detected in the plasma DNA of 55 % of CRC patients compared to 4 % of subjects with adenomas and 7 % of subjects without neoplasia, using a threshold of 3 pg methylated genomic DNA per mL plasma. These results showed that this methylated lncRNA gene could be considered a promising plasma biomarker to use in CRC screening [25].

Shi and co-worker characterized the genome-wide lncRNAs expression profile in plasma from 290 CRC patients compared to cancer-free controls identifying a panel of lncRNAs that might serve as diagnostic tool for CRC. In particular, three of these (XLOC_006844, LOC152578 and XLOC_000303) were differentially present in the plasma of the CRC patients indicating for the first time a promising diagnostic role of these molecules [26].

9.3 Circulating Non-coding RNAs as Prognostic Biomarkers

The association between miRNA expression and CRC prognosis was first described by Xi and colleagues, who demonstrated the association between tumor hsa-miR-200c-3p level and overall survival in colorectal cancer patients [27]. Recently, the key role of miRNAs in cancer has extended from cancer tissues to body fluids, especially plasma and serum, where stable, cell-free miRNA molecules have been detected.

Since 2010, several studies have been published demonstrating the feasibility of cancer prognosis prediction by assessing miRNA levels in plasma or serum (Table 9.2). The very first study was conducted by Pu and colleagues on 103 plasma samples from CRC patients. They found that elevated plasma miR-221-3p levels significantly correlated with shorter survival rate and inversely correlate with p53 expression in cancer, overall indicating that plasma miR-221-3p could be a good prognostic marker in CRC [28]. They analyzed the amount of candidate circulating microRNAs performing RT-qPCR directly on plasma, without RNA extraction.

Subsequently, plasma miR-141-3p amount proved to be an independent prognostic factor for metastatic CRC in two independent cohorts. It was found significantly increased in stage III and in stage IV tumors compared to earlier stage cases, with a sensitivity and specificity in identifying stage IV cancers of 77.1 % and 89.7 % respectively. In addition, it was demonstrated that high plasma miR-141-3p predicted poor survival [29].

Two studies demonstrated that miR-182-5p and miR-378a-3p, whose levels were up-regulated in plasma of patients with colorectal cancer compared to controls, were significantly decreased in the same patients after tumor surgical removal, suggesting a potential role for these miRNAs in patient monitoring during follow-up

Table 9.2 Circulating prognostic microRNAs in colorectal cancer

MicroRNA	Body fluid	Technology	Modulation in poor prognosis group	Effect	Reference
hsa-miR-221-3p	Plasma	Direct amplification from plasma by RT-qPCR	Increase	Shorter survival and reduced p53 expression in CRC.	[28]
hsa-miR-141-3p	Plasma	RT-qPCR	Increase	Advanced stage and shorter survival.	[29]
hsa-miR-183-5p	Plasma	RT-qPCR	Increase	Lymph-node metastases, distant metastases, advanced stage, tumor recurrence, shorter disease-free survival and overall survival. Decrease after surgical tumor removal.	[31]
hsa-miR-182-5p	Plasma	RT-qPCR	Increase	Decrease after surgical tumor removal.	[30]
hsa-miR-378a-3p	Plasma	RT-qPCR	Increase	Decrease after surgical tumor removal.	[18]
hsa-miR-24-3p	Plasma	RT-qPCR	Decrease	Increase after surgical tumor removal.	[32]
hsa-miR-320a	Plasma	RT-qPCR	Decrease	Increase after surgical tumor removal.	[32]
hsa-miR-423-5p	Plasma	RT-qPCR	Decrease	Increase after surgical tumor removal.	[32]
hsa-miR-372-3p	Serum	RT-qPCR	Increase	Shorter overall survival. Decrease after surgical tumor removal.	[36]
hsa-miR-592	Serum	RT-qPCR	Increase	Correlation with distant metastases.	[34]
				Decrease after surgical tumor removal.	
hsa-miR-106a-5p	Serum	TaqMan low-density array and RT-qPCR	Increase	Shorter overall survival. Serum levels decrease after surgical tumor removal.	[37]
hsa-miR-29a-3p	Serum	RT-qPCR	Increase	Early detection of CRC with liver metastases.	[33]
hsa-miR-199a-3p	Serum	Microarray and RT-qPCR	Increase	Correlation with deep wall invasion.	[35]
				Serum levels decrease after surgical tumor removal.	
hsa-miR-200c-3p	Serum	RT-qPCR	Increase	Lymph-node metastases, tumor recurrence, distant metastasis, stage IV, shorter survival.	[15]
hsa-miR-155-5p	Serum	RT-qPCR	Increase	Shorter progression-free and overall survival.	[38]
hsa-miR-17-3p	Serum	TaqMan low-density array and RT-qPCR	Increase	Shorter progression-free and overall survival.	[37]
				Serum levels decrease after surgical tumor removal.	

(continued)

Table 9.2 (continued)

MicroRNA	Body fluid	Technology	Modulation in poor prognosis group	Effect	Reference
hsa-miR-19a-3p	Serum (exosomes)	Microarray and RT-qPCR	Increase	Lymph-node metastases, liver metastases, advanced stage, shorter survival.	[40]
hsa-miR-92a-3p	Serum	RT-qPCR	Increase	Shorter survival.	[39]
hsa-miR-21-5p	Serum	RT-qPCR	Increase	Tumor size, metastases, poor survival.	[41]
				Serum levels decrease after surgical tumor removal.	
hsa-miR-21-5p	Serum	RT-qPCR	Decrease	High local recurrence and increased mortality.	[42]
hsa-miR-145-5p	Serum	TaqMan low-density array and RT-qPCR	Decrease	Increase after surgical tumor removal.	[37]
hsa-miR-218-5p	Serum	RT-qPCR	Decrease	Increase after surgical tumor removal.	[52]
hsa-miR-148a-3p	Serum	MicroRNA array (Applied Biosystems) and RT-qPCR	Decrease	Early relapse after tumor resection.	[53]
hsa-miR-20a-5p, hsa-miR-130, hsa-miR-145-5p, hsa-miR-216a-5p, hsa-miR-372-3p	Serum	TaqMan low-density array and RT-qPCR		Chemosensitivity prediction.	[45]
hsa-miR-19a	Serum	Microarray and RT-qPCR	Increase	Resistance to therapy.	[44]

[18, 30]. Another miRNA that was described upregulated in plasma of CRC patients was miR-183-5p. The authors demonstrated that miR-183-5p amount decreased in 11 post-surgery plasma samples compared to matched pre-surgery to increase again in relapsing patients, although only three relapsing patients were reported in the study [31]. Moreover, higher plasma levels of miR-183-5p were significantly associated with high stage, lymph-node infiltration, distant metastases, tumor recurrence and shorter disease-free and overall survival.

There is only one published study reporting a miRNA down-regulation in plasma of patients affected by CRC. Fang and colleagues demonstrated that plasma miR-24-3p, miR-320a and miR-423-5p levels decreased in patients with both benign colon lesions and colorectal cancer compared to healthy controls. All the three miRNAs increased after tumor surgical removal and patients clinical improvement, suggesting a contribution of these molecules not only to CRC early detection but also to post-surgery patient monitoring [32].

In addition to plasma, several studies were conducted on serum. One of the first studies on the prognostic role of a serum microRNA in CRC identified miR-29a-3p as increased in CRC patients with liver metastasis, providing an area under the curve (AUC) of 0.803 and a 75 % sensitivity and specificity in discriminating metastatic from non-metastatic tumors [33]. Recently, other serum markers of tumor metastasis were identified. MiR-592 was described as over-expressed in both CRC tumor tissues and serum compared to healthy controls and its serum level was further increased in metastatic cancers. Additionally, serum miR-592 decreased in post-surgery patients, suggesting a potential usefulness in patient follow-up [34]. High miR-200c-3p serum levels were associated with high stage, lymph node involvement, distant metastasis and tumor relapse in colorectal cancer patients, thereby suggesting an important role for this microRNA in CRC prognosis definition [15].

Another microRNA up-regulated in serum of CRC patient is miR-199a-3p. This miRNA was identified after an initial microarray screening of 10 matched pre and post-surgery serum samples from CRC-patients and subsequently validated by RT-qPCR on 30 matched samples. In addition, high miR-199a-3p was proved to significantly associate with tumor invasion [35].

Some circulating miRNAs were able to identify patients with good or worse prognosis. MiR-372-3p [36], miR-106a-5p [37] and miR-155-5p [38] up-regulation in serum was associated with shorter disease-free or overall survival. In addition, miR-372-3p increased serum level was found both in CRC and precancerous lesions, to finally drop after surgical removal of tumor [36]. Li et al. performed a screening on 20 paired pre- and post-surgery sera from CRC patients and healthy subjects using TaqMan Low-Density Array (749 miRNAs) for miRNA discovery and RT-qPCR for validation. The authors demonstrated a significant increase of serum miR-106a-5p and miR-17-3p and a decrease of miR-145-5p in pre-operative CRC compared with both healthy controls and post-operative matched samples. The AUC for the three-miRNAs panel in discriminating between pre and post-surgery patients was 0.85 [37]. Moreover, by analyzing stage II and III patients who received chemotherapy, the authors found that miR-17-3p increased levels were correlated with shorter disease-free survival.

MiR-17-3p belongs to miR-17-92 cluster, one of the most studied oncogenic miRNA clusters. Other serum miRNAs from the same cluster were demonstrated to correlate with colorectal cancer prognosis. Liu and colleagues found that miR-92a-3p progressively increased in serum of healthy controls, colorectal adenoma patients and carcinoma patients and high expression in the serum correlated with a lower probability of survival [39]. In another study from Matsumura's group a miRNA microarray analysis was performed on exosomal microRNAs from serum of CRC patients compared with controls. They described an involvement of miR-17-92a cluster in CRC recurrence. Moreover, they demonstrated that miR-19a-3p represented an independent poor prognostic factor in CRC, since high miR-19a-3p level correlated with reduced overall survival and disease-free survival [40]. Altogether, these findings demonstrated that miR-17-92 cluster is a powerful prognostic biomarker in colorectal cancer since the relationship between increased serum levels and poor prognosis was demonstrated in three independent studies.

MiR-21-5p is another well-known oncogenic miRNA, with a role in colorectal cancer. MiR-21-5p was identified as an independent prognostic marker for CRC because its serum levels significantly increased in patients with adenoma and carcinoma compared to healthy controls and miR-21-5p serum up-regulation correlated with tumor size, stage, metastasis and poor survival [41]. In disagreement with these results, Menendez et al. described the association between poor prognosis and low miR-21 serum levels, as reasons behind these opposite and contradictory results could reside on differences in predicted by higher local recurrence and increased risk of mortality [42]. Of course, the reasons behind these opposite and contradictory results could reside on differences in sample processing, miRNA quantification or even data normalization, which are a main source of variation in this kind of studies [9, 43].

9.4 Circulating Non-coding RNAs as Response-to-Therapy Biomarkers

While several studies suggested that circulating miRNAs could be considered prognostic biomarkers in CRC, only few studies demonstrated a role for circulating miRNA as response to therapy predictors (Table 9.2).

In a study published in 2013, serum samples from eight FOLFOX resistant and eight FOLFOX responder patients were analyzed by miRNA microarrays [44]. The miRNAs correlated with treatment resistance were subsequently validated by RT-qPCR on 72 serum samples. The authors described a significant up-regulation of serum miR-19a-3p in FOLFOX resistant patients compared with responders, with an AUC of 0.679. In another study, Zhang and colleagues evaluated the global miRNA expression of pooled serum samples from 253 patients treated with chemotherapy using TaqMan low-density arrays. They found a circulating miRNA signature specific for chemo-resistant and chemo-sensitive patients and they used 17 miRNAs to successfully predict response to chemotherapy. In addition, they used a five-miRNAs signature (miR-20a-5p, miR-130, miR-145-5p, miR-216-5p and miR-372-3p) obtained by RT-qPCR in individual samples to predict the chemo-sensitivity in two independent group of samples, with AUC of 0.841 and 0.918 respectively [45].

9.5 Conclusion

Circulating microRNAs show a great potential as prognostic biomarkers in colorectal cancer, as supported by many studies published in the past few years. However, circulating miRNAs procedures still need to be optimized and standardized high-quality methodologies for cell-free miRNA assessment need to be developed. Several clinical trials including circulating miRNA / ncRNA biomarker assessment are ongoing, but it is necessary for clinical trials to be completed to demonstrate their full potential and independent validations are required.

Once more accurate and standardized practices are established, miRNA / ncRNA quantification in serum or plasma of cancer patients has certainly the potential to become a minimally invasive diagnostic and prognostic tool for colorectal cancer patients.

Acknowledgements The work was supported by funding from the Italian Association for Cancer Research (AIRC) to MF (MFAG 11676) and from the Italian Ministry of Instruction, University and Research FIRB 2011 (Project RBAPIIBYNP) and University of Ferrara (FAR 2012–14) to MN. We thank Miriam Ferracin for her assistance in generating the graphical illustration.

References

1. Siegel R, Ma J, Zou Z, Jemal A. Cancer statistics, 2014. CA Cancer J Clin. 2014;64(1):9–29.
2. Willyard C. Screening: early alert. Nature. 2015;521(7551):S4–5.
3. Arroyo JD, Chevillet JR, Kroh EM, Ruf IK, Pritchard CC, Gibson DF, et al. Argonaute2 complexes carry a population of circulating microRNAs independent of vesicles in human plasma. Proc Natl Acad Sci U S A. 2011;108(12):5003–8.
4. Skog J, Wurdinger T, van Rijn S, Meijer DH, Gainche L, Sena-Esteves M, et al. Glioblastoma microvesicles transport RNA and proteins that promote tumour growth and provide diagnostic biomarkers. Nat Cell Biol. 2008;10(12):1470–6.
5. Vickers KC, Palmisano BT, Shoucri BM, Shamburek RD, Remaley AT. MicroRNAs are transported in plasma and delivered to recipient cells by high-density lipoproteins. Nat Cell Biol. 2011;13(4):423–33.
6. Braicu C, Tomuleasa C, Monroig P, Cucuianu A, Berindan-Neagoe I, Calin GA. Exosomes as divine messengers: are they the Hermes of modern molecular oncology? Cell Death Differ. 2015;22(1):34–45.
7. Schwarzenbach H, Nishida N, Calin GA, Pantel K. Clinical relevance of circulating cell-free microRNAs in cancer. Nat Rev Clin Oncol. 2014;11(3):145–56.
8. Jarry J, Schadendorf D, Greenwood C, Spatz A, van Kempen LC. The validity of circulating microRNAs in oncology: five years of challenges and contradictions. Mol Oncol. 2014;8(4):819–29.
9. Witwer KW. Circulating microRNA biomarker studies: pitfalls and potential solutions. Clin Chem. 2015;61(1):56–63.
10. Balch C, Fang F, Matei DE, Huang TH, Nephew KP. Minireview: epigenetic changes in ovarian cancer. Endocrinology. 2009;150(9):4003–11.
11. Guo HQ, Huang GL, Guo CC, Pu XX, Lin TY. Diagnostic and prognostic value of circulating miR-221 for extranodal natural killer/T-cell lymphoma. Dis Markers. 2010;29(5):251–8.
12. Heredia NJ, Belgrader P, Wang S, Koehler R, Regan J, Cosman AM, et al. Droplet digital PCR quantitation of HER2 expression in FFPE breast cancer samples. Methods. 2013;59(1):S20–3.
13. Kanaan Z, Roberts H, Eichenberger MR, Billeter A, Ocheretner G, Pan J, et al. A plasma microRNA panel for detection of colorectal adenomas: a step toward more precise screening for colorectal cancer. Ann Surg. 2013;258(3):400–8.
14. Belgrader P, Tanner SC, Regan JF, Koehler R, Hindson BJ, Brown AS. Droplet digital PCR measurement of

HER2 copy number alteration in formalin-fixed paraffin-embedded breast carcinoma tissue. Clin Chem. 2013;59(6):991–4.
15. Toiyama Y, Hur K, Tanaka K, Inoue Y, Kusunoki M, Boland CR, et al. Serum miR-200c is a novel prognostic and metastasis-predictive biomarker in patients with colorectal cancer. Ann Surg [Research Support, N.I.H., Extramural Research Support, Non-U.S. Gov't]. 2014;259(4):735–43.
16. Kanaan Z, Rai SN, Eichenberger MR, Roberts H, Keskey B, Pan J, et al. Plasma miR-21: a potential diagnostic marker of colorectal cancer. Ann Surg. 2012;256(3):544–51.
17. Monzo M, Martinez-Rodenas F, Moreno I, Navarro A, Santasusagna S, Macias I, et al. Differential MIR-21 expression in plasma from mesenteric versus peripheral veins: an observational study of disease-free survival in surgically resected colon cancer patients. Medicine (Baltimore). 2015;94(1):e145.
18. Zanutto S, Pizzamiglio S, Ghilotti M, Bertan C, Ravagnani F, Perrone F, et al. Circulating miR-378 in plasma: a reliable, haemolysis-independent biomarker for colorectal cancer. Br J Cancer. 2014;110(4):1001–7.
19. Wang R, Wen H, Xu Y, Chen Q, Luo Y, Lin Y, et al. Circulating microRNAs as a novel class of diagnostic biomarkers in gastrointestinal tumors detection: a meta-analysis based on 42 articles. PLoS One. 2014;9(11):e113401.
20. Han HS, Yun J, Lim SN, Han JH, Lee KH, Kim ST, et al. Downregulation of cell-free miR-198 as a diagnostic biomarker for lung adenocarcinoma-associated malignant pleural effusion. Int J Cancer. 2013;133(3):645–52.
21. Cookson VJ, Bentley MA, Hogan BV, Horgan K, Hayward BE, Hazelwood LD, et al. Circulating microRNA profiles reflect the presence of breast tumours but not the profiles of microRNAs within the tumours. Cell Oncol (Dordr). 2012;35(4):301–8.
22. Ferracin M, Lupini L, Salamon I, Saccenti E, Zanzi MV, Rocchi A, et al. Absolute quantification of cell-free microRNAs in cancer patients. Oncotarget. 2015;6:14545–55.
23. Giraldez MD, Lozano JJ, Ramirez G, Hijona E, Bujanda L, Castells A, et al. Circulating microRNAs as biomarkers of colorectal cancer: results from a genome-wide profiling and validation study. Clin Gastroenterol Hepatol. 2013;11(6):681–8 e3.
24. Zeng W, Tu Y, Zhu Y, Wang Z, Li C, Lao L, et al. Predictive power of circulating miRNAs in detecting colorectal cancer. Tumour Biol. 2015;36(4):2559–67.
25. Pedersen SK, Mitchell SM, Graham LD, McEvoy A, Thomas ML, Baker RT, et al. CAHM, a long non-coding RNA gene hypermethylated in colorectal neoplasia. Epigenetics. 2014;9(8):1071–82.
26. Shi J, Li X, Zhang F, Zhang C, Guan Q, Cao X, et al. Circulating lncRNAs associated with occurrence of colorectal cancer progression. Am J Cancer Res. 2015;5(7):2258–65.
27. Xi Y, Formentini A, Chien M, Weir DB, Russo JJ, Ju J, et al. Prognostic values of microRNAs in colorectal cancer. Biomark Insights. 2006;2:113–21.
28. Pu XX, Huang GL, Guo HQ, Guo CC, Li H, Ye S, et al. Circulating miR-221 directly amplified from plasma is a potential diagnostic and prognostic marker of colorectal cancer and is correlated with p53 expression. J Gastroenterol Hepatol. 2010;25(10):1674–80.
29. Cheng H, Zhang L, Cogdell DE, Zheng H, Schetter AJ, Nykter M, et al. Circulating plasma MiR-141 is a novel biomarker for metastatic colon cancer and predicts poor prognosis. PLoS One. 2011;6(3):e17745.
30. Perilli L, Vicentini C, Agostini M, Pizzini S, Pizzi M, D'Angelo E, et al. Circulating miR-182 is a biomarker of colorectal adenocarcinoma progression. Oncotarget [Research Support, Non-U.S. Gov't]. 2014;5(16):6611–9.
31. Yuan D, Li K, Zhu K, Yan R, Dang C. Plasma miR-183 predicts recurrence and prognosis in patients with colorectal cancer. Cancer Biol Ther [Research Support, Non-U.S. Gov't]. 2015;16(2):268–75.
32. Fang Z, Tang J, Bai Y, Lin H, You H, Jin H, et al. Plasma levels of microRNA-24, microRNA-320a, and microRNA-423-5p are potential biomarkers for colorectal carcinoma. J Exp Clin Cancer Res [Research Support, Non-U.S. Gov't]. 2015;34:86.
33. Wang LG, Gu J. Serum microRNA-29a is a promising novel marker for early detection of colorectal liver metastasis. Cancer Epidemiol. 2012;36(1):e61–7.
34. Liu M, Zhi Q, Wang W, Zhang Q, Fang T, Ma Q. Up-regulation of miR-592 correlates with tumor progression and poor prognosis in patients with colorectal cancer. Biomed Pharmacother [Research Support, Non-U.S. Gov't]. 2015;69:214–20.
35. Nonaka R, Nishimura J, Kagawa Y, Osawa H, Hasegawa J, Murata K, et al. Circulating miR-199a-3p as a novel serum biomarker for colorectal cancer. Oncol Rep [Research Support, Non-U.S. Gov't]. 2014;32(6):2354–8.
36. Yu J, Jin L, Li W, Jiang L, Hu Y, Zhi Q, et al. Serum miR-372 is a diagnostic and prognostic biomarker in patients with early colorectal cancer. Anticancer Agents Med Chem. 2016;16(4):424–31.
37. Li J, Liu Y, Wang C, Deng T, Liang H, Wang Y, et al. Serum miRNA expression profile as a prognostic biomarker of stage II/III colorectal adenocarcinoma. Sci Rep [Research Support, Non-U.S. Gov't]. 2015;5:12921.
38. Lv ZC, Fan YS, Chen HB, Zhao DW. Investigation of microRNA-155 as a serum diagnostic and prognostic biomarker for colorectal cancer. Tumour Biol. 2015;36(3):1619–25.
39. Liu GH, Zhou ZG, Chen R, Wang MJ, Zhou B, Li Y, et al. Serum miR-21 and miR-92a as biomarkers in the diagnosis and prognosis of colorectal cancer. Tumour Biol [Research Support, Non-U.S. Gov't]. 2013;34(4):2175–81.
40. Matsumura T, Sugimachi K, Iinuma H, Takahashi Y, Kurashige J, Sawada G, et al. Exosomal microRNA in

serum is a novel biomarker of recurrence in human colorectal cancer. Br J Cancer [Research Support, Non-U.S. Gov't]. 2015;113(2):275–81.
41. Toiyama Y, Takahashi M, Hur K, Nagasaka T, Tanaka K, Inoue Y, et al. Serum miR-21 as a diagnostic and prognostic biomarker in colorectal cancer. J Natl Cancer Inst [Research Support, N.I.H., Extramural Research Support, Non-U.S. Gov't]. 2013;105(12):849–59.
42. Menendez P, Padilla D, Villarejo P, Palomino T, Nieto P, Menendez JM, et al. Prognostic implications of serum microRNA-21 in colorectal cancer. J Surg Oncol. 2013;108:369–73.
43. Faltejskova P, Bocanek O, Sachlova M, Svoboda M, Kiss I, Vyzula R, et al. Circulating miR-17-3p, miR-29a, miR-92a and miR-135b in serum: evidence against their usage as biomarkers in colorectal cancer. Cancer Biomark. 2012;12(4):199–204.
44. Chen Q, Xia HW, Ge XJ, Zhang YC, Tang QL, Bi F. Serum miR-19a predicts resistance to FOLFOX chemotherapy in advanced colorectal cancer cases. Asian Pac J Cancer Prev [Research Support, Non-U.S. Gov't]. 2013;14(12):7421–6.
45. Zhang J, Zhang K, Bi M, Jiao X, Zhang D, Dong Q. Circulating microRNA expressions in colorectal cancer as predictors of response to chemotherapy. Anticancer Drugs. 2014;25(3):346–52.
46. Ng EK, Chong WW, Jin H, Lam EK, Shin VY, Yu J, et al. Differential expression of microRNAs in plasma of patients with colorectal cancer: a potential marker for colorectal cancer screening. Gut. 2009;58(10):1375–81.
47. Luo X, Stock C, Burwinkel B, Brenner H. Identification and evaluation of plasma MicroRNAs for early detection of colorectal cancer. PLoS One. 2013;8(5):e62880.
48. Huang Z, Huang D, Ni S, Peng Z, Sheng W, Du X. Plasma microRNAs are promising novel biomarkers for early detection of colorectal cancer. Int J Cancer. 2010;127(1):118–26.
49. Wang Q, Huang Z, Ni S, Xiao X, Xu Q, Wang L, et al. Plasma miR-601 and miR-760 are novel biomarkers for the early detection of colorectal cancer. PLoS One. 2012;7(9):e44398.
50. Wang J, Huang SK, Zhao M, Yang M, Zhong JL, Gu YY, et al. Identification of a circulating microRNA signature for colorectal cancer detection. PLoS One. 2014;9(4):e87451.
51. Yong FL, Law CW, Wang CW. Potentiality of a triple microRNA classifier: miR-193a-3p, miR-23a and miR-338-5p for early detection of colorectal cancer. BMC Cancer. 2013;13:280.
52. Yu H, Gao G, Jiang L, Guo L, Lin M, Jiao X, et al. Decreased expression of miR-218 is associated with poor prognosis in patients with colorectal cancer. Int J Clin Exp Pathol [Research Support, Non-U.S. Gov't]. 2013;6(12):2904–11.
53. Tsai HL, Yang IP, Huang CW, Ma CJ, Kuo CH, Lu CY, et al. Clinical significance of microRNA-148a in patients with early relapse of stage II stage and III colorectal cancer after curative resection. Transl Res [Research Support, Non-U.S. Gov't]. 2013;162(4):258–68.

Non-coding RNAs Enabling Prognostic Stratification and Prediction of Therapeutic Response in Colorectal Cancer Patients

10

Samantha O. Perakis, Joseph E. Thomas, and Martin Pichler

Abstract

Colorectal cancer (CRC) is a heterogeneous disease and current treatment options for patients are associated with a wide range of outcomes and tumor responses. Although the traditional TNM staging system continues to serve as a crucial tool for estimating CRC prognosis and for stratification of treatment choices and long-term survival, it remains limited as it relies on macroscopic features and cases of surgical resection, fails to incorporate new molecular data and information, and cannot perfectly predict the variety of outcomes and responses to treatment associated with tumors of the same stage. Although additional histopathologic features have recently been applied in order to better classify individual tumors, the future might incorporate the use of novel molecular and genetic markers in order to maximize therapeutic outcome and to provide accurate prognosis. Such novel biomarkers, in addition to individual patient tumor phenotyping and other validated genetic markers, could facilitate the prediction of risk of progression in CRC patients and help assess overall survival. Recent findings point to the emerging role of non-protein-coding regions of the genome in their contribution to the progression of cancer and tumor formation. Two major subclasses of non-coding RNAs (ncRNAs), microRNAs and long non-coding RNAs, are often dysregulated in CRC and have demonstrated their diagnostic and prognostic potential as biomarkers.

S.O. Perakis (✉)
Division of Oncology, Medical University of Graz, Graz, Austria
e-mail: samantha.perakis@medunigraz.at

J.E. Thomas
Indiana University School of Medicine, Indianapolis, IN, USA
e-mail: thomjoea@iupui.edu

M. Pichler
Department of Experimental Therapeutics, The University of Texas MD Anderson Cancer Center, Houston, TX, USA
e-mail: martin.pichler@medunigraz.at

O. Slaby, G.A. Calin (eds.), *Non-coding RNAs in Colorectal Cancer*, Advances in Experimental Medicine and Biology 937, DOI 10.1007/978-3-319-42059-2_10

These ncRNAs are promising molecular classifiers and could assist in the stratification of patients into appropriate risk groups to guide therapeutic decisions and their expression patterns could help determine prognosis and predict therapeutic options in CRC.

Keywords
MicroRNAs • Long non-coding RNAs • Colorectal cancer • Prognosis • Therapeutic response

10.1 Introduction

Colorectal cancer (CRC) remains a significant global health burden and is the third most common cause of cancer-related mortality worldwide, accounting for about 608,000 deaths annually, which corresponds to approximately half of its incidence [1–4]. Although there has been a dramatic decline in the incidence of CRC in the past 10 years due to the more widespread access of screening methods and the development of new chemotherapeutic drugs, incidence still remains high and CRC survival rates have not substantially improved [5–7]. It has been reported that about 20 % of CRC patients initially present with metastatic disease and up to 35 % of these individuals will later develop metastases in the later stages of progression [2, 6]. If detected early, the removal of early-stage cancer and precancerous lesions is possible, rendering CRC a potentially curable disease. However, once the disease has spread to distant sites (metastasized), therapeutic options diminish and the majority of patients can only be treated with palliative options with the sole objective of prolonging life and improving quality of life [8]. This dismal patient outlook addresses the need for a better understanding of the origins and biological nature of CRC in order to better develop effective preventative and diagnostic tools. It would be of great value to be able to efficiently identify the correct patient population that may benefit from more thorough screening methods for monitoring disease recurrence or to determine which patients might be candidates for adjuvant chemotherapy [9]. CRC patient outcomes are heterogeneous during early stages, with survival rates for stage II patients averaging between 72 and 88 % and 40–71 % for stage III patients [10]. Because approximately 25 % of patients with early-stage disease are confronted with recurrence, there is a clear and present demand for new markers to allow for the stratification of high-risk patients and to identify those who should be subjected to stricter monitoring and potential systemic treatments [11–14]. Improvements in preventative methods, early detection and treatment options can help tremendously in combating this malignancy, but such strategies require a comprehensive understanding of CRC at both the genetic and molecular levels.

10.1.1 Current Prognosis and Screening Methods in CRC

The currently applied method in clinical practice for prognosis of CRC and for guiding therapeutic decisions is the tumor-node-metastasis-system (TNM), which is based on primary tumor attributes, the presence and extent of the involvement of the lymph nodes in addition to the presence of distant metastases [15]. Although it remains a powerful tool for the prediction of late stages, it is less informative when applying to intermediate stages of disease [16, 17]. For example, while TNM stage III patients tend to benefit from adjuvant therapy following surgical resection, it remains unclear if such treatment would be advantageous for TNM stage II patients, especially since multiple clinical trials have offered conflicting results [18–22]. Furthermore, current TNM guidelines do not advocate that adjuvant chemotherapy be administered to early-stage patients, yet 20–30 % of these stage I and II patients will succumb to CRC within 5 years, thus provoking the question of whether or not these patients might have survived had the appropriate therapy been selected and adapted accordingly in advance [23]. The TNM staging system also remains limited as it relies on macroscopic features and cases of surgical resection, fails to incorporate new molecular data and information, and cannot perfectly predict the variety of outcomes and responses to treatment associated with tumors of the same stage [24, 25]. Although additional histopathologic features have recently been applied in order to better classify individual tumors and further clinico-pathological prognostic factors have been applied, i.e. tumor grade, perineural spread and vascular invasion, they add no particular value in terms of stratifying patients according to their specific treatment requirements [26]. These limitations have fueled great interest in the search for new prognostic factors which could ensure that select patients are neither forced to undergo needless chemotherapy nor left untreated on the basis of an initial inaccurate assessment. Given the significance of precise prognostic prediction throughout all stages of disease, it is understood that the discovery of such novel biomarkers will be fundamental for the diagnosis, treatment and prognosis of CRC.

Much like the standard screening methods, current therapies for the treatment of CRC patients also have bottlenecks in terms of efficacy and improving patient outcome. Advances in defining the underlying molecular mechanisms of metastatic CRC (mCRC) have led to the development of the drugs cetuximab and panitumumab, both monoclonal antibodies which selectively target the epidermal growth factor receptor (*EGFR*) extracellular domain. Although these novel agents have greatly improved available therapeutic options as well as clinical outcome for mCRC patients, it has been reported that only generally 10–20 % of these patients clinically benefit from anti-EGFR therapy [27–29]. Furthermore, even though it has been determined that mutations of the *KRAS* gene serve as negative predictors of the effectiveness of EGFR-targeted agents in patients with mCRC, *KRAS* and *NRAS* mutations only account for approximately 50 % of non-responders [30–33]. Patients also encounter underlying issues regarding the use of current chemotherapeutic agents. Failure of chemotherapy is a result of resistance to such chemotherapeutic agents as 5-FU, oxaliplatin and irinotecan, which in turn leads to both cancer relapse and poor prognosis for the patient [14]. These statistics demonstrate the urgent need to be able to predict treatment response to EGFR-targeted therapy as well as chemotherapeutic treatment plans in the subgroups of patients with mCRC, thereby preventing patients from being needlessly exposed to ineffective therapies.

10.1.2 ncRNAs as Biomarkers in CRC

The development of biomarkers could assist in disease management by providing means of early cancer detection as well as improving stratification of patients and their response to therapy, thereby resulting in a more positive outlook of patient prognosis [34]. This has fueled the search for a more sensitive set of biomarkers. Due to the biological heterogeneity of colorectal cancer, only a subset of patients tend to benefit from particular treatments; for this reason, being able to prospectively filter out patients who will most likely respond to a targeted therapy would be of immense clinical significance. Such predictive markers could help guide the choice of therapy. For example, patients with concentrations of a particular marker associated with resistance could in turn be prescribed an alternative and more beneficial treatment fitting to their condition. Additionally, predictive markers could help define not only the optimal drug dosage but perhaps also assess toxicity issues, in turn potentially decreasing costs associated with health care and enhancing the patient's quality of life [35, 36]. In contrast, potential colorectal cancer prognostic markers could predict the natural course of the malignancy specific to the individual and determine which patients might have a more desirable or more negative disease outcome [37]. Ideally, such cancer biomarkers would be involved in fundamental cell functions such as cell proliferation, differentiation, invasion or metastatic progression (Tables 10.1 and 10.2).

Table 10.1 Current key prognostic miRNAs and expression trends in colorectal cancer

miRNA(s)	Patients in cohort (n)	Expression/outcome	P-value	HR	Reference
miR-31	12	Increased in stage IV tissue compared to stage II	0.028		[43]
	98	Increased in CRC compared to normal tissue; expression positively related to advanced TNM stage, deeper invasion	0.001; 0.026, 0.024		[63]
	29	Increased in tumors	0.0006		[67]
miR-106a	363	Overexpressed in CRC tissue	0.001		[68]
	28	Low expression levels indicates better clinical outcome	<0.05		[69]
	50	High expression predicts short DFS and OS in stage II patients	0.008 (DFS) 0.049 (OS)	2.91 (DFS) 2.25 (OS)	[70]
miR-21	50	High expression associated with short DFS and OS in stage II patients	0.015 (DFS) 0.029 (OS)	2.68 (DFS) 2.47 (OS)	[70]
	88	Increased expression in CRC compared to adjacent normal tissue	<0.0001		[72]
	84 (test cohort) 113 (validation cohort)	High expression in tumors associated with poor survival prognosis independent of tumor staging	0.008	2.7	[110]
	15	Expression associated with clinical progression	0.008		[95]
	156	High expression associated with worse OS and DFS	0.001 (OS) 0.007 (DFS)	0.335 (OS) 0.316 (DFS)	[122]
	113	High expression associated with poor prognosis	0.0005	3.0	[125]
	129	High expression associated with shorter DFS	0.004	1.28	[124]
	46	High expression associated with shorter DFI	0.0026		[123]
miR-675	20	Significantly increased in CRC compared to non-cancerous tissue	0.019		[76]
miR-92a	88	Significantly increased in CRC compared to normal tissue	<0.0001		[72]
	37	Elevated expression in advanced adenomas compared to normal controls	<0.0001		[148]

(continued)

Table 10.1 (continued)

miRNA(s)	Patients in cohort (n)	Expression/outcome	P-value	HR	Reference
miR-135a	15	Increased expression associated with clinical progression	0.032		[95]
	43	Tumors showed increased expression compared to normal colonic epithelium	<0.0001		[73]
miR-135b	43	Tumors showed increased expression compared to normal colonic epithelium	<0.0001		[73]
	125	Increased expression in CRC tumor tissue; associated with higher pre-operative serum levels of CEA, CA19-9	<0.001; 0.0338 (CEA), 0.0360 (CA19-9)	0.33; 0.42 (CEA), 0.41(CA19-9)	[118]
miR-143	20	Decreased expression in tumors	0.003		[81]
	77	Low expression is independent prognostic factor of cancer-specific survival in KRAS WT patients	<0.031	1.86	[85]
	34	Increased expression associated with shorter PFS in patients with KRAS mutated tumor	0.04	1.59	[153]
miR-148b	96	Decreased expression in CRC tissue	<0.0001		[90]
miR-345	31	Decreased expression in CRC tissue associated with higher rate of lymph node metastases, worse histological type	0.037, 0.040		[91]
miR-17-92 cluster	55	All 6 miRNAs overexpressed during colorectal adenoma to adenocarcinoma progression	miR-17: 0.001 (FC 2.6)		[105]
			miR-18a: 0.04 (FC 2.4)		
			miR-19a: <0.001 (FC 3.4)		
			miR-20a: 0.001 (FC 2.6)		
			miR-19b-1: 0.021 (FC 1.6)		
			miR-92a-1: <0.001 (FC 4.5)		
miR-21-5p, miR-20a-5p, miR-103a-5p, miR-106b-5p, miR-143-5p, miR-215	138	High-risk patients have greater likelihood of recurrence and lower 5-year DFS	<0.0001	4.24	[107]

(continued)

Table 10.1 (continued)

miRNA(s)	Patients in cohort (n)	Expression/outcome	P-value	HR	Reference
miR-17	185	Elevated expression in tumors indicates shorter OS	0.002	2.41	[130]
	48	High expression leads to reduced OS, associated with risk of death	0.007	2.67	[128]
miR-215	34	Decreased expression in stage II and III colon tumors but high expression levels associated with poor OS	0.025	3.516	[138]
	107	Decreased expression in CRC tissue	<0.001		[137]
miR-16	143	Decreased expression is independent prognostic factor indicating lower 5-year OS	0.018	1.67	[146]
	126	Decreased expression associated with shorter DFS, OS	0.01 (DFS, OS)	2.598 (DFS) 2.912 (OS)	[24]
miR-15a	126	Decreased expression associated with shorter DFS, OS	0.01 (DFS, OS)	2.782 (DFS) 3.016 (OS)	[24]
miR-29a	110	High expression associated with longer DFS in stage II patients	0.0043	0.194	[147]
miR-93	77	Expression decreased in early relapse patients	<0.0001 (FC 6.1)		[152]
miR-155	156	High expression associated with shorter DFS, OS	0.023 (DFS) 0.014 (OS)	0.387 (DFS) 0.427 (OS)	[122]
	109	Increased expression compared to normal mucosa	0.005 (FC 2.3)		[156]
miR-148a	273	Low expression associated with shorter DFS and poorer OS	0.017 (DFS) 0.014 (OS)	1.83 (DFS) 1.93 (OS)	[133]
miR-141	258	Increased expression in plasma correlates with poor survival	0.016	2.4	[92]
miR-320	37	Low expression correlated with probability of RFS in stage II patients	0.002	6.6	[102]
miR-498	37	Low expression correlated with probability of RFS in stage II patients	0.03	11.5	[102]
miR-556	50	High expression associated with short DFS in stage II patients	0.018	2.0	[70]
miR-200b	34	Increased expression correlates with better PFS in patients with KRAS-mutated tumors	0.01	5.6	[153]

HR hazard ratio, *DFS* disease-free survival, *DFI* disease-free interval, *FC* fold change, *PFS* progression-free survival, *RFS* recurrence-free survival

Table 10.2 Current prognostic lncRNAs in colorectal cancer

lncRNA	CRC patients (n)	Outcome	P-value	HR/RR	Reference
PVT-1	Total: 164	↑expression shows poorer prognosis	0.0101	2.532	[158]
	High expression: 131	Independent indicator for OS	0.016		
	Low expression: 33				
91H	Total: 72	↑ expression shows poorer prognosis	<0.001	3.66	[159]
	High expression: 42	Independent indicator for OS	0.001		
	Low expression: 30				
MALAT-1	Total: 146	↑ expression shows poorer prognosis	0.003	3.968; 2.863	[160]
	High expression: 73	Independent indicator for OS; DFS	0.002; <0.001		
	Low expression: 73				
HOTAIR	Total: 100	↑ expression shows poorer prognosis	0.0046	5.62	[162]
	High expression: 20	Independent indicator for OS	0.008		
	Low expression 80				
PCAT-1	Total: 108	↑ expression shows poorer prognosis	<0.001	3.12	[163]
	High expression: 58	Independent indicator for OS	0.007		
	Low expression: 50				
NEAT1	Total: 239	↑ expression shows poorer prognosis	<0.001	1.7; 1.8	[164]
	High expression: 110	Independent indicator for OS; DFS	0.005; 0.001		
	Low expression: 129				
ncRAN long variant	Total: 81	↓ expression shows poorer prognosis	0.014	0.192-3.872	[165]
	High expression: 49	Independent indicator for OS	0.024		
	Low expression: 32				
ncRAN short variant	Total: 81	↓ expression shows poorer prognosis	0.02		[165]
	High expression: 46				
	Low expression: 35				
LOC285194	Total: 81	↓ expression shows poorer prognosis	0.01	0.337	[166]
	High expression: 33	Independent indicator for DSS	0.034		
	Low expression: 48				
GAS5	Total: 66	↓ expression shows poorer prognosis	<0.001	0.036	[167]
	High expression: 33	Independent indicator for OS	0.034		
	Low expression: 33				
MEG3	Total: 62	↓ expression shows poorer prognosis	<0.001	0.133	[168]
	High expression: 31	Independent indicator for OS	0.049		
	Low expression: 31				

HR hazard ratio, *RR* relative risk

Although it is without question that genetic and epigenetic aberrations play a central role in the development of human disease, recent findings in the past decade have shed light on the role of non-protein-coding genomic regions in the formation of cancer. These regions often pertain to so-called non-coding RNAs (ncRNAs). Identification of the proteins and pathways associated with CRC and regulated by these ncRNAs could provide new opportunities for refining prognostic and diagnostic applications and could potentially facilitate improved patient stratification [38]. NcRNAs fit under various classifications; such categories include members of the PIWI-interacting RNA (piRNA) family, the small nucleolar RNA (snoRNA) family, the large intragenic noncoding RNA (lincRNA) family and the long noncoding RNA (lncRNA) family, but the most well-characterized and extensively studied are microRNAs (miRNAs), which have been shown to perform essential functions during tissue development, cell differentiation and proliferation, as well as survival [39–41]. Numerous recent publications have focused on the significance of miRNAs in the development, classification, diagnosis and prognosis of CRC [42–47]. Dysregulated miRNAs typically witnessed in cancer could serve as biomarkers and could potentially be used in combination biomarker panels in order to enhance current diagnosis and prognosis methods of CRC patients, thereby increasing both sensitivity and specificity.

10.2 miRNAs in CRC

10.2.1 miRNA Expression in Normal vs. Diseased Tissue and Phenotypic Classification

It has also been proposed that miRNA profiling could greatly contribute to the diagnostic and prognostic classification of human cancers. Whole-genome miRNA profiling has demonstrated that the expression of miRNA is drastically different in most types of cancer, as it is often tissue-specific, and that expression of miRNAs offers information about the pathophysiological state of a person [48–58]. Indeed, many studies have validated unique CRC microRNA profiles which could accurately distinguish between malignant tissues and benign colorectal mucosa [59–62]. Many individual miRNAs with abnormal expression in CRC have already been described in the literature as the search for both diagnostic and prognostic biomarkers continues among researchers and clinicians. Classifying the regulatory role of these miRNAs has proven difficult, nevertheless, as they are known to act on multiple mRNA targets and since their expressional status is specific to the type of cancer and tissue; however, the trend in the field has been to categorize expression as overexpressed, implicating an oncogenic role of the miRNA, or as underexpressed, suggesting that these miRNAs act as tumor suppressors.

There are numerous examples of overexpressed miRNAs in CRC. Perhaps the most well-established and well-studied oncogenic miRNA in this particular disease is miR-31. It is one of the most upregulated miRNAs in colorectal neoplasms and has been associated with tumor lymph node metastasis stage, in particular the pT stage and deep tumor invasion processes, and thereby implicates advanced disease stage [14, 43, 63]. Interestingly, low expression of miR-31 is mainly observed in poorly differentiated tumors [64–67]. Expression measurements of another miRNA, miR-106a, have demonstrated that it is also one such miRNA highly expressed in metastatic colorectal cancer cells and is known to play a role in both migration and invasion [68–70]. Numerous recent reports have also identified the upregulation of miR-21 and its association with metastasis, particularly distant metastasis involving the liver, as well as how its upregulation correlates with reduced expression of the gene encoding the tumor suppressor protein *PDCD4* [70–74]. Further miRNAs, such as miR-92a, miR-96, miR-135a, miR-135b, and miR-183 have been found to be significantly higher in CRC tissues when compared to adjacent normal tissue; miR-135a and miR-135b upregulation in particular is associated with dowregulation of the adenomatous polyposis coli (*APC*) gene, a loss of function which triggers a chain of events involving molecular and histological changes [72, 73, 75]. Furthermore, upregulation of miR-675 in CRC tissue promotes aggressive tumor cell growth and has been shown to regulate the cell cycle by targeting retinoblastoma (*RB*), a known tumor suppressor [76].

There are several examples of downregulated tumor suppressor miRNAs in CRC, although these occurrences are less common in comparison to the overexpressed cases. Perhaps the most well-known are those of the let-7 family, which include let-7a, let-7b, let- 7c, let-7d, let-7e, let-7f, let-7 g, let-7i, and miR-98, which are known to target the kirsten rat sarcoma viral oncogene homolog (*KRAS*) [77–79]. MiR-143 is another such example that is often downregulated in colorectal neoplasms, particularly at stages of liver invasion, and its reduced expression promotes the invasion and migration of malignant cells and correlates with the aggressive mucinous phenotype [80–89]. Further studies have demonstrated that significantly downregulated expression of miR-148b in CRC tissues directly correlates with tumor size [90]. miR-345 is also significantly downregulated in over 50 % of colorectal neoplasms and its low expression could serve as an indicator for both lymph node metastasis and unfavorable histological classifications [91].

Further evidence which demonstrates the advanced phenotypic classification of miRNA expression patterns can be seen in tumors harboring *KRAS* mutations, which occur in 35–45 % of CRC cases; these tumors have shown altered miRNA expression patterns as well [57, 93, 94]. One study found that KRAS-mutated CRC cell lines exhibited overexpression of miR-9, miR-95, miR-148a, miR-190, and miR-372 when compared to human normal, healthy colon cells [43]. Furthermore, another group examining the let-7 family members showed that let-7a expression is upregulated in metastatic CRC harboring *KRAS* mutations when compared to normal mucosa and non-metastatic disease [94].

There has even been evidence which suggests that such miRNA expression signatures might provide more accurate subtype classification than their protein-coding RNA counterparts and conventional cytology approaches [95, 96]. Liu et al. demonstrated this by showing how miRNA profiles could identify tumors of unknown origin with more success than when applying mRNA profiles [97]. This proves to be very promising, as it can often be very difficult to locate the origin of the tumor in cases with many growing metastases. Given the fact that it can often be challenging to obtain tumor tissue for analysis, miRNAs have thus been recognized as an attractive source of information for accurate diagnosis and prognosis and as predictors of tumors. For these reasons, miRNA expression profiles are currently being used to classify tumors based on both tissue type and disease stage [98–101]. Rosenfeld et al. utilized one such pattern in a blind test set to accurately predict the origin of tumor tissue in 86 % of cases, of which 77 % were metastatic cases [52]. Moreover, since various phenotypic subtypes of CRC can be discriminated using gene expression data from microarray platforms, miRNA expression patterns may likewise help classify these important subgroups, which include microsatellite instability (MSI), *TP53* status, and *KRAS* mutation status. Because these subgroups vary in terms of therapeutic response, projected survival also tends to vary and this classification of subgroups could prove important. miRNAs, for example, are expressed differently in microsatellite stable (MSS) and MSI tumors and these patterns can therefore accurately classify a tumor as MSS or MSI [56, 102, 103]. Several miRNAs investigated for any potential association with CRC MSI status were indeed shown to be differentially expressed in CRC tissue when compared to normal mucosa [43, 56, 104, 110]. In addition, some studies have demonstrated that higher expression levels of the miR-17-92 cluster, comprised of six miRNAs (miR-17, miR-18a, miR-19a, miR-20a, miR-19b-1, and miR-92a-1), are associated with MSS colorectal tumors, transformation of colonic epithelium, as well as progression of adenoma to carcinoma [46, 56, 105]. Another study by Schepeler et al. verified that expression levels of miR-320 and miR-498 were significantly lower in MSS in comparison to normal tissue, whereas upregulation of miR-20a and miR-92 correlated with survival with low probability of recurrence [102].

10.2.2 Prediction of Recurrence and Clinical Outcome

Characteristic miRNA signatures have already been revealed by expression profiling analyses that can predict the clinical outcomes of CRC cases [106]. Zhang et al. for example, developed and validated a 6-miRNA signature that was able to improve the prediction of disease recurrence in stage II colon cancer patients following resection. This prognostic tool successfully categorized patients as either high-risk or low-risk and better predicted patient survival in comparison to mismatch repair status and other applied clinicopathological risk factors [107]. In another investigation, Cheng et al. was able to generate a 3-miRNA signature using ANN analysis and an RT-qPCR-based microarray platform to predict tumor status in stage II CRC cases with 100 % accuracy. This distinct miRNA signature included miR-139-5p, miR-31, and miR-17-92 and was able to distinguish between normal and tumor tissues [92]. This demonstrated the potential of these three biomarkers to more effectively stratify patients with an increased risk of disease recurrence to help guide adjuvant therapy. Numerous reports have also examined the relation of miRNA expression to disease-free survival (DFS) and patient overall survival (OS). Perhaps the most extensively studied case involves miR-21 expression. The main targets of miR-21 in CRC cells include nuclear factor 1 b-type (*NFIB*), Sprouty2 (*SPRY2*) and programmed cell death protein 4 (*PDCD4*), a tumor suppressor protein which is responsible for inhibiting neoplastic transformation and thereby invasion and intravasation as well. For these reasons, *PDCD4* suppression, as a result of miR-21 expression, is associated with poor prognosis for CRC patients [108, 109]. In addition, another study demonstrated that tumors which showed upregulation of miR-21 were associated with poorer prognosis regarding survival and poor therapeutic outcome [110]. This may perhaps be explained by the fact that miR-21 expression dictates epithelial invasion and expression increases as the disease advances; high expression levels thus correlate with decreased recurrence-free survival and shorter OS [80, 95, 111–122]. A report by Yu et al. also confirmed the correlation between high miR-21 expression and clinical CRC stage, i.e. LNM and distant metastasis [114]. There have been several international and extensive studies validating the robustness of miR-21 as an early-stage biomarker for the identification of individuals with a high risk of cancer progression who currently show no signs of metastasis or advanced disease. The first report involved an American cohort of 84 CRC patients evaluated via microarray analyses and a Chinese cohort of 113 Chinese patients evaluated via qRT-PCR [110]. Both cohorts exhibited elevated miR-21 expression levels in cases in which the tumor was associated with worse survival prognosis and therapeutic outcome. Moreover, this association was also significant in TNM stage II CRC patients and these conclusions were drawn independent of any staging or clinical characteristics, thus truly demonstrating the prognostic and predictive potential of this biomarker. Since the conduction of this study, the relation of increased miR-21 expression to poor survival outcomes has been further validated by three additional research groups, including 156 CRC patients in Japan, 46 in the Czech Republic and 130 patients in Denmark [122–124]. These studies provide incredible evidence of miR-21 expression as a potent prognostic classifier for CRC, especially given the diverse ethnic nature of the populations investigated and the variety of technical approaches used to generate the data [9]. Further strengthening this evidence, a study by Schetter et al. in 2009 applied the predictive power of this expression data in combination with a classifier involving genes related to inflammation, enabling a significant improvement in stratifying patients according to risk and cancer-specific death [125].

A variety of studies have demonstrated the use of a combination of miRNAs to determine how their expression correlates with survival. Several of these such studies have been able to verify that the expression of three individual miRNAs, miR-17, miR-135a and miR135b, is associated with poor survival, as they all contribute to tumor cell proliferation, growth and progression of the cell cycle by targeting their respective tumor suppressors [80, 95, 118, 126–132]. As a result, expression of these particular miRNAs could be used as a prognostic marker for predicting clinical stage, liver metastasis, DFS and OS. Conversely, low expression of miR-148a and miR-215 has been associated with a significantly shorter DFS and OS and could potentially supplement the predictive capabilities of overexpressed miR-17, miR-135a and miR135b [118, 133–138]. Moreover, the expression level of miR-215 could be used as an independent predictive marker for relapse.

Many recent studies have focused on unveiling prognostic miRNA tools to be able to predict overall patient outcome. As in other types of cancer, miRNAs encoded by the miR-15a/16 cluster are often also either deleted or downregulated in CRC, indicating that they play a role in tumor suppression [139–144]. Ma et al. was able to demonstrate that miR-16 overexpression, for example, harnessed the intrinsic apoptosis pathway to inhibit CRC cell proliferation and induce apoptosis, whereas Qian et al. demonstrated that downregulation of miR-16 predicted poor prognosis [145, 146]. Furthermore, another study associated the low expression of both individual and combined miR-15a and miR-16 with advanced TNM stage, poor histological grade and positive lymph node metastasis. This provides important implications for the use of aberrant miR-15a and miR-16 expression status for the improved stratification of CRC patients with aggressive tumors and for also determining which patients will have worse prognosis following surgery [24]. Another study by Weissman-Brenner et al. suggested the use of miR-29a as a prognostic tool for stage II CRC patients who underwent resection [147]. Levels of miR-29a expression were significantly higher in patients without recurrence within three years of surgery of their primary tumor when compared to those who experience recurrence in this time frame. Furthermore, upregulation of miR-29a was associated with decreased risk of recurrence and improved duration of DFS. This study compared the expression of miR-29a in the surgically resected tissue of CRC patients who did and did not experience recurrence to successfully confirm the prognostic value of this particular miRNA.

There have been numerous other reports identifying aberrantly expressed miRNAs in CRC which were related to poor survival or which portended good prognosis. Skog et al. for example, identified the differential expression of 37 miRNAs, among which 5 highly expressed miRNAs (miR-20a, miR-21, miR-106a, miR-181b and miR-203) were associated with poor survival in the evaluated cohort [105]. In addition, it has been suggested that miR-15b, miR-21, miR-181b, miR-191, and miR-200c are also linked to the development and progression of CRC and are therefore putative prognostic CRC markers [148, 149]. Other miRNAs such as miR-92 and miR-17-3p, both which are encoded by the miR-17-92 cluster, have similarly been implicated in the progression of tumor cell growth and suppression of apoptosis and it has been estimated that miR-92 could serve as an early predictive marker of CRC with a sensitivity and specificity of 89 % and 70 %, respectively [150, 151]. In contrast to this example of tumor cell proliferation regarding the miR-17-92 cluster, a somewhat recent study by Yang et al. reported the significant differential expression of miR-93 in CRC patients and demonstrated that this particular miRNA could inhibit tumorigenesis and reduce the recurrence of CRC [152].

10.2.3 Prediction of Response and Resistance to Treatment

miRNA profiling can further be applied to identify treatment-resistant cancers and to predict the response of CRC patients to various treatment regimens, as miRNA expression is closely related to the efficacy of therapy. Kong et al. conducted a study in which they linked the deregulation of four miRNAs to CRC resistance to EGFR-targeted agents [57]. Ragusa et al. compared the expression of 667 miRNAs in the Caco-2 and HCT-116 CRC cell lines, which are sensitive to cetuximab and resistant to cetuximab, respectively, and found approximately 20 miRNAs in each cell line that were differentially expressed [149]. This would suggest a potential signature for the prediction of a successful therapeutic response of CRC patients undergoing anti-EGFR treatment. One study also demonstrated the predictive prowess of a multi-miRNA-based classifier derived from the LASSO Cox regression model which could determine disease-free survival as well as benefit from adjuvant chemotherapy in stage II CRC patients who had already been subjected to surgery [47]. This signature was tested in two internal patient cohorts as well as validated in an independent patient group to confirm its prognostic and predictive accuracy. This study thus highlighted the existence of an miRNA panel with promising potential for supplementing current approaches regarding prognosis and prediction of response in CRC. Mekenkemp et al. proposed that upregulation of miR-200b expression could be useful for predicting which patients harboring a *KRAS* mutation would most likely benefit from EGFR-targeted therapy, a topic which is still highly debated in the clinic [153]. Although there is evidence that miR-143 might potentially target *KRAS* and that miR-143 expression levels have demonstrated prognostic power in *KRAS* wild-type CRC patients, it has unfortunately not been proven that it could serve as a predictive marker for anti-EGFR treatment [85]. Several investigations conducted showed that increased expression of the oncogenic miR-155, which is known to promote the proliferation, migration and invasion of tumor cells, has been correlated with increased chemoresistance and therefore poor prognosis [122, 154–156]. Another example of association with poor prognosis, which was demonstrated by several groups, showed that down-regulation of miR-148a is not only associated with a significantly shorter DFS, but also indicates poor therapeutic response and OS [133–135]. Expression levels of miR-148a could therefore serve as markers of disease progression and have important implications in predicting response to chemotherapy regimens. As the list of potentially predictive miRNAs regarding treatment successes or failures continues to grow, extensive research on the reliability of each miRNA predictor will need to be performed before being implemented in the clinical setting.

10.3 LncRNAs in CRC

Long non-coding RNAs (LncRNAs) have shown promise in several therapeutic applications involving cancer. Evidence has shown that lncRNAs may play an important role as biomarkers for predicting prognosis with respect to survival, recurrence, and chemotherapeutic drug response in multiple cancer types, including CRC. MiRNAs are a similar class of molecules, and have been shown to circulate throughout the body in plasma, serum, and other fluids, making their utility as biomarkers much more accessible. It is plausible that lncRNAs share this characteristic, and their circulating levels may be easily accessed for therapeutic uses. In addition, research has shown that lncRNAs may be secreted in exosomes from cells, making them specific and easily accessible biomarkers [157]. This subsection details the current knowledge and information regarding the prognostic use of lncRNAs in CRC.

10.3.1 Up-Regulated lncRNAs in CRC and Influence on Prognosis

Several studies have conducted analyses of tissue-based expression levels of certain lncRNAs and applying classical methods of statistics (i.e. Kaplan-Meier survival curves and Cox proportional models) indicated the significance of lncRNA expression levels on CRC patient survival. The up-regulation of a number of lncRNAs in tumor tissue which display oncogene-like properties have shown to be associated with poorer prognoses in many such patients. These lncRNAs therefore have the potential to serve as biomarkers for prognostic purposes in CRC. Further research and replicative studies will help determine the usability of these molecules in clinical applications.

PVT-1 A study involving 164 CRC patient samples indicated that those with a higher expression of the lncRNA PVT-1 showed a poorer prognosis compared to patients with lower levels of PVT-1 ($P=0.0101$). PVT-1 was also shown to act as a significant independent factor for predicting overall survival in these patients (Hazard ratio: 2.532, $P=0.016$). It was further shown that PVT-1 was associated with apoptosis, likely through regulation of TGF-beta signaling, proliferation, and invasion of CRC cells. An association between elevated expression of PVT-1 in CRC patients and lymph node metastasis and venous invasion was also described in this study [158].

91H 91H is another example of a lncRNA that has been found to be significantly elevated in CRC cells compared to healthy samples ($P<0.001$). A study involving 72 patients with CRC showed that higher levels of 91H were predictive of a poorer prognosis ($P<0.001$), and 91H holds true as an independent factor for predicting overall survival in CRC patients (Hazard ratio: 3.66, $P=0.001$). This study has shown that repression of 91H expression in CRC cells leads to a reduced cell proliferation, migration, and invasion, indicating its potential role in CRC progression [159].

MALAT-1 A univariate analysis investigating the prognostic potential of the well-known lncRNA MALAT1 in CRC has shown that higher levels of this molecule were significantly associated with both decreased overall survival ($P = 0.003$) and disease free survival ($P=0.001$). Multivariate analysis further showed that MALAT-1 was able to serve as an independent prognostic indicator of both overall survival (Hazard ratio: 3.968, $P=0.002$) and disease-free survival (HR: 2.863, $P<0.001$). MALAT-1 has been implicated in reducing apoptosis, promoting EMT, and enhancing invasion in cervical, bladder, and lung cancers, respectively. Increased levels of MALAT-1 have also been found to be associated with increased colony formation and invasiveness in CRC cells, yet the exact biological mechanisms of this lncRNA's pathological effects in CRC are still under investigation [160].

HOTAIR A study by Kogo et al. has shown that an increased level of the lncRNA HOTAIR within CRC tissue is associated with a poorer prognosis ($P=0.0046$). An elevated expression of this lncRNA was found significantly associated with an increased number of liver metastases ($P=0.006$), and increased levels of HOTAIR were found correlated with increased invasiveness of CRC cells. Furthermore, studies have shown that HOTAIR can serve as an independent prognostic factor in CRC (Relative risk: 5.62, $P=0.008$). Interestingly, CRC cells showed significantly lowered levels of HOTAIR expression compared to healthy controls ($P=0.002$). This points to the complexity of lncRNAs and the pleomorphic effects they may exert. Studies have shown that HOTAIR may be involved in the widespread targeting of polycomb repressive complex 2 (PRC2), affecting the expression of numerous tumor-suppressive genes and oncogenic genes [161]. A more recent study has looked at the differential expression of HOTAIR in CRC patient peripheral blood samples. Kaplan-Meier analysis has found that an increased level of HOTAIR was again correlated with decreased survival and poorer prognosis ($P=0.008$) [162].

PCAT-1 Increased expression of the lncRNA PCAT-1 has been found to be significantly associated with a decreased survival in CRC patients ($P<0.001$) as well as being able to serve as an independent prognostic factor (Hazard ratio=3.12, $P=0.007$) shown through multivariate analysis. The increased expression of PCAT-1 was also correlated with distant metastasis in this study, indicating a potential role of this lncRNA in the promotion of metastases. In prostate cancer, PCAT-1 has been shown to exert some of its pathological effects through its association with PRC2; however no such association has yet been discovered in CRC [163].

NEAT1 In a study involving 239 CRC clinical samples, it was found that increased expression of the lncRNA NEAT1 was associated with a decreased overall ($P<0.001$) and disease-free

(P<0.001) survival. NEAT1 was furthermore indicated as an independent prognostic factor in CRC by multivariate analyses for both disease-free (Adjusted hazard ratio = 1.8) and overall (Adjusted hazard ratio = 1.7) survival. The pathogenic effects of this lncRNA have been thought to involve the regulation of genes associated with CRC differentiation, invasion, and metastasis [164].

CCAT2 Increased levels of the lncRNA CCAT2 in CRC cells have recently been shown to be associated with a higher rate of migration, invasion, and metastasis. While no survival or prognostic analyses have yet been conducted for this lncRNA in CRC, elevated CCAT2 levels were found to be associated with decreased disease-free survival in breast cancer patients. Further research with CCAT2 is required to verify its prognostic capabilities with regards to CRC, but the current research indicates a strong possibility for its potential use as a prognostic marker. CCAT2 is thought to induce its pathological effects through its association with WNT signaling and its upregulation of MYC [174].

10.3.2 Down-Regulated lncRNAs in CRC and Influence on Prognosis

Decreased levels of lncRNAs with tumor-suppressor-like features have also shown prognostic value with regards to CRC patients. These molecules also have the potential ability to serve as clinical biomarkers for CRC patients.

ncRAN Kaplan-Meier analysis has shown that lowered levels of both the long and short variants of the lncRNA ncRAN are associated with a decreased overall survival in CRC patients (P=0.014; P=0.020). Decreased levels of this lncRNA were shown to be associated with CRC cells that were more metastatic. Furthermore, ncRAN was found to act as a tumor suppressor in CRC cells by inhibiting invasion and migration. Multivariate analysis has further shown that the long variant of this lncRNA can serve as an independent prognostic indicator for CRC patients (P=0.024) [165].

LOC295194 CRC patients with a decreased level of LOC295194, another lncRNA, have been found to have a poorer prognosis compared to patients with higher levels of this lncRNA (P=0.010). Levels of LOC295194 have also shown predictive capabilities with regards to disease-specific survival in CRC patients (P=0.019). This lncRNA has further shown to be able to serve as an independent prognostic factor for disease-specific survival in CRC patients via multivariate analysis (P=0.034). LOC295194 is thought to exert its tumor-suppressive effects by inhibiting tumor growth and metastasis. It was found that lower levels of this molecule were associated with an increased tumor size, increased incidence of metastases, and a higher TNM stage [166].

GAS5 Studies have found that GAS5 acts as a lncRNA tumor-suppressor in CRC cells. It has been shown that decreased levels of this lncRNA are associated with a poorer prognosis in CRC patients, and multivariate analyses have shown that this lncRNA can serve as an independent prognostic marker for CRC (P=0.034). GAS5 is thought to regulate CRC cell proliferation, and decreased levels have been found to be associated with decreased differentiation, an increased tumor size, and an increased TNM stage [167].

MEG3 Recent studies have also indicated that decreased levels of the lncRNA MEG3 are associated with a poorer prognosis in CRC (P<0.001). In addition, multivariate analyses have indicated that this lncRNA can serve as an independent predictor of prognosis in CRC patients (P=0.049). MEG3 potentially decreases CRC cell proliferation and metastasis, and decreased levels have been shown to be related to increased tumor stage, decreased tumor differentiation, and increased tumor depth of invasion [168].

10.3.3 Large Scale Data Analyses Involving lncRNAs in CRC Prognosis

Research techniques have also utilized data mining of microarray gene expression compilations in search of lncRNAs with potential prognostic capabilities. A study utilizing such a technique has found a 6-lncRNA profile that is prognostic for CRC survival, regardless of TNM staging factors or prior chemotherapeutic treatment [169]. Meta-analyses of numerous literary and scholarly sources have also helped elucidate the usefulness of lncRNAs in CRC prognosis. The association between elevated levels of MALAT-1 in CRC patients and a poorer prognosis has been shown through this method of research [170]. Additionally, a meta-analysis of HOTAIR's involvement in CRC (as well as several other cancers) has shown that this lncRNA may serve as a biomarker for lymph node metastasis, which may contribute to a decreased survival in patients [171]. Furthermore, lncRNA databases have been developed to assist researchers and educators with their understanding of recent lncRNA developments. Several of these tools currently exist, and studies have been conducted to determine those with the most utility and reliability. One of the largest databases, termed Noncode, contains about 200,000 lncRNAs. Another database that manually maintains a collection of functional lncRNAs from published literature is lncRNAdb. Numerous other databases contain lncRNA-specific information, as well as more comprehensive information relating to all types of non-coding RNAs. Some additional popular databases include lncRNABase, ChIPBase, LNCipedia, MONOCLdb, lncRNome, and NRED.

10.3.4 Predictive Ability of lncRNAs to Drug Response in CRC

Due to the pleiotropic and widespread effects lncRNAs have in CRC, it has been hypothesized that altered levels of these molecules may serve as predictive factors for chemotherapeutic drug response. A related class of molecules, miRNAs, have shown numerous associations with drug response in CRC, so it is likely that many such associations also exist with lncRNAs. As of now, however, only a handful of studies have suggested such a relationship. One study has indicated that the lncRNA snaR has a contributing role in the development of resistance to 5-FU in CRC cells. Decreased levels of this molecule have been shown to be associated with a lowered sensitivity to 5-FU. The exact roles of snaR are not yet known, but it is hypothesized to regulate some form of cellular growth [172]. While these findings are still preliminary, future research is promising. One study utilizing microarray analysis techniques demonstrated that 2662 lncRNAs were differentially expressed between parental CRC cells and those that were resistant to 5-FU. In addition, studies have shown that altered levels of the lncRNAs UCA1 and HOTAIR are associated with chemotherapeutic resistance to cisplatin in bladder cancer and lung adenocarcinoma, respectively. Oxaliplatin, which is commonly used in the treatment of CRC, is a chemotherapeutic drug in the same class as cisplatin. It is therefore plausible that these lncRNAs may also play a role in the development of resistance to oxaliplatin in CRC patients. Studies such as these highlight the importance of lncRNAs in the development of drug resistance and entice further investigations in this field [173].

10.4 Future Work

Research involving the prognostic capabilities of miRNAs and lnCRNAs in CRC is still preliminary in its advances. Although there has been extensive miRNA research in the past decade, only a handful of studies detailing the association between lncRNA molecules and CRC have been published. The findings for both miRNAs and lncRNAs, however, are in need of replication and further investigation. Furthermore, novel ncRNAs with respect to CRC pathology and prognosis will continue to be elucidated, and the growing knowledge base will likely prove advantageous in the development of novel prognostic tools for the treatment of CRC. While this is indeed an exciting new field in cancer research, a great deal more work is required before these discoveries can realistically be put into clinical practice.

References

1. Vladimír Janout HK. Epidemiology of colorectal cancer. Biomed Papers. 2001;145(1):5–10.
2. Siegel R, DeSantis C, Jemal A. Colorectal cancer statistics, 2014. CA Cancer J Clin. 2014;64(2):104–17.
3. World Cancer Research Fund and American Institute for Cancer Research. Food, Nutrition, Physical Activity, and the Prevention of Cancer: A Global Perspective. Washington, D.C.: 2007.
4. Ferlay J, Shin HR, Bray F, Forman D, Mathers C, Parkin DM. Estimates of worldwide burden of cancer in 2008: GLOBOCAN 2008. Int J Cancer. 2010;127:2893–917.
5. Muratore A, Zorzi D, Bouzari H, Amisano M, Massucco P, Sperti E, et al. Asymptomatic colorectal cancer with un-resectable liver metastases: immediate colorectal resection or up-front systemic chemotherapy? Ann Surg Oncol. 2007;14(2):766–70.
6. Van Cutsem E, Nordlinger B, Adam R, Köhne CH, Pozzo C, Poston G, et al. Towards a pan-European consensus on the treatment of patients with colorectal liver metastases. Eur J Cancer. 2015;42(14):2212–21.
7. Cheng HY, Zhang L, Cogdell D, Zheng H, Schetter A, Nykter M, et al. Circulating plasma MiR-141 is a novel biomarker for metastatic colon cancer and predicts poor prognosis. PLoS One. 2011;6:e17745.
8. Pichler M, Winter E, Stotz M, Eberhard K, Samonigg H, Lax S, et al. Down-regulation of KRAS-interacting miRNA-143 predicts poor prognosis but not response to EGFR-targeted agents in colorectal cancer. Br J Cancer. 2012;106:1826–32.
9. Schetter AJ, Okayama H, Harris CC. The role of microRNAs in colorectal cancer. Cancer J. 2012;18:244–52.
10. Gunderson LL, Jessup JM, Sargent DJ, Greene FL, Stewart AK. Revised TN categorization for colon cancer based on national survival outcomes data. J Clin Oncol. 2010;28:264–71.
11. Poste G. Bring on the biomarkers. Nature. 2011;469:2.
12. Poste G, Carbone DP, Parkinson DR, Verweij J, Hewitt SM, Jessup JM. Leveling the playing field: bringing development of biomarkers and molecular diagnostics up to the standards for drug development. Clin Cancer Res. 2012;18:1515–23.
13. Henry NL, Hayes DF. Cancer biomarkers. Mol Oncol. 2012;6:140–6.
14. Xuan Y, Yang H, Zhao L, Lau WB, Lau B, Ren N, et al. MicroRNAs in colorectal cancer: small molecules with big functions. Cancer Lett. 2015;360:89–105.
15. Thorsteinsson M, Jess P. The clinical significance of circulating tumor cells in non-metastatic colorectal cancer – a review. Eur J Surg Oncol. 2011;37:1–7.
16. Klump B, Nehls O, Okech T, Hsieh CJ, Gaco V, Gittinger FS, et al. Molecular lesions in colorectal cancer: impact on prognosis? Original data and review of the literature. Int J Colorectal Dis. 2004;19:23–42.
17. Wallner M, Herbst A, Behrens A, Crispin A, Stieber P, Goke B, et al. Methylation of serum DNA is an independent prognostic marker in colorectal cancer. Clin Cancer Res. 2006;12:7347–52.
18. Chua YJ, Zalcberg JR. Progress and challenges in the adjuvant treatment of stage II and III colon cancers. Expert Rev Anticancer Ther. 2008;8(4):595–604.
19. Wolpin BM, Mayer RJ. Systemic treatment of colorectal cancer. Gastroenterology. 2008;134(5):1296–310.
20. Guerra A, Borda F, Javier Jimenez F, Martinez-Penuela JM, Larrinaga B. Multivariate analysis of prognostic factors in resected colorectal cancer: a new prognostic index. Eur J Gastroenterol Hepatol. 1998;10:51–8.
21. Steinberg SM, Barwick KW, Stablein DM. Importance of tumor pathology and morphology in patients with surgically resected colon cancer. Findings from the Gastrointestinal Tumor Study Group. Cancer. 1986;58:1340–5.
22. Wolters U, Stutzer H, Keller HW, Schroder U, Pichlmaier H. Colorectal cancer-a multivariate analysis of prognostic factors. Eur J Surg Oncol. 1996;22:592–7.
23. Smits KM, Paranjape T, Nallur S, Wouters KA, Weijenberg MP, Schouten LJ, et al. A let-7 microRNA SNP in the KRAS 3′UTR is prognostic in early-stage colorectal cancer. Clin Cancer Res. 2011;17(24):7723–31.
24. Xiao G, Tang H, Wei W, Li J, Ji L, Ge J. Aberrant Expression of MicroRNA-15a and MicroRNA-16 Synergistically Associates with Tumor Progression and Prognosis in Patients with Colorectal Cancer. Gastroenterol Res Pract. 2014;8.
25. Erstad DJ, Tumusiime G, Cusack J,Jr. Prognostic and Predictive Biomarkers in Colorectal Cancer: Implications for the Clinical Surgeon. Annals of surgical oncology. 2015;22(11).
26. Compton CC, Fielding LP, Burgart LJ, Conley B, Cooper HS, Hamilton SR, et al. Prognostic factors in colorectal cancer. College of American Pathologists Consensus Statement 1999. Arch Pathol Lab Med. 2000;124:979–94.
27. Cunningham D, Humblet Y, Siena S, Khayat D, Bleiberg H, Santoro A, et al. Cetuximab monotherapy and cetuximab plus irinotecan in irinotecanrefractory metastatic colorectal cancer. N Engl J Med. 2004;351(4):337–45.

28. Douillard JY, Siena S, Cassidy J, Tabernero J, Burkes R, Barugel M, et al. Randomized, phase III trial of panitumumab with infusional fluorouracil, leucovorin, and oxaliplatin (FOLFOX4) versus FOLFOX4 alone as first-line treatment in patients with previously untreated metastatic colorectal cancer: the PRIME study. J Clin Oncol. 2010;28(31):4697–705.
29. Peeters M, Price TJ, Cervantes A, Sobrero AF, Ducreux M, Hotko Y, et al. Randomized phase III study of panitumumab with fluorouracil, leucovorin, and irinotecan (FOLFIRI) compared with FOLFIRI alone as second-line treatment in patients with metastatic colorectal cancer. J Clin Oncol. 2010;28(31):4706–13.
30. Amado RG, Wolf M, Peeters M, Van Cutsem E, Siena S, Freeman DJ, et al. Wild-type KRAS is required for panitumumab efficacy in patients with metastatic colorectal cancer. J Clin Oncol. 2008;26(10):1626–34.
31. Karapetis CS, Khambata-Ford S, Jonker DJ, O'Callaghan CJ, Tu D, Tebbutt NC, et al. K-ras mutations and benefit from cetuximab in advanced colorecal cancer. N Engl J Med. 2008;359(17):1757–65.
32. Lievre A, Bachet JB, Boige V, Cayre A, Le Corre D, Buc E, et al. KRAS mutations as an independent prognostic factor in patients with advanced colorectal cancer treated with cetuximab. J Clin Oncol. 2008;26(3):374–9.
33. Bardelli A, Siena S. Molecular mechanisms of resistance to cetuximab and panitumumab in colorectal cancer. J Clin Oncol. 2010;28(7):1254–61.
34. Madhavan D, Zucknick M, Wallwiener M, Cu K, Modugno C, Scharpff M, et al. Circulating miRNAs as surrogate markers for circulating tumor cells and prognostic markers in metastatic breast cancer. Clin Cancer Res. 2010;18:5972–82.
35. Duffy MJ, O'Donovan N, Crown J. Use of molecular markers for predicting therapy response in cancer patients. Cancer Treat Rev. 2011;37:151–9.
36. Huang RS, Ratain MJ. Pharmacogenetics and pharmacogenomics of anticancer agents. CA Cancer J Clin. 2009;59:42–55.
37. Sawyers CL. The cancer biomarker problem. Nature. 2008;452:548–52.
38. Hrdlickova B, de Almeida RC, Borek Z, Withoff S. Genetic variation in the non-coding genome: involvement of micro-RNAs and long non-coding RNAs in disease. Biochim Biophys Acta. 1842;2014:1910–22.
39. Esteller M. Non-coding RNAs in human disease. Nat Rev Genet. 2011;12:861–74.
40. Ling H, Fabbri M, Calin GA. MicroRNAs and other non-coding RNAs as targets for anticancer drug development. Nat Rev Drug Discov. 2013;12:847–65.
41. Garzon R, Marcucci G, Croce CM. Targeting microRNAs in cancer: rationale, strategies and challenges. Nat Rev Drug Discov. 2010;9(10):775–89.
42. Michael MZ, O'Connor SM, van Holst Pellekaan NG, Young GP, James RJ. Reduced accumulation of specific microRNAs in colorectal neoplasia. Mol Cancer Res. 2003;1:882–91.
43. Bandres E, Cubedo E, Agirre X, Malumbres R, Zarate R, Ramirez N, et al. Identification by real-time PCR of 13 mature microRNAs differentially expressed in colorectal cancer and nontumoral tissues. Mol Cancer. 2006;5:29.
44. Cummins JM, He Y, Leary RJ, Pagliarini R, Diaz Jr LA, Sjoblom T, et al. The colorectal microRNAome. Proc Natl Acad Sci U S A. 2006;103:3687–92.
45. Volinia S, Calin GA, Liu CG, Ambs S, Cimmino A, Petrocca F, et al. A microRNA expression signature of human solid tumors defines cancer gene targets. Proc Natl Acad Sci U S A. 2006;103:2257–61.
46. Monzo M, Navarro A, Bandres E, Artells R, Moreno I, Gel B, et al. Overlapping expression of microRNAs in human embryonic colon and colorectal cancer. Cell Res. 2008;18:823–33.
47. Earle JS, Luthra R, Romans A, Abraham R, Ensor J, Yao H, et al. Association of microRNA expression with microsatellite instability status in colorectal adenocarcinoma. J Mol Diagn. 2010;12:433–40.
48. Calin GA, Croce CM. MicroRNA-cancer connection: the beginning of a new tale. Cancer Res. 2006;66:7390–4.
49. Calin GA, Dumitru CD, Shimizu M, Bichi R, Zupo S, Noch E, et al. Frequent deletions and down-regulation of micro-RNA genes miR15 and miR16 at 13q14 in chronic lymphocytic leukemia. Proc Natl Acad Sci U S A. 2002;99:15524–9.
50. Visone R, Rassenti LZ, Veronese A, Taccioli C, Costinean S, Aguda BD, et al. Karyotype-specific microRNA signature in chronic lymphocytic leukemia. Blood. 2009;114:3872–9.
51. Lebanony D, Benjamin H, Gilad S, Ezagouri M, Dov A, Ashkenzai K, et al. Diagnostic assay based on hsa-miR-205 expression distinguishes squamous from nonsquamous non-small-cell lung carcinoma. J Clin Oncol. 2009;27:2030–7.
52. Rosenfeld N, Aharonov R, Meiri E, Rosenwald S, Spector Y, Zepeniuk M, et al. MicroRNAs accurately identify cancer tissue origin. Nat Biotechnol. 2008;26:462–9.
53. Zhu W, Liu X, He J, Chen D, Hunag Y, Zhang YK. Overexpression of members of the microRNA-183 family is a risk factor for lung cancer: a case control study. BMC Cancer. 2011;11:393.
54. Blenkiron C, Goldstein LD, Thorne NP, Spiteri I, Suet-Feung C, Dunning M, et al. MicroRNA expression profiling of human breast cancer identifies new markers of tumor subtype. Genome Biol. 2007;8:R214.
55. Rothe F, Ignatiadis M, Chaboteaux C, Haibe-Kains B, Kheddoumi N, Majjaj S, et al. Global microRNA expression profi ling identifi es MiR-210 associated with tumor proliferation, invasion and poor clinical outcome in breast cancer. PLoS One. 2011;6:e20980.

56. Lanza G, Ferracin M, Gafa R, Veronese A, Spizzo R, Pichiorri F, et al. mRNA/microRNA gene expression profile in microsatellite unstable colorectal cancer. Mol Cancer. 2007;6:54.
57. Mosakhani N, Sarhadi VK, Borze I, Karjalainen-Lindsberg ML, Sundström J, Ristamäki R, et al. MicroRNA profiling differentiates colorectal cancer according to KRAS status. Genes Chromosomes Cancer. 2012;51:1–9.
58. Lu J, Getz G, Miska EA, Alvarez-Saavedra E, Lamb J, Peck D, et al. MicroRNA expression profiles classify human cancers. Nature. 2005;435:834–8.
59. Khvorova A, Reynolds A, Jayasena SD. Functional siRNAs and miRNAs exhibit strand bias. Cell. 2003;115:209–16.
60. Blahna MT, Hata A. Regulation of miRNA biogenesis as an integrated component of growth factor signaling. Curr Opin Cell Biol. 2013;25:233–40.
61. Sun X, Jiao X, Pestell TG, Fan C, Qin S, Mirabelli E, et al. MicroRNAs and cancer stem cells: the sword and the shield. Oncogene. 2013. doi:10.1038/onc.2013.492 [Epub ahead of print].
62. Davis BN, Hilyard AC, Lagna G, Hata A. SMAD proteins control DROSHA-mediated microRNA maturation. Nature. 2008;454:56–61.
63. Wang CJ, Zhou ZG, Wang L, Yang L, Zhou B, Gu J, et al. Clinicopathological significance of microRNA-31, −143 and −145 expression in colorectal cancer. Dis Markers. 2009;26(1):27–34.
64. Krichevsky AM, Gabriely G. miR-21: a small multi-faceted RNA. J Cell Mol Med. 2009;13:39–53.
65. Luo XY, Burwinkel B, Tao S, Brenner H. MicroRNA signatures: novel biomarker for colorectal cancer? Cancer Epidemiol Biomarkers Prev. 2011;20:1272–86.
66. Asangani IA, Rasheed SA, Nikolova DA, Leupold JH, Colburn NH, Post S, et al. MicroRNA-21 (miR-21) post-transcriptionally downregulates tumor suppressor Pdcd4 and stimulates invasion, intravasation and metastasis in colorectal cancer. Oncogene. 2008;27:2128–36.
67. Slaby O, Svoboda M, Fabian P, Smerdova T, Knoflickova D, Bednarikova M, et al. Altered expression of miR-21, miR-31, miR-143 and miR-145 is related to clinicopathologic features of colorectal cancer. Oncology. 2007;72(5–6):397–402.
68. Catela Ivkovic T, Aralica G, Cacev T, Loncar B, Kapitanovic S. miR-106a overexpression and pRB downregulation in sporadic colorectal cancer. Exp Mol Pathol. 2013;94:148–54.
69. Feng B, Dong TT, Wang LL, Zhou HM, Zhao HC, Dong F, et al. Colorectal cancer migration and invasion initiated by microRNA-106a. PLoS One. 2012;7:e43452.
70. Bullock MD, Pickard K, Mitter R, Sayan AE, Primrose JN, Ivan C, et al. Stratifying risk of recurrence in stage II colorectal cancer using deregulated stromal and epithelial microRNAs. Oncotarget. 2015;6(9):7262–79.
71. Lopez-Camarillo C, Marchat LA, Arechaga-Ocampo E, Perez-Plasencia C, Del Moral-Hernandez O, Castaneda-Ortiz EJ, et al. MetastamiRs: non-coding microRNAs driving cancer invasion and metastasis. Int J Mol Sci. 2012;13:1347–79.
72. Wu CW, Ng SSM, Dong YJ, Ng SC, Leung WW, Lee CW, et al. Detection of miR-92a and miR-21 in stool samples as potential screening biomarkers for colorectal cancer and polyps. Gut. 2012;61:739–45.
73. Nagel R, le Sage C, Diosdado B, van der Waal M, Oude Vrielink JA, Bolijn A, et al. Regulation of the adenomatous polyposis coli gene by the miR-135 family in colorectal cancer. Cancer Res. 2008;68:5795–802.
74. Yamada A, Horimatsu T, Okugawa Y, Nishida N, Honjo H, Ida H, et al. Serum miR-21, miR-29a, and miR-125b are promising biomarkers for the early detection of colorectal neoplasia. Clin Cancer Res. 2015;21(18):4234–42.
75. Fodde R. The APC, gene in colorectal cancer. Eur J Cancer. 2002;38(7):867–71.
76. Tsang WP, Ng EK, Ng SS, Jin H, Yu J, Sung JJ, et al. Oncofetal H19-derived miR-675 regulates tumor suppressor RB in human colorectal cancer. Carcinogenesis. 2010;31(3):350–8.
77. Koga Y, Yasunaga M, Takahashi A, Kuroda J, Moriya Y, Akasu T, et al. MicroRNA expression profiling of exfoliated colonocytes isolated from feces for colorectal cancer screening. Cancer Prev Res. 2010;3:1435–42.
78. Jérôme T, Laurie P, Louis B, Pierre C. Enjoy the silence: the story of let-7 microRNA and cancer. Curr Genomics. 2007;8:229–33.
79. Thornton JE, Gregory RI. How does Lin28 let-7 control development and disease? Trends Cell Biol. 2012;22:474–82.
80. Drebber U, Lay M, Wedemeyer I, Vallböhmer D, Bollschweiler E, Brabender J, et al. Altered levels of the onco-microRNA 21 and the tumor-supressor microRNAs 143 and 145 in advanced rectal cancer indicate successful neoadjuvant chemoradiotherapy. Int J Oncol. 2011;39:409–15.
81. Chang KH, Miller N, Kheirelseid EA, Lemetre C, Ball GR, Smith MJ, et al. MicroRNA signature analysis in colorectal cancer: identification of expression profiles in stage II tumors associated with aggressive disease. Int J Colorectal Dis. 2011;26:1415–22.
82. Kahlert C, Klupp F, Brand K, Lasitschka F, Diederichs S, Kirchberg J, et al. Invasion front-specific expression and prognostic significance of microRNA in colorectal liver metastases. Cancer Sci. 2011;102:1799–807.
83. Kamatani A, Nakagawa Y, Akao Y, Maruyama N, Nagasaka M, Shibata T, et al. Downregulation of anti-oncomirs miR-143/145 cluster occurs before APC gene aberration in the development of colorectal tumors. Med Mol Morphol. 2013;46:166–71.

84. Gregersen LH, Jacobsen A, Frankel LB, Wen J, Krogh A, Lund AH. MicroRNA-143 down-regulates Hexokinase 2 in colon cancer cells. BMC Cancer. 2012;12:232.
85. Pichler M, Winter E, Stotz M, Eberhard K, Samonigg H, Lax S, et al. Down-regulation of KRAS-interacting miRNA- 143 predicts poor prognosis but not response to EGFR-targeted agents in colorectal cancer. Br J Cancer. 2012;106:1826–32.
86. Zhang Y, Wang Z, Chen M, Peng L, Wang X, Ma Q, et al. MicroRNA-143 targets MACC1 to inhibit cell invasion and migration in colorectal cancer. Mol Cancer. 2012;11:23.
87. Li JM, Zhao RH, Li ST, Xie CX, Jiang HH, Ding WJ, et al. Down-regulation of fecal miR-143 and miR-145 as potential markers for colorectal cancer. Saudi Med J. 2012;33:24–9.
88. Chen DT, Hernandez JM, Shibata D, McCarthy SM, Humphries LA, Clark W, et al. Complementary strand microRNAs mediate acquisition of metastatic potential in colonic adenocarcinoma. J Gastrointest Surg. 2012;16:905–12.
89. Qian X, Yu J, Yin Y, He J, Wang L, Li Q, et al. MicroRNA-143 inhibits tumor growth and angiogenesis and sensitizes chemosensitivity to oxaliplatin in colorectal cancers. Cell Cycle. 2013;12:1385–94.
90. Song Y, Xu Y, Wang Z, Chen Y, Yue Z, Gao P, et al. MicroRNA-148b suppresses cell growth by targeting cholecystokinin-2 receptor in colorectal cancer. Int J Cancer. 2012;131:1042–51.
91. Tang JT, Wang JL, Du W, Hong J, Zhao SL, Wang YC, et al. MicroRNA 345, a methylation-sensitive microRNA is involved in cell proliferation and invasion in human colorectal cancer. Carcinogenesis. 2011;32:1207–15.
92. Cheng H, Zhang L, Cogdell DE, Zheng H, Schetter AJ, Nykter M, et al. Circulating plasma MiR-141 is a novel biomarker for metastatic colon cancer and predicts poor prognosis. PLoS One. 2011;6:e17745.
93. Brink M, de Goeij AF, Weijenberg MP, Roemen GM, Lentjes MH, Pachen MM, et al. K-ras oncogene mutations in sporadic colorectal cancer in The Netherlands Cohort Study. Carcinogenesis. 2003;24:703–10.
94. Baldus SE, Schaefer KL, Engers R, Hartleb D, Stoecklein NH, Gabbert HE. Prevalence and heterogeneity of KRAS, BRAF, and PIK3CA mutations in primary colorectal adenocarcinomas and their corresponding metastases. Clin Cancer Res. 2010;16:790–9.
95. Vickers MM, Bar J, Gorn-Hondermann I, Yarom N, Daneshmand M, Hanson JE, et al. Stage-dependent differential expression of microRNAs in colorectal cancer: potential role as markers of metastatic disease. Clin Exp Metastasis. 2012;29:123–32.
96. Gusev Y, Brackett DJ. MicroRNA expression profiling in cancer from a bioinformatics prospective. Expert Rev Mol Diagn. 2007;7:787–92.
97. Karim BO, Ali SZ, Landolfi JA, Mann JF, Liu G, Christian A, et al. Cytomorphologic differentiation of benign and malignant mammary tumors in fine needle aspirate specimens from irradiated female Sprague–Dawley rats. Vet Clin Pathol. 2008;37:229–36.
98. Liu A, Tetzlaff MT, Vanbelle P, Elder D, Feldman M, Tobias JW, et al. MicroRNA expression profiling outperforms mRNA expression profiling in formalin-fixed paraffin-embedded tissues. Int J Clin Exp Pathol. 2009;2:519–27.
99. Ryan BM, Robles AI, Harris CC. Genetic variation in microRNA networks: the implications for cancer research. Nat Rev Cancer. 2010;10:389–402.
100. Calin GA, Croce CM. MicroRNA signatures in human cancers. Nat Rev Cancer. 2006;6:857–66.
101. Esquela-Kerscher A, Slack FJ. Oncomirs - microRNAs with a role in cancer. Nat Rev Cancer. 2006;6:259–69.
102. Schepeler T, Reinert JT, Ostenfeld MS, Christensen LL, Silahtaroglu AN, Dyrskjøt L, et al. Diagnostic and prognostic microRNAs in stage II colon cancer. Cancer Res. 2008;68(15):6416–24.
103. Sarver AL, French AJ, Borralho PM, Thayanithy V, Oberg AL, Silverstein KA, et al. Human colon cancer profiles show differential microRNA expression depending on mismatch repair status and are characteristic of undifferentiated proliferative states. BMC Cancer. 2009;9:401.
104. Akao Y, Nakagawa Y, Naoe T. MicroRNA-143 and −145 in colon cancer. DNA Cell Biol. 2007;26:311–20.
105. Diosdado B, van de Wiel MA, Terhaar Sive Droste JS, Mongera S, Postma C, Meijerink WJ, et al. MiR-17-92 cluster is associated with 13q gain and c-myc expression during colorectal adenoma to adenocarcinoma progression. Br J Cancer. 2009;101(4):707–14.
106. Skog J, Würdinger T, van Rijn S, Meijer DH, Gainche L, Sena-Esteves M, et al. Glioblastoma microvesicles transport RNA 32 and proteins that promote tumour growth and provide diagnostic biomarkers. Nat Cell Biol. 2008;10:1470–6.
107. Zhang JX, Song W, Chen ZH, Wei JH, Liao YJ, Lei J, et al. Prognostic and predictive value of a microRNA signature in stage II colon cancer: a microRNA expression analysis. Lancet Oncol. 2013;14:1295–306.
108. Iorio MV, Croce CM. MicroRNAs in cancer: small molecules with a huge impact. J Clin Oncol. 2009;27:5848–56.
109. Leupold JH, Yang HS, Colburn NH, Asangani I, Post S, Allgayer H. Tumor suppressor Pdcd4 inhibits invasion/intravasation and regulates urokinase receptor (u-PAR) gene expression via Sp-transcription factors. Oncogene. 2007;26:4550–62.
110. Schetter AJ, Leung SY, Sohn JJ, Zanetti KA, Bowman ED, Yanaihara N, et al. MicroRNA expres-

sion profiles associated with prognosis and therapeutic outcome in colon adenocarcinoma. JAMA. 2008;299:425–36.
111. Bullock MD, Pickard KM, Nielsen BS, Sayan AE, Jenei V, Mellone M, et al. Pleiotropic actions of miR-21 highlight the critical role of deregulated stromal microRNAs during colorectal cancer progression. Cell Death Dis. 2013;4:e684.
112. Toiyama Y, Takahashi M, Hur K, Nagasaka T, Tanaka K, Inoue Y, et al. Serum miR-21 as a diagnostic and prognostic biomarker in colorectal cancer. J Natl Cancer Inst. 2013;105:849–59.
113. Yu Y, Sarkar FH, Majumdar AP. Down-regulation of miR-21 induces differentiation of chemoresistant colon cancer cells and enhances susceptibility to therapeutic regimens. Trans Oncol. 2013;6:180–6.
114. Yu Y, Kanwar SS, Patel BB, Oh PS, Nautiyal J, Sarkar FH, et al. MicroRNA-21 induces stemness by downregulating transforming growth factor beta receptor 2 (TGFβR2) in colon cancer cells. Carcinogenesis. 2012;33:68–76.
115. Xiong B, Cheng Y, Ma L, Zhang C. MiR-21 regulates biological behavior through the PTEN/PI-3 K/Akt signaling pathway in human colorectal cancer cells. Int J Oncol. 2013;42:219–28.
116. Schee K, Boye K, Abrahamsen TW, Fodstad Ø, Flatmark K. Clinical relevance of microRNA miR-21, miR-31, miR-92a, miR-101, miR-106a and miR-145 in colorectal cancer. BMC Cancer. 2012;12:505.
117. Kanaan Z, Rai SN, Eichenberger MR, Roberts H, Keskey B, Pan J, Galandiuk S. Plasma miR-21: a potential diagnostic marker of colorectal cancer. Ann Surg. 2012;256:544–51.
118. Faltejskova P, Svoboda M, Srutova K, Mlcochova J, Besse A, Nekvindova J, et al. Identification and functional screening of microRNAs highly deregulated in colorectal cancer. J Cell Mol Med. 2012;16(11):2655–66.
119. Yamashita S, Yamamoto H, Mimori K, Nishida N, Takahashi H, Haraguchi N, et al. MicroRNA-372 is associated with poor prognosis in colorectal cancer. Oncology. 2012;82:205–12.
120. Liu K, Li G, Fan C, Zhou X, Wu B, Li J. Increased expression of microRNA-21 and its association with chemotherapeutic response in human colorectal cancer. J Int Med Res. 2011;39:2288–95.
121. Horiuchi A, Iinuma H, Akahane T, Shimada R, Watanabe T. Prognostic significance of PDCD4 expression and association with microRNA-21 in each Dukes' stage of colorectal cancer patients. Oncol Rep. 2012;27:1384–92.
122. Shibuya H, Iinuma H, Shimada R, Horiuchi A, Watanabe T. Clinicopathological and prognostic value of microRNA-21 and microRNA-155 in colorectal cancer. Oncology. 2010;79:313–20.
123. Kulda V, Pesta M, Topolcan O, Liska V, Treska V, Sutnar A, et al. Relevance of miR-21 and miR-143 expression in tissue samples of colorectal carcinoma and its liver metastases. Cancer Genet Cytogenet. 2010;200(2):154–60.
124. Nielsen BS, Jørgensen S, Fog JU, Søkilde R, Christensen IJ, Hansen U, et al. High levels of microRNA-21 in the stroma of colorectal cancers predict short disease-free survival in stage II colon cancer patients. Clin Exp Metastasis. 2011;28(1):27–38.
125. Schetter AJ, Nguyen GH, Bowman ED, Mathe EA, Yuen ST, Hawkes JE, et al. Association of inflammation-related and microRNA gene expression with cancer-specific mortality of colon adenocarcinoma. Clin Cancer Res. 2009;15(18):5878–87.
126. Zhang J, Xiao Z, Lai D, Sun J, He C, Chu Z, et al. miR-21, miR-17 and miR-19a induced by phosphatase of regenerating liver-3 promote the proliferation and metastasis of colon cancer. Br J Cancer. 2012;107:352–9.
127. Nishida N, Nagahara M, Sato T, Mimori K, Sudo T, Tanaka F, et al. Microarray analysis of colorectal cancer stromal tissue reveals upregulation of two oncogenic miRNA clusters. Clin Cancer Res. 2012;18:3054–70.
128. Yu G, Tang JQ, Tian ML, Li H, Wang X, Wu T, et al. Prognostic values of the miR-17-92 cluster and its paralogs in colon cancer. J Surg Oncol. 2012;106:232–7.
129. Slaby O, Sachlova M, Brezkova V, Hezova R, Kovarikova A, Bischofová S, et al. Identification of microRNAs regulated by isothiocyanates and association of polymorphisms inside their target sites with risk of sporadic colorectal cancer. Nutr Cancer. 2013;65:247–54.
130. Ma Y, Zhang P, Wang F, Zhang H, Yang Y, Shi C, et al. Elevated oncofoetal miR-17- 5p expression regulates colorectal cancer progression by repressing its target gene P130. Nat Commun. 2012;3:1291.
131. Xu XM, Qian JC, Deng ZL, Cai Z, Tang T, Wang P, et al. Expression of miR-21, miR-31, miR-96 and miR-135b is correlated with the clinical parameters of colorectal cancer. Oncol Lett. 2012;4:339–45.
132. Gaedcke J, Grade M, Camps J, Søkilde R, Kaczkowski B, Schetter AJ, et al. The rectal cancer microRNAome--microRNA expression in rectal cancer and matched normal mucosa. Clin Cancer Res. 2012;18:4919–30.
133. Takahashi M, Cuatrecasas M, Balaguer F, Hur K, Toiyama Y, Castells A, et al. The clinical significance of MiR-148a as a predictive biomarker in patients with advanced colorectal cancer. PLoS One. 2012;7:e46684.
134. Kalimutho M, Di Cecilia S, Del Vecchio BG, Roviello F, Sileri P, Cretella M, et al. Epigenetically silenced miR-34b/c as a novel faecal-based screening marker for colorectal cancer. Br J Cancer. 2011;104:1770–8.
135. Zhang H, Li Y, Huang Q, Ren X, Hu H, Sheng H, et al. MiR-148a promotes apoptosis by targeting Bcl-2 in colorectal cancer. Cell Death Differ. 2011;18(11):1702–10.

136. Li S, Gao J, Gu J, Yuan J, Hua D, Shen L. MicroRNA-215 inhibits relapse of colorectal cancer patients following radical surgery. Med Oncol. 2013;30:549.
137. Chiang Y, Song Y, Wang Z, Liu Z, Gao P, Liang J, et al. microRNA-192, −194 and −215 are frequently downregulated in colorectal cancer. Exp Ther Med. 2012;3:560–6.
138. Karaayvaz M, Pal T, Song B, Zhang C, Georgakopoulos P, Mehmood S, et al. Prognostic significance of miR-215 in colon cancer. Clin Colorectal Cancer. 2011;10(4):340–7.
139. Calin GA, Cimmino A, Fabbri M, Ferracin M, Wojcik SE, Shimizu M, et al. MiR-15a and miR-16-1 cluster functions in human leukemia. Proc Natl Acad Sci U S A. 2008;105(13):5166–71.
140. Aqeilan RI, Calin GA, Croce CM. MiR-15a and miR-16-1 in cancer: discovery, function and future perspectives. Cell Death Differ. 2010;17(2):215–20.
141. Luo Q, Li X, Li J, Kong X, Zhang J, Chen L, et al. MiR-15a is underexpressed and inhibits the cell cycle by targeting CCNE1 in breast cancer. Int J Oncol. 2013;43(4):1212–8.
142. Huang YH, Lin KH, Chen HC, Chang ML, Hsu CW, Lai MW, et al. Identification of postoperative prognostic microRNA predictors in hepatocellular carcinoma. PLoS One. 2012;7(5):e37188.
143. Musumeci M, Coppola V, Addario A, Patrizii M, Maugeri-Sacca M, Memeo L, et al. Control of tumor and microenvironment cross-talk by miR-15a and miR-16 in prostate cancer. Oncogene. 2011;30:4231–42.
144. Bandi N, Vassella E. miR-34a and miR-15a/16 are co-regulated in non-small cell lung cancer and control cell cycle progression in a synergistic and Rb-dependent manner. Mol Cancer. 2011;10:55.
145. Ma Q, Wang XF, Li ZF, Li BF, Ma FF, Peng LF, et al. microRNA-16 represses colorectal cancer cell growth in vitro by regulating the p53/survivin signaling pathway. Oncol Rep. 2013;29(4):1652–8.
146. Qian J, Jiang BF, Li MF, Chen JF, Fang M. Prognostic significance of microRNA-16 expression in human colorectal cancer. World J Surg. 2013;12:2944–9.
147. Weissmann-Brenner A, Kushnir M, Yanai GL, Aharonov R, Gibori H, Purim O, et al. Tumor microRNA-29a expression and the risk of recurrence in stage II colon cancer. Int J Oncol. 2012;40:2097–103.
148. Huang ZH, Huang D, Ni SJA, Peng ZL, Sheng WQ, Du X. Plasma microRNAs are promising novel biomarkers for early detection of colorectal cancer. Int J Cancer. 2010;127:118–26.
149. Ragusa M, Majorana AF, Statello LF, Maugeri MF, Salito LF, Barbagallo DF, et al. Specific alterations of microRNA transcriptome and global network structure in colorectal carcinoma after cetuximab treatment. Mol Cancer Ther. 2010;9:3396–409.
150. Ng EK, Chong WW, Jin H, Lam EK, Shin VY, Yu J, et al. Differential expression of microRNAs in plasma of patients with colorectal cancer: a potential marker for colorectal cancer screening. Gut. 2009;58:1375–81.
151. Ruzzo A, Graziano FF, Vincenzi BF, Canestrari EF, Perrone GF, Galluccio NF, et al. High let-7a microRNA levels in KRAS-mutated colorectal carcinomas may rescue anti-EGFR therapy effects in patients with chemotherapy-refractory metastatic disease. Oncologist. 2012;17:823–9.
152. Yang IP, Tsai HL, Hou MF, Chen KC, Tsai PC, Huang SW, et al. MicroRNA-93 inhibits tumor growth and early relapse of human colorectal cancer by affecting genes involved in the cell cycle. Carcinogenesis. 2012;33:1522–30.
153. Mekenkamp LJ, Tol J, Dijkstra JR, de Krijger I, Vink-Borger E, Teerenstra S, et al. Beyond KRAS mutation status: influence of KRAS copy number status and microRNAs on clinical outcome to cetuximab in metastatic colorectal cancer patients. BMC Cancer. 2012;12:292.
154. Shibutani M, Noda E, Maeda K, Nagahara H, Ohtani H, Hirakawa K. Low expression of claudin-1 and presence of poorly-differentiated tumor clusters correlate with poor prognosis in colorectal cancer. Anticancer Res. 2013;33:3301–6.
155. Pu J, Bai D, Yang X, Lu X, Xu L, Lu J. Adrenaline promotes cell proliferation and increases chemoresistance in colon cancer HT29 cells through induction of miR-155. Biochem Biophys Res Commun. 2012;428:210–5.
156. Wang M, Zhang P, Li Y, Liu G, Zhou B, Zhan L, et al. The quantitative analysis by stem-loop real-time PCR revealed the microRNA-34a, microRNA-155 and microRNA-200c overexpression in human colorectal cancer. Med Oncol. 2012;29:3113–8.
157. Gezer U, Özgür E, Cetinkaya M, Isin M, Dalay N. Long non-coding RNAs with low expression levels in cells are enriched in secreted exosomes. Cell Biol Int. 2014;38(9):1076–9. doi:10.1002/cbin.10301.
158. Takahashi Y, Sawada G, Kurashige J, Uchi R, Matsumura T, Ueo H, et al. Amplification of PVT-1 is involved in poor prognosis via apoptosis inhibition in colorectal cancers. Br J Cancer. 2014;110(1):164–71. doi:10.1038/bjc.2013.698.
159. Deng Q, He B, Gao T, Pan Y, Sun H, Xu Y, et al. Up-regulation of 91H promotes tumor metastasis and predicts poor prognosis for patients with colorectal cancer. PLoS One. 2014;9(7):e103022. doi:10.1371/journal.pone.0103022.
160. Zheng HT, Shi DB, Wang YW, Li XX, Xu Y, Tripathi P, et al. High expression of lncRNA MALAT1 suggests a biomarker of poor prognosis in colorectal cancer. Int J Clin Exp Pathol. 2014;7(6):3174–81.
161. Kogo R, Shimamura T, Mimori K, Kawahara K, Imoto S, Sudo T, et al. Long noncoding RNA HOTAIR regulates polycomb-dependent chromatin modification and is associated with poor prognosis

in colorectal cancers. Cancer Res. 2011;71(20):6320–6. doi:10.1158/0008-5472.CAN-11-1021.
162. Svoboda M, Slyskova J, Schneiderova M, Makovicky P, Bielik L, Levy M, et al. HOTAIR long non-coding RNA is a negative prognostic factor not only in primary tumors, but also in the blood of colorectal cancer patients. Carcinogenesis. 2014;35(7):1510–5. doi:10.1093/carcin/bgu055.
163. Ge X, Chen Y, Liao X, Liu D, Li F, Ruan H, et al. Overexpression of long noncoding RNA PCAT-1 is a novel biomarker of poor prognosis in patients with colorectal cancer. Med Oncol. 2013;30(2):588. doi:10.1007/s12032-013-0588-6.
164. Li Y, Li Y, Chen W, He F, Tan Z, Zheng J, et al. NEAT expression is associated with tumor recurrence and unfavorable prognosis in colorectal cancer. Oncotarget. 2015 Aug 7.
165. Qi P, Xu MD, Ni SJ, Shen XH, Wei P, Huang D, et al. Down-regulation of ncRAN, a long non-coding RNA, contributes to colorectal cancer cell migration and invasion and predicts poor overall survival for colorectal cancer patients. Mol Carcinog. 2015;54(9):742–50. doi:10.1002/mc.22137.
166. Qi P, Xu MD, Ni SJ, Huang D, Wei P, Tan C, et al. Low expression of LOC285194 is associated with poor prognosis in colorectal cancer. J Transl Med. 2013;11:122. doi:10.1186/1479-5876-11-122.
167. Yin D, He X, Zhang E, Kong R, De W, Zhang Z. Long noncoding RNA GAS5 affects cell proliferation and predicts a poor prognosis in patients with colorectal cancer. Med Oncol. 2014;31(11):253. doi:10.1007/s12032-014-0253-8.
168. Yin DD, Liu ZJ, Zhang E, Kong R, Zhang ZH, Guo RH. Decreased expression of long noncoding RNA MEG3 affects cell proliferation and predicts a poor prognosis in patients with colorectal cancer. Tumour Biol. 2015;36(6):4851–9. doi:10.1007/s13277-015-3139-2.
169. Hu Y, Chen HY, Yu CY, Xu J, Wang JL, Qian J, et al. A long non-coding RNA signature to improve prognosis prediction of colorectal cancer. Oncotarget. 2014;5(8):2230–42.
170. Zhu L, Liu J, Ma S, Zhang S. Long noncoding RNA MALAT-1 can predict metastasis and a poor prognosis: a meta-analysis. Pathol Oncol Res. 2015;21(4):1259–64. doi:10.1007/s12253-015-9960-5.
171. Cai B, Wu Z, Liao K, Zhang S. Long noncoding RNA HOTAIR can serve as a common molecular marker for lymph node metastasis: a meta-analysis. Tumour Biol. 2014;35(9):8445–50. doi:10.1007/s13277-014-2311-4.
172. Lee H, Kim C, Ku JL, Kim W, Yoon SK, Kuh HJ, et al. A long non-coding RNA snaR contributes to 5-fluorouracil resistance in human colon cancer cells. Mol Cells. 2014;37(7):540–6. doi:10.14348/molcells.2014.0151.
173. Xiong W, Jiang YX, Ai YQ, Liu S, Wu XR, Cui JG, et al. Microarray analysis of long non-coding RNA expression profile associated with 5-fluorouracil-based chemoradiation resistance in colorectal cancer cells. Asian Pac J Cancer Prev. 2015;16(8):3395–402.
174. Ling H, Spizzo R, Atlasi Y, Nicoloso M, Shimizu M, Redis RS, et al. CCAT2, a novel noncoding RNA mapping to 8q24, underlies metastatic progression and chromosomal instability in colon cancer. Genome Res. 2013;23(9):1446–61. doi:10.1101/gr.152942.112.

Part III

Non-coding RNAs: Therapeutic Targets and Colorectal Cancer Therapeutics

Involvement of Non-coding RNAs in Chemo- and Radioresistance of Colorectal Cancer

11

Daniele Fanale, Marta Castiglia, Viviana Bazan, and Antonio Russo

Abstract

Despite recent progress in understanding the cancer signaling pathways and in developing new therapeutic strategies, however, the resistance of colorectal cancer (CRC) cells to chemo- and radiotherapy represents the main hurdle to the successful treatment, leading to tumor recurrence and, consequently, a poor prognosis. Therefore, overcoming drug and radiation resistance, enhancing drug and radiation sensitivity of CRC cells, and improving the efficacy of chemo- and radiotherapy have an important significance in the treatment of CRC. The identification of new molecular biomarkers which can predict therapy response and prognosis is one of the most significant aims in pharmacogenomics and cancer research.

Recent studies showed that non-coding RNAs (ncRNAs), such as microRNAs (miRNAs) and long non-coding RNAs (lncRNAs), may play important roles in the regulation of chemo- and radioresistance of CRC, by controlling several signaling pathways, including cell cycle, proliferation, apoptosis and DNA damage repair. Recent data have demonstrated that selective modulation of the ncRNA activity can improve the response to chemo- and radiotherapy, providing an innovative anti-tumor approach based on a ncRNA-related gene therapy. Therefore, ncRNAs could not only be useful as predictive and prognostic biomarkers but also serve as targets for the development of novel therapeutic strategies to overcome drug and radiation resistance in CRC. In this chapter, we discuss the involvement of ncRNAs in chemo- and radiotherapy resistance of CRC, highlighting the impact of these molecules in prediction of the treatment

D. Fanale • M. Castiglia • V. Bazan
A. Russo (✉)
Department of Surgical, Oncological and Oral Sciences, Section of Medical Oncology, University of Palermo, Via del Vespro 129, 90127 Palermo, Italy
e-mail: fandan@libero.it; martacastiglia@gmail.com; viviana.bazan@unipa.it; antonio.russo@usa.net

O. Slaby, G.A. Calin (eds.), *Non-coding RNAs in Colorectal Cancer*, Advances in Experimental Medicine and Biology 937, DOI 10.1007/978-3-319-42059-2_11

response and modification of the therapy, and describing possible intracellular pathways involved in these processes.

Keywords
Chemoresistance • Chemotherapy • miRNAs • Non-coding RNA • Predictive biomarkers • Radioresistance • Radiotherapy • Targeted therapy • Therapy response

11.1 Introduction

Colorectal cancer (CRC) is one of the most common malignant tumors worldwide accounting for the second leading cause of cancer-related death in the Western Europe countries and the third in the United States [1, 2]. Improvements in early screening strategies, the emergence of new therapies and recent progress in understanding the genetic and molecular basis of CRC have greatly reduced death rates [3, 4]. Although, in the last few years, early detection methods and innovative therapeutic strategies have been developed in order to prolong survival and improve patient life quality, surgical resection remains the most successful treatment option [5]. However, a large proportion of CRC patients develops unresectable distant metastatic lesions, which can be detected early at diagnosis or at a later stage [6]. For this reason, surgery alone appears to be inadequate and insufficient in eradicating the disease and improving prognosis. Therefore, along with surgery, chemotherapy, targeted therapy and radiotherapy are the preferred treatments for CRC [7]. However, the development and selection of cancer cells resistant to chemo- and radiotherapy is one of the major issues for the clinical management of CRC patients, leading to tumor recurrence and, consequently, an unfavorable prognosis [8, 9]. Therefore, implementing appropriate strategies able to overcome the resistance that patients may develop during chemo- or radiotherapy is the main goal of clinical research [10]. Understanding the molecular mechanisms responsible for the occurrence of therapy resistance and identifying new targets to improve efficacy of therapeutic treatment might help oncologists to promote the development of personalized approaches for cancer cure [11, 12]. The identification of new predictive and prognostic biomarkers could represent an important tool to select patients who may benefit from a specific treatment and a crucial step toward a tailored therapy [13, 14].

In recent years, a large number of molecular and genetic alterations related to tumor cell proliferation and survival, and therapy response were found as potential biomarkers for clinical use, thank to advances in the field of genomics, biotechnology and molecular pathology [15]. Also, several evidence showed that response to treatment can be affected by epigenetic mechanisms involving gene expression regulation [16].

Recent progress in the field of transcriptomics highlighted the functional relevance in human cancer of non-coding RNAs (ncRNAs), including microRNAs (miRNAs) and long non-coding RNAs (lncRNAs), that seem to be involved in transcriptional and post-transcriptional regulation of gene expression, modulation of protein activity and genomic imprinting [17–20]. Experimental evidence suggested that dysregulation of specific ncRNAs may be involved in the tumor initiation, progression, metastatic processes and acquisition of tumor resistance to therapy [21–24]. Therefore, ncRNAs could not only be useful as predictive and/or prognostic biomarkers for CRC [25–27], but also serve as targets for the development of novel ncRNA-based therapeutic strategies to overcome drug resistance and radioresistance [28–30].

In this chapter, we will discuss the involvement of ncRNAs in resistance to chemo- and radiotherapy of CRC, highlighting the impact of these molecules in prediction of the treatment response and modification of the therapy, and describing possible intracellular pathways involved in these processes.

11.2 Non-coding RNAs Involved in Drug Resistance of CRC

Aberrant expression of ncRNAs has been reported in several types of human cancer, including CRC, suggesting a potential role in cancer pathogenesis [17]. Recent studies showed that ncRNAs, especially miRNAs and lncRNAs, may play important roles in the regulation of chemoresistance of CRC, by controlling several signaling pathways, including cell cycle, proliferation, apoptosis and DNA damage repair or other key cellular signaling pathways [31]. Furthermore, the expression of drug targets and genes involved in drug metabolism or transport may be regulated by ncRNAs [32]. Additionally, the inactivation of oncogenic miRNAs, called oncomiRs, inhibiting the expression of target tumor suppressor genes, and activation of tumor suppressor miRNAs, called anti-oncomiRs, inhibiting the expression of oncogenes, may be important mechanisms that regulate the expression of specific genes able to restore drug sensitivity [33]. Therefore, targeting of selected ncRNAs could be an innovative therapeutic strategy in order to develop a suitable anticancer therapy able to abolish drug resistance of cancer cells or eradicate cells that are usually resistant to conventional and targeted therapies [29]. Since selective modulation of the ncRNA activity may improve the response to therapy in CRC, the characterization of ncRNA expression profiles could help us to increase our knowledge about the molecular biology of CRC offering the possibility to identify new prognostic and/or predictive markers which could be used as new therapeutic targets [34]. Identifying new ncRNAs as predictive biomarkers of response to therapies could improve efficacy of therapeutic treatment and allow the development of individualized and more tailored treatment regimens for CRC patients. Recent advances in microarray technology, and the ongoing development of new targeted therapies have opened up new roads to fight drug resistance.

11.2.1 Chemotherapy and Resistance Mechanisms

The management of CRC patients involves screening, staging, and treatment with surgery, chemotherapy, novel targeted agents and/or radiation. However, the primary antitumor drug treatment for both resectable and advanced CRC remains the conventional chemotherapy.

The first important drug, classified as antimetabolite, belonging to the fluoropyrimidine family, was 5-fluorouracil (5-FU), which exerts its anti-cancer activity through the inhibition of RNA synthesis and function, block of thymidylate synthase (TS) activity, and incorporation into DNA, causing DNA strand breaks. Initially, fluoropyrimidine-based adjuvant chemotherapy has been used to improve survival and reduce the risk of tumor recurrence [35]. Afterwards, thanks to the entry into clinical practice of cytotoxic agents such as irinotecan (CPT-11) and oxaliplatin (L-OHP), the overall survival (OS), progression-free survival (PFS) and response rate (RR) are improved [36]. With the advent of new targeted therapies, nowadays, many therapeutic regimens involving several drugs used in combination or as monotherapy were approved for the treatment of unresectable mCRC. The addition of targeted agents to conventional chemotherapy regimens has led to a considerable improvement in survival of mCRC patients [37, 38].

Despite the development of new and different therapeutic strategies, current therapies are not always able to totally eradicate the disease due to the occurrence of resistance. In fact, a lack of response to anticancer therapy and frequent relapse were observed in a relevant percentage of CRC patients. The resistance of CRC cells to chemotherapy is the main hurdle to the successful treatment, by reducing the effectiveness of anticancer therapies, causing tumor recurrence and, consequently, determining an unfavorable prognosis [39]. Tumors may be intrinsically insensitive to therapeutic treatment prior to therapy (intrinsic or primary resistance), or, after being initially sensitive to therapy, may develop a resistance acquired after treatment (acquired or secondary resistance). Acquired resistance not only makes tumors resistant to originally used drugs, but may also cause cross-resistance to other drugs with different mechanisms of action. Therefore, one of the most compelling challenges of the current cancer research is to identify the mechanisms underlying the resistance and implement strategies to circumvent the resistance to therapy, increase chemosensitivity of CRC cells, and improve the effectiveness of chemotherapy [11]. In recent years, several pharmacogenomic studies were carried out in order to identify new molecular biomarkers which could predict therapy response, improving the ability of clinicians to determine the most effective therapeutic treatment for CRC patients.

Drug resistance is a complex and multifactorial event involving several major mechanisms and factors such as the pharmacokinetic profile of the drug, reduced drug uptake, drug efflux/inactivation, modifications of membrane lipids, alterations in drug target, reactivation of the targeted pathway, hyperactivation of alternative pathways, cross-talk with the microenvironment, activation of detoxification, apoptosis inhibition, drug-induced DNA damage repair, and alterations in cell cycle checkpoints. Therefore, drug resistance mechanisms can limit drug accumulation within cancer cells or affect the tumor microenvironment. Additionally, the intra-tumor heterogeneity of most tumors can restrict therapy response [40, 41]. In some cases, the resistance to chemotherapy agents which induce DNA damage either indirectly (e.g., 5-FU and CPT-11) or directly (e.g., L-OHP) may depend on the enhanced ability of cancer cells to repair damaged DNA due to alterations in repair pathways, such as nucleotide excision-repair (NER), mismatch-repair (MMR), base excision-repair (BER), non-homologous end-joining (NHEJ) and homologous-recombination (HR) [40]. Following DNA damage, some signaling pathways are triggered within the cell to arrest the cell cycle and allow the DNA repair. If DNA damage is not repaired completely, the cell will undergo apoptosis. DNA damage induced by platinum-based agents is repaired using NER as major repair system [42].

Experimental evidence showed that alterations in drug metabolism and targets, and variations in expression levels of genes involved in apoptosis can cause CRC resistance to 5-FU [43]. Thymidylate synthase (TS), encoded by *TYMS* gene, is the pivotal molecular target of 5-FU and a predictive biomarker of response to 5-FU-based chemotherapy. TS is a enzyme involved in DNA replication and repair processes through *de novo* synthesis of thymidylate. 5-FU mainly exerts its anticancer activity via formation of a ternary complex, consisting of the active metabolite fluorodeoxyuridine monophosphate (FdUMP), TS and folate cofactor, that causes repression of TS enzyme activity. TS overexpression is considered the main acquired resistance mechanism to 5-FU, and a potential predictive and prognostic biomarker [44]. Different mechanisms can determinate increased expression levels of TS, including gene amplification, transcription and/or translation upregulation, and epigenetic modifications [45]. Some *TYMS* mutations have been also correlated with the CRC resistance to 5-FU, as they have generated structural alterations in protein, reducing its binding affinity for FdUMP in preclinical models [46].

Other molecular changes associated with CRC resistance to 5-FU are the overexpression of dihydropyrimidine dehydrogenase (DPD), a key enzyme involved in the catabolism of 5-FU, and low expression levels of 5-FU activating enzymes such as uridine monophosphate kinase (UMPK) [47] and orotate phosphoribosyl transferase (OPRT) [48].

Irinotecan inhibits the activity of DNA topoisomerase I (topo-1), an enzyme involved in DNA replication and repair, by means of its active metabolite SN-38, which allows the irreversible binding of topo-1 to DNA, thus inducing DNA damage, G2 arrest and apoptosis [49]. A positive correlation between topo-1 activity and sensitivity to CPT-11/SN-38 was observed in human CRC cells, while the formation of topo-1/DNA complexes seems to be a predictive factor of response in CRC xenografts. Moreover, a reduced affinity for SN-38 was shown by topo-1 mutants, suggesting that *Top-1* mutations can affect CPT-

11 sensitivity [39]. The key enzyme involved in CPT-11 detoxification process is uridine diphosphogluronysltransferase 1A1 (UGT1A1), which inactivates SN-38 via glucuronidation. An UGT1A1-induced increase of drug clearance may enhance CRC resistance to CPT-11. For this reason, irinotecan-based therapies may be more effective for patients harboring a silenced *UGT1A1* gene in the primary tumor and patients with an active *UGT1A1* gene in normal cells [50].

A pivotal mediator of CPT-11 sensitivity is the cell cycle control gene *p16*, which is methylated in CRC. Crea et al. [51] have observed that *p16* methylation renders CRC cells more resistant to irinotecan-induced cell cycle arrest. Loss of *MLH1* gene has been shown to be associated with a greater irinotecan sensitivity in CRC cell lines, as it increases SN-38-induced apoptosis due to defective DNA repair processes, suggesting that MMR deficiency may be a predictive factor of CPT-11 response in advanced CRC [52].

Oxaliplatin is a third-generation platinum compound, derivative of cisplatin, which exerts its anticancer activity, by binding to DNA and forming GG intrastrand cross-links and DNA adducts, that inhibit DNA synthesis and trigger apoptosis [53]. However, drug efficacy is restricted by development of resistance mechanisms that lead to a decreased drug accumulation or reduced DNA-Pt adduct formation. Several mechanisms may be responsible for resistance to platinum compounds, including increased cellular efflux, reduced cellular uptake, suppression of DNA adducts through reaction with glutathione or other metallothioneins, increases in the NER pathway [43]. Cellular defense mechanisms may remove DNA adducts (e.g., NER and BER) or inhibit their formation (e.g., glutathione-S-transferase). MMR deficiency is not correlated with L-OHP resistance, as MMR proteins are unable to bind DNA adducts formed by oxaliplatin [54]. Conversely, ERCC1-mediated NER seems to be the main pathway implicated in oxaliplatin processing and platinum drug-induced DNA damage repair. In fact, increased expression levels of some pivotal components of the NER machinery, such as ERCC1 and XPA, are associated with CRC resistance to L-OHP [55]. Recent findings suggested that other genes and multiple pathways may be involved in the development of oxaliplatin resistance, including PI3K/Akt activation, pyruvate kinase M2 (PKM2) down-regulation and altered mitochondria-mediated apoptosis [56].

11.2.2 Impact of ncRNAs in Resistance to Conventional Chemotherapy

Chemotherapy sensitivity or resistance may be affected by epigenetic mechanisms that mainly involve variations in intracellular miRNAs expression levels. Accumulating evidence demonstrated that different miRNAs and lncRNAs are involved in the acquisition of CRC cell resistance to conventional agents such as 5-FU, oxaliplatin and irinotecan [29] (Table 11.1).

Most of findings regarding the role of miRNAs in drug resistance mostly relies on preclinical *in vitro* models. These studies suggested that 5-FU resistance can be mediated by many miRNAs, including miR-10b, miR-19a/b, miR-20a, miR-21, miR-22, miR-23a, miR-31, miR-34a, miR-122, miR-129, miR-140, miR-143, miR-145, miR-148a, miR-192/215, miR-200 family, miR-224, miR-497, miR-519c and miR-520g. Furthermore, other miRNAs have been shown to mediate the irinotecan resistance (miR-21, miR-451 and miR-519c) and oxaliplatin resistance (miR-27b, miR-20a, miR-21, miR-133a, miR-143, miR-153, miR-181b, miR-196a, miR-203, miR-222, miR-297, miR-520g, miR-625-3p and miR-1915) [57] (Table 11.1). Nishida et al. [58] have showed that miR-10b is an independent prognostic marker for survival in CRC and its expression can be associated with chemosensitivity to 5-FU in HCT-116 CRC cell lines. High miR-10b expression levels may confer 5-FU chemoresistance, suppressing drug-induced apoptosis through direct inhibition of the pro-apoptotic *BIM* gene, a BH3-only Bcl-2 family member [58].

In vitro studies carried out by Kurokawa and colleagues [59] revealed that miR-19b and miR-

Table 11.1 ncRNAs involved in resistance to conventional chemotherapy

ncRNA	Expression	Drugs	Targets	References
Let-7g/miR-181b	↑	S-1	*RAS, cyclin D, C-myc, E2F, cytochrome C*	[102]
miR-10b	↑	5-FU	*BIM*	[58]
miR-19b	↑	5-FU	*SFPQ, MYBL2*	[59]
miR-20a	↑	5-FU, Oxaliplatin	*BNIP2*	[64]
miR-21	↑	5-FU, Irinotecan, Oxaliplatin	*hMSH2/hMSH6, PDCD4*	[59, 63, 66–70]
miR-22	↓	5-FU	*BTG1*	[72]
miR-23a	↑	5-FU	*APAF-1*	[73]
miR-31	↑	5-FU	N/S	[74]
miR-34a	↓	5-FU	*Sirt1, E2F3, c-Kit, LDHA*	[75–78]
miR-122	↓	5-FU	*PKM2*	[82]
miR-129	↓	5-FU	*Bcl-2, TYMS, E2F3*	[83]
miR-133a	↓	Oxaliplatin	*RFFL*	[104]
miR-140	↑	5-FU	*HDAC4*	[84]
miR-141/200c	↓	Oxaliplatin	*ZEB1*	[108]
miR-143	↓	5-FU	*Bcl-2, NF-kB, ERK5*	[86]
miR-143	↓	Oxaliplatin	*IGF1-R*	[87]
miR-145	↓	5-FU	*Fli-1, RAD18*	[78–81]
miR-148a	↓	5-FU + Oxaliplatin	N/S	[88]
miR-153	↑	Oxaliplatin	*FOXO3a*	[105]
miR-192/215	↑	5-FU	*TYMS, DHFR*	[89, 91, 92]
miR-196a	↓	Oxaliplatin	*HoxA7, HoxB8, HoxC8, HoxD8*	[107]
miR-200 cluster	↓	5-FU	EMT-related genes	[93–95]
miR-203	↓	5-FU	*TYMS*	[96]
miR-203	↑	Oxaliplatin	*ATM*	[97]
miR-224	↑	5-FU	N/S	[98]
miR-222	↓	Oxaliplatin	*ADAM17*	[109]
miR-297	↓	Oxaliplatin	*ABCC2*	[110]
miR-451	↓	Irinotecan	*ABCB1*	[103]
miR-497	↓	5-FU	*IGF1-R*	[99]
miR-519c	↓	5-FU, Irinotecan	*ABCG2, HuR*	[100]
miR-520g	↑	5-FU, Oxaliplatin	*p21*	[101]
miR-625-3p/27b/181b	↑	Oxaliplatin	N/S	[112]
miR-1915	↓	Oxaliplatin	*Bcl-2*	[111]
snaR	↓	5-FU	N/S	[113]
BACE1AS	↓	5-FU	N/S	[113]

↑ Up-regulated and ↓ down-regulated ncRNAs in chemoresistance. 5-FU = 5-fluorouracil. *N/S* target not specified

21 were up-regulated in 5FU-resistant DLD-1 cells. Although miR-19b is encoded by the *miR-17-92* cluster and its activation is dependent on the accumulation of the transcription factor E2F1 in the G1 phase of the cell cycle, no alteration in cell cycle profile in response to 5-FU treatment was found. Further validation experiments confirmed data from computational analysis showing that, after transfection of miR-19b, the *SFPQ* and *MYBL2* genes, involved in cell cycle regulation, were putative targets implicated in 5-FU resistance [59]. SFPQ (splicing factor proline and glutamate-rich) is involved in mRNA processing and keeping sister chromatid interac-

tion during cell cycle [60]. The loss of its function determines abnormal accumulation of cells in the cell cycle S phase [61]. MYBL2 (v-Myb myeloblastosis viral oncogene homolog-like2) is a transcription factor directly regulated by E2F and greatly induced during the G1 to S-phase progression of cell cycle [62]. Likewise, Rossi et al. showed that miR-19a (a paralogue of miR-19b) and miR-21 were overexpressed in HCT-119 and HT29 cells in response to 5-FU [63].

Up-regulation of miR-20a has been shown to be associated with chemoresistance to 5-FU and oxaliplatin in SW620 and SW480 CRC adenocarcinoma cells, by inhibiting apoptosis through targeting of the pro-apoptotic *BNIP2* gene and down-regulation of its expression [64]. BNIP2 is a member of the BH3-only Bcl-2 protein family whose pro-apoptotic activity depends on the caspase-mediated cleavage [65]. Conversely, miR-20a knockdown has determined increased cancer cell sensitivity to chemotherapeutic agents, indicating that this miRNA may be a therapeutic target for drug resistance in CRC [64].

Several papers reported that miR-21 oncomiR is a potential mediator of the CRC chemoresistance to 5-FU, irinotecan and oxaliplatin, through modulation of different molecular mechanisms. Overexpression of miR-21 has been detected in many types of cancer, including CRC, and was correlated with down-regulation of several tumor suppressor target genes such as *p21*, *PDCD4*, *TIMP3*, *TGFBR2*, *PTEN*, *RECK*, *TPM1*, *RhoB*, *Bax* [66]. Faltejskova et al. [67] have analyzed the effects of miR-21 knockdown on apoptosis, cell cycle, viability and chemosensitivity of DLD1 cells and found that miR-21 silencing alone does not affect the cell viability, except when it is in combination with therapeutic agents such as 5-FU, oxaliplatin and irinotecan, an event in which it determines decreased cell viability. In addition, suppression of miR-21 has produced a increase in apoptosis rate only when combined with 5-FU, but not with L-OHP and CPT-11. Also, no change in cell cycle distribution was detected following the anti-miR-21 transfection in combination with 5-FU, L-OHP and CPT-11, even if a higher number of DLD-1 cells in S-phase was observed [67]. Valeri et al. [68] showed that miR-21 induces 5-FU chemoresistance by down-regulating the expression of hMSH2 (human mutS homolog 2), a core MMR component, thus resulting in a reduction of 5-FU-induced G2/M damage arrest and apoptosis. High expression levels of miR-21 detected in 5-FU-resistant CRC cells and xenografts are likely to increase mutation rates, generating defects in MMR system and, thereby, enhancing tumor progression [68]. Afterwards, Deng and collaborators [69] have confirmed the previously obtained results, demonstrating that miR-21 overexpression was associated with 5-FU chemoresistance also in HT-29 colon cancer cells, through targeting of *hMSH2*, indirect decrease of TP and DPD expression, inhibition of apoptosis and increased invasion and cell proliferation. Contrariwise, miR-21 knockdown has reversed these effects, restoring the HT-29 chemosensitivity to 5-FU [69]. Further, Yu et al. [70] have reported that miR-21 silencing induces differentiation of cancer stem/stem-like cells (CSCs/CSLCs)-enriched chemoresistant HCT-116 and HT-29 cells, by decreasing the ability to form colonospheres and T-cell factor/lymphoid enhancer factor (TCF/LEF) activity, increasing the expression of proapoptotic *PDCD4* target gene, and consequently enhancing cancer cell chemosensitivity to combined therapeutic regimens containing 5-FU and L-OHP [70].

Since inhibition of autophagy by hydroxychloroquine (HCQ) and 3-methyladenine was shown to promote 5-FU-induced apoptosis in CRC cells [71], Zhang et al. [72] investigated the opportunity to target the switch between autophagy and apoptosis in order to overcome chemoresistance. Using preclinical *in vitro* and *in vivo* models, they suggested that miR-22 may modulate CRC chemosensitivity to 5-FU, by inducing apoptosis and inhibiting autophagy pathway, one of the most important mechanisms of chemotherapy resistance supporting the tumor cell survival. Up-regulation of miR-22 may inhibit autophagy by down-regulating the expression of its target gene *BTG1* (B-cell translocation gene 1), which, in turn, may suppress miR-22-induced inhibition of autophagy [72].

Recently, miR-23a silencing was associated with enhanced chemosensitivity to 5-FU in HCT116 and HT29 CRC cells, through increased expression of its target gene *APAF-1* and activation of the 5-FU-induced mitochondria-mediated apoptosis. In fact, anti-miR-23a overexpression caused activation of the caspases 3, 7 and 9, whereas miR-23a up-regulation reversed these effects, inhibiting 5-FU-induced apoptosis [73].

Wang and colleagues [74] showed that miR-31 suppression enhanced 5-FU chemosensitivity at an early stage in HCT-116 cells, inhibited proliferation partly in combination with 5-FU through an apoptotic mechanism, decreased migration but increased invasive capacity. However, little is known on the biological functions of miR-31 due to lack of knowledge about its target genes in CRC [74].

Several studies reported that miR-34a is one of the most down-regulated miRNAs in CRC cells chemoresistant to 5-FU. Akao et al. [75] have observed that 5-FU-resistant DLD-1 cells revealed an increase in growth and a probable inhibition of apoptosis determined by activation of the PI3K/Akt signaling pathway, low expression levels of miR-34a, and increased expression of Sirt1 and E2F3. *Sirt-1* is a target gene of miR-34a and up-regulation of its expression was correlated with 5-FU chemoresistance, whereas suppression of this gene caused enhanced 5-FU chemosensitivity in 5-FU-resistant cells. Conversely, transfection of 5-FU-resistant DLD-1 cells with ectopic miR-34a induced cell growth arrest and significantly reduced the 5-FU chemoresistance, by down-regulating the expression of *Sirt1* and *E2F3* [75]. In addition, Siemens et al. [76] have demonstrated that p53-induced up-regulation of miR-34 mediated repression of *c-Kit* by p53 via direct targeting of *c-Kit* mRNA, determining a higher CRC cell sensitivity to 5-FU, and leading to a decrease in Erk signaling and transformation, induced by c-Kit down-regulation, and inhibition of stem cell factor (SCF)-induced invasion/migration [76]. Recently, Li et al. [77] have indicated another mechanism by which miR-34a up-regulation may render 5-FU-resistant CRC cells sensitive to 5-FU, through direct repression of the lactate dehydrogenase A (LDHA) expression, resulting in inhibition of glucose metabolism. In a recent work, Akao and other authors also showed that DLD-1 cell chemoresistance to 5-FU was significantly correlated with the intra- and extracellular levels of miR-34a and miR-145, and was caused by increased secretion of both miRNAs via microvesicles that reduced their intracellular levels [78]. Furthermore, miR-145 was shown to inhibit cell proliferation and sensitize LS174T colon cancer cells to 5-FU-induced apoptosis, through targeting and down-regulation of *Fli-1* oncogene, resulting in Rb up-regulation and Bcl-2 down-regulation [79]. A further recently discovered mechanism by which miR-145 may reverse 5-FU resistance in CRC cells is the direct targeting of *RAD18* gene, encoding a DNA damage-activated E3 ubiquitin ligase involved in DNA damage repair process. The repression of RAD18 expression by miR-145 increases DNA damage, enhancing effectiveness of 5-FU [80]. Recently, Findlay et al. [81] demonstrated that SNAI2 (Slug), a protein involved in epithelial-mesenchymal transition (EMT) process, may mediate 5-FU resistance by repressing the activity of miR-145 promoter and thus miR-145 expression in CRC cells.

A significant increase in glucose metabolism was associated with miR-122 down-regulation in 5-FU-resistant CRC cells. Indeed, overexpression of miR-122 in 5-FU-resistant cells allowed to overcome the 5-FU resistance, through inhibition of glycolysis by directly targeting *PKM2* both *in vitro* and *in vivo*, thus restoring the sensitivity to drug [82].

Karaayvaz and colleagues [83] showed that miR-129 is a key mediator of 5-FU-induced cell death involved in CRC chemosensitivity to 5-FU. Preclinical *in vitro* and *in vivo* models revealed that ectopic expression of miR-129 triggered the intrinsic apoptotic pathway through direct targeting and down-regulation of *Bcl-2*, inhibited cell proliferation, induced cell cycle arrest, and enhanced 5-FU cytotoxicity in CRC cells. In addition, miR-129 has been shown to exert a synergistic effect in restoring the 5-FU chemosensitivity, suppressing also the expression of E2F3 and 5-FU target enzyme TS [83].

Experimental evidence indicated that miR-140 induced 5-FU chemoresistance in HCT116 cells by suppressing one of most important target genes such as *HDAC4* [84]. Since HDAC4 has been shown to promote growth of colon cancer cells by repressing p21 [85], ectopic expression of miR-140 induced an increase in p53 and p21 expression levels in wt-p53 HCT116 cells, inhibiting cell proliferation through G1 and G2 cell cycle arrest. However, null-p53 HCT116 cells did not shown increased expression of p21, suggesting that miR-140 exerts its functions in a p53-dependent manner. Therefore, targeting of miR-140 might be an effective strategy to overcome 5-FU resistance in CRC [84]. Contrariwise, miR-143-overexpressing HCT116 cells exposed to 5-FU exhibited reduced viability and increased cell death, indicating that miR-143 enhanced sensitivity to 5-FU, by down-regulating the expression of target genes, such as *ERK5*, *NF-kB*, *Bcl-2*, involved in the regulation of cell proliferation, death and chemotherapy response. Since direct activation of NF-kB by ERK5 promotes cell cycle progression through G2-M, a decreased expression of both proteins may cause reduced cell growth and greater response to 5-FU. Additionally, miR-143-induced reduction of expression of anti-apoptotic protein Bcl-2 may confer to miR-143 a putative pro-apoptotic role [86]. In the same way, though with a different mechanism, miR-143 has been shown to confer a greater chemosensitivity to oxaliplatin in CRC cells, by directly inhibiting its target gene *IFG1-R*, thus resulting in suppression of cell proliferation and tumor growth, and increased induction of apoptosis through caspase-3 activation [87].

A recent study performed on a large cohort of specimens revealed that miR-148a could be an potential predictive biomarker of therapy response in stage IV CRC patients treated with combined therapeutic regimen containing 5-FU and L-OHP. Indeed, they found that decreased miR-148a expression in these patients may be correlated with poor survival and unfavorable response to 5-FU and L-OHP through a mechanism not yet elucidated, whereas it is linked to a poor outcome in stage III patients treated alone with 5-FU [88].

Since TYMS is a target for chemotherapeutic drugs such as 5-FU and its transcriptional and translational regulation was shown to affect cell chemosensitivity, Boni et al. [89] examined the effects of down-regulation of its expression mediated by miR-192 and miR-215 in CRC cell lines, reporting that both miRNAs could be potential predictive biomarkers of 5-FU resistance. Overexpression of miR-192 and miR-215 decreased cell proliferation, blocking cell cycle progression into the S phase and thereby reducing 5-FU sensitivity. Also, miR-192/-215-mediated regulation of cell cycle was dependent in part on p53 status, as it was associated with p21 and p27 induction [89]. Since it has previously been reported that miR-192 down-regulates also the expression of DHFR, a key enzyme of folate metabolism involved in DNA synthesis and targeted by antifolate-based chemotherapy [90], it has been speculated that miR-192 and miR-215 via targeting of *TYMS* and *DHFR* may modulate p53 activity, altering the cell cycle and influencing therapy response in CRC [89]. Furthermore, a recent work confirmed, using miRNA microarray analysis, that miR-215 could potentially predict response to adjuvant chemotherapy in stage II CRC patients [91]. In addition, miR-215 was reported to be a novel potential biomarker of chemoresistance to DHFR inhibitor methotrexate (MTX) and TS inhibitor Tomudex (TDX) in CRC cells, by inducing cell cycle G2-arrest through inhibition of DTL (denticleless protein homolog) target, a key E3 ubiquitin ligase required for cell cycle control [92].

Recently, several studies showed that miR-200 family members may mediate 5-FU chemosensitivity in CRC cells, by down-regulating the expression of proteins, including ZEB1 and ZEB2, involved in regulation of EMT, and preventing also the suppression of E-cadherin synthesis, necessary for intercellular adhesion [93–95].

Divergent results were reported concerning the miR-203 role in conferring chemoresistance to 5-FU or oxaliplatin in CRC. Li et al. [96] found that miR-203 was down-regulated in 5-FU-resistant cells, whereas the inhibitory effects of 5-FU on tumor growth were enhanced by miR-203 overexpression in preclinical *in vivo* models, increasing 5-FU chemosensitivity via targeting of *TYMS*. Conversely, Zhou et al. [97] showed that miR-203 overexpression induced acquired chemoresistance to oxaliplatin in CRC cell lines via negative regulation of expression of ataxia telangiectasia mutated (ATM), a kinase involved in DNA damage response pathway. Furthermore, mutations in the 3′ untranslated region (3′UTR) of the *ATM* mRNA that prevent the binding with miR-203 have been shown to suppress the inhibitory effect of miR-203 on ATM [97].

In a recent work [98], miR-224 silencing was correlated with a greater chemosensitivity to 5-FU-based chemotherapy in CRC cell lines, leading to alterations in cell proliferation, invasion and EMT phenotype. Interestingly, CRC cells harboring *KRAS* and *BRAF* mutations were more sensitive to 5-FU than wt-*KRAS* and wt-*BRAF* cells [98].

Guo and colleagues [99] showed that miR-497 may increase sensitivity of CRC cell to 5-FU, favoring drug-induced apoptosis through targeting and down-regulation of the expression of insulin-like growth factor 1 receptor (IGF1-R), and inhibition of Akt activation. Conversely, miR-497 down-regulation caused by gene copy number reduction determines increased IGF1-R expression and activation of the IGF-1/IGF-1R and PI3K/Akt signaling pathways, inducing chemoresistance to 5-FU [99].

Experimental evidence reported that CRC resistance to 5-FU and irinotecan may be caused by overexpression of the MDR transporter ABCG2, as both agents are substrates for ABCG2. To et al. [100] have observed that CRC patients resistant to adjuvant chemotherapy exhibited high expression levels of ABCG2 and mRNA binding protein HuR, but concomitantly reduced miR-519c expression, instead patients respondent to therapy showed the reverse situation. Since ABCG2 and HuR are known to be targets of miR-519c, the CRC chemosensitivity to 5-FU and irinotecan seems to be dependent on the induced-miR-519c inhibition of ABCG2 and HuR expression [100].

Recent findings revealed that miR-520g may confer chemoresistance to 5-FU and oxaliplatin in CRC cells, by inhibiting 5-FU- or L-OHP-induced apoptosis through down-regulation of p21 expression. Furthermore, CRC xenograft models showed a decrease in 5-FU-mediated suppression of tumor growth following the ectopic expression of miR-520g. In addition, p53 has been shown to inhibit miR-520g expression, whereas the loss of p53 function caused an increase in expression levels of miR-520g, suggesting an pivotal role of the p53/miR-520g/p21 signaling axis in therapy response. For this reason, miR-520g could be a potential therapeutic target to overcome drug resistance in CRC patients [101].

Let-7g and miR-181b were found down-regulated in tumor tissue specimens from CRC patients respondent to treatment with S-1, an analogue of 5-FU. Therefore, overexpression of both miRNAs was associated with chemoresistance to S-1, by inhibiting several predicted target genes, including *RAS*, *cyclin D*, *c-myc*, *E2F* and *cytochrome C* [102].

The anticancer therapy failure often is due to the presence of a small sub-population of cells within tumor, called cancer stem cells (CSCs), responsible for the onset, growth and progression of tumor as well as resistance to cytotoxic agents. Therefore, CSCs play a key role in determining therapy response in many tumors [31]. Bitarte et al. [103] showed that reduced expression levels of miR-451 were associated with chemoresistance of CRC stem cells to irinotecan-based first-line therapy. Conversely, miR-451 up-regulation has been shown to induce chemosensitivity to irinotecan via inhibition of expression of the ATP-binding cassette drug transporter ABCB1. Moreover, the authors found that miR-451 down-regulation causes an increase in expression of its target gene, macrophage migration inhibitory factor (*MIF*), responsible, in turn, for the expression of cyclooxygenase-2 (COX-2). Then, COX-2-mediated activation of Wnt pathway promotes CSC growth [103].

Recent findings revealed that miR-133a overexpression may confer chemosensitivity to oxaliplatin in CRC cells, by increasing apoptosis and suppressing cancer cell proliferation through a mechanism which probably involves p53 and the ring finger and FYVE-like domain containing E3-ubiquitin protein ligase (RFFL). Indeed, it has been demonstrated that miR-133a increases p53 expression and up-regulates the p53/p21 pathway, by directly targeting RFFL, a negative regulator of p53 [104].

Functional studies showed that increased miR-153 expression induced platinum-based chemotherapy resistance both *in vitro* and *in vivo*, by directly repressing the expression of the Forkhead transcription factor FOXO3a and mediating anti-apoptotic effects through reduced caspase-3 activation, up-regulation of anti-apoptotic genes and down-regulation of pro-apoptotic genes such as *PUMA* and *Bim* [105]. FOXO3a is a member of the subfamily of FOXO transcription factors, involved in apoptosis, differentiation, DNA damage response, and chemoresistance [106].

Schimanski and colleagues [107] reported that high expression levels of miR-196a increased chemosensitivity of CRC cells towards platinum-derived agents but not towards 5-FU or irinotecan, via inhibition of the *HoxA7*, *HoxB8*, *HoxC8*, and *HoxD8* target genes, involved in regulation of oncogenesis, embryogenesis and organogenesis. Since miR-196a has not been shown to affect cell proliferation or apoptosis, the exact chemosensitivity mechanism is still to be elucidated [107].

Recent evidence showed that the expression of miR-141 and miR-200c was significantly down-regulated in SW620 CRC cells with acquired resistance to oxaliplatin, but not in cells resistant to 5-FU and irinotecan. In addition, L-OHP-resistant cells exhibited EMT phenotype and increased expression of ZEB1. Therefore, acquisition of L-OHP resistance may be suppressed by up-regulation of miR-141 and miR-200c which block EMT, by inhibiting ZEB1 [108].

Preclinical *in vitro* models suggested that miR-222 down-regulation is correlated with a novel MDR mechanism mediated by up-regulation of ADAM17 (a disintegrin and metallopeptidase domain 17), conferring chemoresistance to oxaliplatin in CRC cells. Conversely, ectopic expression of miR-222 in resistant cells reduced the ADAM17 expression via direct targeting of its mRNA, sensitizing these cells to L-OHP-induced apoptosis [109]. Furthermore, Xu et al. [110] showed that miR-297 down-regulation was involved in a MDR mechanism conferring L-OHP resistance mediated by up-regulation of the *ABCC2* gene encoding the MDR-associated protein 2 (MPR-2). Likewise, ectopic expression of miR-222 in multidrug-resistant CRC cells reduced the MRP-2 transporter levels through targeting of *ABCC2*, rendering cells sensitive to L-OHP-induced apoptosis [110]. Also miR-1915 has been also shown to modulate MDR via inhibition of Bcl-2, by enhancing sensitivity of CRC cells to L-OHP-induced apoptosis [111].

In a recent work, Rasmussen et al. [112] observed that up-regulation of miR-625-3p, miR-181b and miR-27b was associated with chemoresistance to L-OHP-based first-line therapy (XELOX/FOLFOX) in CRC cells. In particular, miR-625-3p has not shown to be a prognostic biomarker but only a response biomarker closely associated with resistance to XELOX treatment, through a mechanism yet unidentified, since to date there is not a validate target gene for miR-625 and little is known about its functions [112].

Lastly, although the functional link between lncRNAs and the acquisition of drug resistance is yet unclear, recent *in vitro* studies aimed to identify lncRNAs involved in resistance to chemotherapy revealed that down-regulation of the lncRNA snaR (small NF90-associated RNA) may contribute to confer 5-FU resistance in CRC cells, increasing viability and inhibiting cancer cell death without altering the cell cycle distribution. Furthermore, other lncRNAs have been shown to be differentially expressed in 5-FU-resistant CRC cells, including BACE1AS (down-regulated) [113].

11.2.3 Targeted Therapies and Resistance Mechanisms

Recent advances in understanding molecular mechanisms driving tumors led to the development of new therapeutic modalities targeting selectively specific molecular pathways, by improving the prognosis of patients with advanced CRC. New biological agents mainly target two different pathways in mCRC: tumor growth mediated by proangiogenic factors such as vascular endothelial growth factor (VEGF), and cell proliferation triggered by epidermal growth factor receptor (EGFR) [114]. Currently, targeted therapies approved for mCRC include three drug groups: (1) monoclonal antibodies (mAbs) targeting VEGF (bevacizumab) and EGFR (cetuximab and panitumumab); (2) recombinant fusion proteins targeting angiogenic factors, including VEGF (aflibercept); (3) molecules that inhibit tyrosine kinase receptors (TKIs) located on the cancer cell membrane (e.g., regorafenib) [115, 116]. Several studies suggested that the first-line treatment for mCRC patients should include an oxaliplatin- or irinotecan-based standard chemotherapy in combination with a targeted agent such as anti-EGFR or anti-VEGF mAb, taking into account the RAS mutation status [117]. Indeed, about 30–40 % of mCRC patients harbors *KRAS* activating mutations that induce the constitutive activation of the downstream signaling pathway RAF-MEK-ERK, causing lack of response to anti-EGFR therapies [114]. Although novel targeted agents have improved prognosis and clinical outcomes of mCRC patients, the emergence of therapeutic resistance was frequently observed upon treatment with these agents.

Recent findings from preclinical and clinical studies have revealed that several mechanisms may be involved in the event of a failure or poor response to antiangiogenic therapy, including recruitment of bone marrow stromal cells, enhanced pericyte coverage of tumor blood vessels to support vasculature, hypoxia resistance, activation of alternative signaling pathways, increased expression of other angiogenic factors and cytokines [118]. Recently, Weickhardt et al. [119] showed that VEGF-D could be a predictive marker of resistance to bevacizumab, as mCRC patients resistant to bevacizumab-containing therapeutic regimens exhibited increased expression levels of VEFG-D. Moreover, several studies have demonstrated that bevacizumab resistance may be associated with a increase in expression levels of VEGF-C, P1GF, VEGFR-1, soluble VEGFR-2, thrombospondin-1 (THBS1), angiopoietins (Ang-2 and Tie-2), suggesting a possible involvement of these factors in tumor resistance [120].

Experimental evidence showed that *KRAS* mutations are negative predictive factors of response to anti-EGFR mAbs, as mCRC patients with alterations in codon 12 or 13 of *KRAS* exon 2 respond little to cetuximab or panitumumab, and thus should not be considered for monoclonal therapy. However, not all patients with *K-RAS wild-type* mCRCs are sensitive to anti-EGFR-based therapies and, conversely, not all responders bear *K-RAS wild-type* tumors [121]. Furthermore, there are different opinions about the possible predictive role of *BRAF* gene whose mutations are mutually exclusive to *KRAS* mutations [114]. Recently, multiple studies suggested that activating mutations in *KRAS* exon 2, *BRAF*-p.V600E, *PI3KCA*-exon 9, and loss of *PTEN* expression were correlated negatively with response to anti-EGFR mAbs in mCRC patients. Therefore, CRCs showing no alterations in these genes are defined as quadruple negative and exhibit a greater chance of responding to anti-EGFR treatment [122].

11.2.4 Impact of ncRNAs in Resistance to Novel Targeted Agents

Recent evidence revealed that some miRNAs may be involved in the acquisition of CRC cell resistance to novel targeted agents such as bevacizumab and cetuximab [123] (Table 11.2). However, there are still few studies on the topic. Since *KRAS* mutations are negative predictive biomarkers of response to anti-EGFR monoclonal therapy in CRC, miRNA-mediated

Table 11.2 ncRNAs involved in resistance to novel targeted agents in CRC

ncRNA	Expression	Agent	Targets	References
Let-7 family	↓	Cetuximab, Panitumumab	*KRAS*	[125–131]
miR-7	↓	Cetuximab	EGFR, RAF1, *ERK1/2*, *AKT*	[132]
miR-17* (miR-17-3p)	↑	Cetuximab	N/S	[125]
miR-31	↑	Cetuximab	N/S	[133]
cir-miR-126	↑	Bevacizumab + XELOX	N/S	[137]
miR-146b-3p	↑	Cetuximab	*IL1A*	[125]
miR-199a-5p/375	↑	Cetuximab	*PHLPP1*	[134]
miR-486-5p	↑	Cetuximab	*ARHGAP5*, ST5, DOCK3, *TOB1*, *PIK3R1*	[125]

↑ Up-regulated and ↓ down-regulated ncRNAs in resistance to targeted therapies. *N/S* target not specified

mechanisms which induce post-transcriptional down-regulation of mutated *KRAS* may improve the effectiveness of therapy in mCRC patients [124]. The members of the Let-7 family are miRNAs involved in response to anti-EGFR agents, by targeting *KRAS*. Indeed, high expression levels of Let-7 may enhance sensitivity of mCRC to anti-EGFR treatments, by inhibiting the expression of mutated *KRAS*, and increase survival benefits for patients. Additionally, other genes involved in regulation of cell cycle, such as *Myc* and *Bcl-2*, may be modulated by Let-7 [125]. The analysis of Let-7a expression levels in mCRC patients treated with cetuximab and irinotecan showed that Let-7a expression was correlated with response to therapy and OS in both *KRAS* mutant and *KRAS wild-type* individuals [126]. Ragusa et al. [125] suggested that down-regulation of Let-7b and Let-7e and up-regulation of miR-17* (known also as miR-17-3p) may be used as potential predictive markers of cetuximab resistance, although, to date, no clinical data confirmed these findings. Furthermore, the authors have observed high expression levels of miR-146b-3p and miR-486-5p in *KRAS*-mutated CRC patients when compared with *wild-type KRAS*, suggesting that these miRNAs may be involved in EGFR pathway and predict the cetuximab response. Upregulation of miR-146b-3p and miR-486-5p could be determined by constitutive activation of *KRAS* signaling. *PIK3R1*, which has been shown to be a target gene of miR-486-5p, was up-regulated in responsive patients after cetuximab treatment [125].

Recently, another study identified a signature consisting of the cluster Let-7c/miR-99a/miR-125b that could be useful for predict sensitivity of mCRC patients to EGFR-targeting agents. Patients harboring *KRAS wild-type* exhibited high-intensity signatures correlated with a significantly longer PFS. Therefore, the Let-7c/miR-99a/miR-125b signature could help to improve the selection of *KRAS wild-type* mCRC patients for anti-EGFR therapy [127]. Experimental evidence also demonstrated that high expression levels of Let-7g were associated with a good prognosis in rectal cancer patients treated with chemoradiotherapy, indicating that Let-7g could be used as predictive biomarker of chemoradiosensitivity [128].

Several studies revealed that the presence of the LCS6 polymorphism in the binding site for Let-7 of the 3′-UTR region of *KRAS* may be a predictive biomarker of response to anti-EGFR monotherapy in *wt-KRAS* and *wt-BRAF* mCRC patients, also correlating with improved outcomes in early stages of CRC. Furthermore, combination therapy with conventional chemotherapy agents did not give additional benefits [129–131].

Recently, Suto et al. [132] showed that miR-7 may mediate sensitivity to cetuximab in CRC cell lines via EGFR regulation, by targeting *EGFR* and *RAF-1* and suppressing the ERK1/2 and pAKT expression, thereby resulting in the inhibition of the cell proliferation. Ectopic expression of miR-7 induced cetuximab sensitivity in cetuximab-resistant HCT116 and SW480 cells harboring *KRAS* mutations, and HT29 cells harboring a *BRAF* mutation [132].

A recent study found that miR-31-5p/3p was up-regulated in mCRC patients who did not respond to cetuximab therapy, thus it could be used as a biomarker able to predict specifically cetuximab resistance. Moreover, no association between miR-31-5p/3p expression levels and response to panitumumab was detected [133].

Recently, Mussnich et al. [134] showed that up-regulation of miR-199a-5p and miR-375 determined cetuximab resistance in CRC cells, via targeting of tumor-suppressor gene *PHLPP1* (PH domain and leucine-rich repeat protein phosphatase 1) which negatively regulates the AKT pathway.

Finally, miR-126 has been supposed to be a putative tumor suppressor involved in the regulation of angiogenesis, a process targeted by bevacizumab. Previously, Hansen et al. [135] have observed that high expression levels of miR-126 were detected in primary tumors from mCRC patients who responded to XELOX, conferring chemosensitivity to first-line XELOX treatment. Conversely, miR-126 down-regulation in primary tumors was correlated with resistance to XELOX, causing reduced integrity of tumor vessels and increased interstitial pressure [135]. These results were confirmed by another study that demonstrated a correlation between high expression of miR-126 and a longer PFS in mCRC patients [136]. Lastly, a recent finding revealed that high levels of circulating miR-126 (cir-miRNA-126) were associated with bevacizumab resistance and lack of benefits in mCRC patients treated with bevacizumab plus XELOX. For this reason, cir-miR-126 could become, in future, a potential predictive biomarker for the resistance to anti-angiogenic therapies [137].

11.3 Non-coding RNAs Involved in Radioresistance of CRC

Since radiation therapy appears to affect epigenetic patterns, by causing a state of genetic instability and inducing apoptosis in cancer cells, other important factors that may influence the response/resistance to chemoradiotherapy (CRT) are ncRNAs, such as miRNAs and lncRNAs. Their aberrant alterations have been widely investigated in many tumors, including CRC, and seem to have an important role in therapy response, by affecting radiation sensitivity of cancer cells [138].

11.3.1 Radiotherapy and Radiation Resistance

Radiotherapy is a localized treatment by means of ionizing radiations mainly used in combination with chemotherapy, preferentially in rectal cancer [139]. CRT can be used either prior or after surgery. CRT exerts its action inducing DNA damage mainly through irradiation or production of chemicals radicals. The concomitant administration of chemotherapeutic agents may serve as a radiosensitizer [140, 141]. Preoperative radiotherapy (neoadjuvant) is used in rectal cancer patients to reduce the risk of cancer recurrence after surgery, but also to determine tumor shrinkage favoring the complete removal of the mass. In addition, since about 15 % of all rectal cancer patients is diagnosed with unresectable disease and is not amenable for primary surgical resection, neo-adjuvant CRT aimed at tumor shrinkage may be an effective option [142].

Even though CRT represents an effective treatment against colorectal cancer, it has been shown that not all patients experience the same response rate. In the era of the “precision medicine” it is fundamental to choose the right patient for the appropriate treatment, thus the identification of a predictive biomarker for CRT could further improve survival for rectal cancer patients [143].

Recently, increasing evidences showed that cancer stem cells (CSCs) may be also responsible for resistance to different therapies, including CRT. CSCs are a heterogeneous cancer cell population, able to differentiate and to self-renew, that shows resistance to radiations [144]. It has been shown that radiosensitivity/radioresistance of CSCs is related to both extrinsic properties, which include signals from the extracellular environment, and intrinsic properties including DNA repair, cell cycle status and survival pathways. Upon radiation, CSCs seem to protect themselves, by increasing the ability to repair DNA damage through several mechanisms, including double-strand break repair [145], MMR [146], NER [147] and BER [148]. An additional property of CSCs associated with radioresistance is their capacity to remain in a quiescent state [149, 150].

11.3.2 Impact of ncRNAs in Radioresistance

Non-coding RNAs are important determinants that may influence the response/resistance to radiotherapy (Table 11.3). RNA expression profiling has revealed that miRNA deregulation in CRC tissues influences the activity of signaling pathways that may be associated with prognosis and response to CRT.

MiRNA-622 and miRNA-630 demonstrated a remarkable efficacy in predicting pathological complete response (pCR) [151]. These miRNAs regulate genes and signaling pathways involved in cell repair following CRT. In particular, miRNA-630 reduces ability of cells to repair DNA damage after cisplatin-based chemotherapy in non-small cell lung cancer thus providing a possible explanation for the benefit seen in the patient cohort receiving oxaliplatin-based CRT, which, however, may not be transferable to the more standard 5-FU-based neoadjuvant treatment [151]. However, there are some conflicting data reported by Ma et al. in rectal cancer cell lines [152]. In this recent paper it is reported that miR-622 is significantly up-regulated in CRC cell lines exposed to ionizing radiations. Interestingly, this over-expression is maintained and persisted stably in surviving cells treated with continuous low-dose radiation, providing an evidence that miR-622 induces radioresistance *in vitro*. In the same work it was reported that miR-622 inhibits Rb by directly targeting *RB1*-3′UTR, and miR-622-induced radioresistance may be reversed by overexpressing Rb [152]. Thus, miR-622 may be a radioresistance biomarker. On the contrary, miR-630 has been found to positively correlate with radiosensitivity in CRC cell lines [153]. The levels of miR-630 are also significantly decreased after repeated ionic radiation confirming the possible role of miR-630 in regulating pathway fundamental for radiosensitivity. The main targets of miR-630 are *BCL2L2* and *TP53RK* that are both involved in cell survival and apoptosis inhibition. Therefore, miR-630 may be considered a radiosensitivity biomarker, because its up-regulation negatively influences the expression of BCL2L2 and TP53RK, leading to the activation of apoptotic mechanisms [153].

Another radioresistance biomarker may be miR-100 as recently reported by Yang et al. [154]. In this study 33 differentially expressed miRNAs were identified in CRC cell lines through miRNA sequencing. Of these miRNAs, miR-100 shows a lower expression in CRC tissue than in normal tissue. Furthermore, miR-100 is

Table 11.3 ncRNAs involved in resistance/sensitivity to CRT

miRNA	Expression	Targets	Effects	References
miR-100	↓	N/S	Radioresistance	[154]
miR-106b	↑	*PTEN*, *p21*	Radioresistance	[155]
miR-622	↑	*Rb*	Radioresistance	[151, 152]
miR-630	↑	*BCL2L2*, *TP53RK*	Radiosensitivity	[151, 153]
lincRNA-p21	↓	*β-catenin*	Radiosensitivity	[156]

↑ Up-regulated and ↓ down-regulated ncRNAs in resistance/sensitivity to CRT. *N/S* target not specified down-regulated after X-ray irradiation of CCL-244 cells and seems to be involved in radioresistance, therefore, up-regulation of miR-100 restores radiation sensitivity [154].

MiR-106b overexpression can determine radioresistance both *in vitro* and *in vivo* by inhibiting apoptosis and promoting cell proliferation. To confirm this hypothesis Zheng et al. [155] have demonstrated that knock-down of miR-106b in CRC cell lines re-constitutes the radiosensitivity. The main factors affected by the overexpression of miR-106b seem to be PTEN and p21. Indeed, upon miR-106b up-regulation a reduction in PTEN and p21 expression is observed. The down-regulation of these proteins leads to the activation of AKT determining cell survival and proliferation [155].

Finally, the long intergenic non-coding RNA-p21 (lincRNA-p21) seems to be a interesting radiosensitivity biomarker, being involved in the regulation of the β-catenin pathway. In particular, the authors observed low expression levels of lincRNA-p21 in CRC cell lines and tissue samples and, concomitantly, increased levels of β-catenin. The expression of lincRNA-p21 increases following radiation exposure and enforced expression of the lincRNA enhances the CRC sensitivity to radiotherapy, by promoting cell apoptosis. These preliminary data indicates that lincRNA-p21 could serve as radiosensitivity marker in CRC patients [156].

11.4 Conclusions and Future Perspectives

Resistance to conventional chemotherapy may be caused by several mechanisms that partially overlap with those implicated in resistance to targeted therapies, including modifications of drug targets, reduced drug uptake, drug inactivation, increased drug elimination, etc. In addition, radioresistance is mainly related to the aptitude of CSCs to protect themselves, by increasing the ability to repair DNA damage induced by radiations.

Today, several studies focused on the use of ncRNAs, especially miRNAs and lncRNAs, as possible therapeutic targets in chemo- and radioresistance of CRC, by investigating their aberrant expression in several chemo- and radioresistant tumors, in order to suppress this dysregulation by means of ncRNA-based drugs (antagomiRs and miRNA mimics) and overcome the resistance. The identification of new potential molecular mechanisms involved in chemo- and radioresistance could be an important clinical tool to select CRC patients who may benefit from individualized therapies. Recent progress in the development of miRNA-based anti-cancer therapeutic approaches provided interesting results. However, in spite of encouraging obtained results, the introduction of ncRNAs in clinical practice seems to be still far. The development of new therapeutic approaches concerning the possible use of ncRNAs as potential targets has raised some doubts. Since miRNA targeting is sequence-specific, whereas gene silencing requires only a partial complementarity between miRNA and mRNA, a possible issue is to preserve target specificity, avoiding that a specific miRNA-based therapy may induce unexpected gene alterations. Another restricting factor is to obtain a high therapeutic effectiveness in relation to the number of cells that must be targeted and extent of target gene modulation. Hopefully, in the near future, specific ncRNA signatures could offer new insights about the possible mechanisms of chemo- and radioresistance that CRC patients may develop before starting therapy, whereas the modulation of expression of specific ncRNAs might provide a new tool to overcome acquired resistance. In conclusion, the identification of candidate ncRNAs able to modulate the resistance in CRC and the study of their molecular mechanisms could serve for designing novel and targeted ncRNA-based therapeutic strategies to improve the clinical outcome of CRC patients. However, further investigations are needed to specifically assess these approaches in CRC patients who do not respond to chemo- and radiotherapy.

References

1. Siegel RL, Miller KD, Jemal A. Cancer statistics, 2015. CA Cancer J Clin. 2015;65(1):5–29.
2. Torre LA, Bray F, Siegel RL, Ferlay J, Lortet-Tieulent J, Jemal A. Global cancer statistics, 2012. CA Cancer J Clin. 2015;65(2):87–108.

3. Edwards BK, Ward E, Kohler BA, Eheman C, Zauber AG, Anderson RN, et al. Annual report to the nation on the status of cancer, 1975–2006, featuring colorectal cancer trends and impact of interventions (risk factors, screening, and treatment) to reduce future rates. Cancer. 2010;116(3):544–73.
4. Zhang X, Li J. Era of universal testing of microsatellite instability in colorectal cancer. World J Gastrointest Oncol. 2013;5(2):12–9.
5. Caruso S, Bazan V, Rolfo C, Insalaco L, Fanale D, Bronte G, et al. MicroRNAs in colorectal cancer stem cells: new regulators of cancer stemness? Oncogenesis. 2012;1:e32.
6. Bronte G, Silvestris N, Castiglia M, Galvano A, Passiglia F, Sortino G, et al. New findings on primary and acquired resistance to anti-EGFR therapy in metastatic colorectal cancer: do all roads lead to RAS? Oncotarget. 2015;6(28):24780–96.
7. Dienstmann R, Salazar R, Tabernero J. Personalizing colon cancer adjuvant therapy: selecting optimal treatments for individual patients. J Clin Oncol. 2015;33(16):1787–96.
8. Marin JJ, Sanchez de Medina F, Castano B, Bujanda L, Romero MR, Martinez-Augustin O, et al. Chemoprevention, chemotherapy, and chemoresistance in colorectal cancer. Drug Metab Rev. 2012;44(2):148–72.
9. Skvortsova I, Debbage P, Kumar V, Skvortsov S. Radiation resistance: cancer stem cells (CSCs) and their enigmatic pro-survival signaling. Semin Cancer Biol. 2015;35:39–44.
10. Sierra JR, Cepero V, Giordano S. Molecular mechanisms of acquired resistance to tyrosine kinase targeted therapy. Mol Cancer. 2010;9:75.
11. Ramos P, Bentires-Alj M. Mechanism-based cancer therapy: resistance to therapy, therapy for resistance. Oncogene. 2015;34(28):3617–26.
12. Passiglia F, Bronte G, Castiglia M, Listi A, Calo V, Toia F, et al. Prognostic and predictive biomarkers for targeted therapy in NSCLC: for whom the bell tolls? Expert Opin Biol Ther. 2015;15(11):1553–66.
13. Rizzo S, Bronte G, Fanale D, Corsini L, Silvestris N, Santini D, et al. Prognostic vs predictive molecular biomarkers in colorectal cancer: is KRAS and BRAF wild type status required for anti-EGFR therapy? Cancer Treat Rev. 2010;36(3):S56–61.
14. Rolfo C, Fanale D, Hong DS, Tsimberidou AM, Piha-Paul SA, Pauwels P, et al. Impact of microRNAs in resistance to chemotherapy and novel targeted agents in non-small cell lung cancer. Curr Pharm Biotechnol. 2014;15(5):475–85.
15. Sawyers CL. The cancer biomarker problem. Nature. 2008;452(7187):548–52.
16. Boni V, Zarate R, Villa JC, Bandres E, Gomez MA, Maiello E, et al. Role of primary miRNA polymorphic variants in metastatic colon cancer patients treated with 5-fluorouracil and irinotecan. Pharmacogenomics J. 2011;11(6):429–36.
17. Huang T, Alvarez A, Hu B, Cheng SY. Noncoding RNAs in cancer and cancer stem cells. Chin J Cancer. 2013;32(11):582–93.
18. Ma L, Bajic VB, Zhang Z. On the classification of long non-coding RNAs. RNA Biol. 2013;10(6):925–33.
19. Sana J, Faltejskova P, Svoboda M, Slaby O. Novel classes of non-coding RNAs and cancer. J Transl Med. 2012;10:103.
20. Nie L, Wu HJ, Hsu JM, Chang SS, Labaff AM, Li CW, et al. Long non-coding RNAs: versatile master regulators of gene expression and crucial players in cancer. Am J Transl Res. 2012;4(2):127–50.
21. Prensner JR, Chinnaiyan AM. The emergence of lncRNAs in cancer biology. Cancer Discov. 2011;1(5):391–407.
22. Wu Y, Lu W, Xu J, Shi Y, Zhang H, Xia D. Prognostic value of long non-coding RNA MALAT1 in cancer patients. Tumour Biol. 2015;37(1):897–903.
23. Tokarz P, Blasiak J. The role of microRNA in metastatic colorectal cancer and its significance in cancer prognosis and treatment. Acta Biochim Pol. 2012;59(4):467–74.
24. Arndt GM, Dossey L, Cullen LM, Lai A, Druker R, Eisbacher M, et al. Characterization of global microRNA expression reveals oncogenic potential of miR-145 in metastatic colorectal cancer. BMC Cancer. 2009;9:374.
25. Zheng HT, Shi DB, Wang YW, Li XX, Xu Y, Tripathi P, et al. High expression of lncRNA MALAT1 suggests a biomarker of poor prognosis in colorectal cancer. Int J Clin Exp Pathol. 2014;7(6):3174–81.
26. Wang LL, Du LT, Li J, Liu YM, Qu AL, Yang YM, et al. Decreased expression of miR-133a correlates with poor prognosis in colorectal cancer patients. World J Gastroenterol. 2014;20(32):11340–6.
27. Corsini LR, Bronte G, Terrasi M, Amodeo V, Fanale D, Fiorentino E, et al. The role of microRNAs in cancer: diagnostic and prognostic biomarkers and targets of therapies. Expert Opin Ther Targets. 2012;16(2):S103–9.
28. Luo X, Burwinkel B, Tao S, Brenner H. MicroRNA signatures: novel biomarker for colorectal cancer? Cancer Epidemiol Biomarkers Prev. 2011;20(7):1272–86.
29. Hummel R, Hussey DJ, Haier J. MicroRNAs: predictors and modifiers of chemo- and radiotherapy in different tumour types. Eur J Cancer. 2010;46(2):298–311.
30. Kitade Y, Akao Y. MicroRNAs and their therapeutic potential for human diseases: microRNAs, miR-143 and -145, function as anti-oncomirs and the application of chemically modified miR-143 as an anti-cancer drug. J Pharmacol Sci. 2010;114(3):276–80.
31. Raza U, Zhang JD, Sahin O. MicroRNAs: master regulators of drug resistance, stemness, and metastasis. J Mol Med (Berl). 2014;92(4):321–36.
32. Zheng T, Wang J, Chen X, Liu L. Role of microRNA in anticancer drug resistance. Int J Cancer. 2010;126(1):2–10.

33. Donzelli S, Mori F, Biagioni F, Bellissimo T, Pulito C, Muti P, et al. MicroRNAs: short non-coding players in cancer chemoresistance. Mol Cell Ther. 2014;2:16.
34. Neelakandan K, Babu P, Nair S. Emerging roles for modulation of microRNA signatures in cancer chemoprevention. Curr Cancer Drug Targets. 2012;12(6):716–40.
35. Gill S, Loprinzi CL, Sargent DJ, Thome SD, Alberts SR, Haller DG, et al. Pooled analysis of fluorouracil-based adjuvant therapy for stage II and III colon cancer: who benefits and by how much? J Clin Oncol. 2004;22(10):1797–806.
36. Lombardi L, Morelli F, Cinieri S, Santini D, Silvestris N, Fazio N, et al. Adjuvant colon cancer chemotherapy: where we are and where we'll go. Cancer Treat Rev. 2010;36(3):S34–41.
37. Cunningham D, Humblet Y, Siena S, Khayat D, Bleiberg H, Santoro A, et al. Cetuximab monotherapy and cetuximab plus irinotecan in irinotecan-refractory metastatic colorectal cancer. N Engl J Med. 2004;351(4):337–45.
38. Saltz LB, Clarke S, Diaz-Rubio E, Scheithauer W, Figer A, Wong R, et al. Bevacizumab in combination with oxaliplatin-based chemotherapy as first-line therapy in metastatic colorectal cancer: a randomized phase III study. J Clin Oncol. 2008;26(12):2013–9.
39. Longley DB, Allen WL, Johnston PG. Drug resistance, predictive markers and pharmacogenomics in colorectal cancer. Biochim Biophys Acta. 2006;1766(2):184–96.
40. Rodrigues AS, Dinis J, Gromicho M, Martins C, Laires A, Rueff J. Genomics and cancer drug resistance. Curr Pharm Biotechnol. 2012;13(5):651–73. doi:BSP/CPB/E-Pub/0000153-13-7 [pii].
41. Longley DB, Johnston PG. Molecular mechanisms of drug resistance. J Pathol. 2005;205(2):275–92.
42. Al-Ejeh F, Kumar R, Wiegmans A, Lakhani SR, Brown MP, Khanna KK. Harnessing the complexity of DNA-damage response pathways to improve cancer treatment outcomes. Oncogene. 2010;29(46):6085–98.
43. Crea F, Nobili S, Paolicchi E, Perrone G, Napoli C, Landini I, et al. Epigenetics and chemoresistance in colorectal cancer: an opportunity for treatment tailoring and novel therapeutic strategies. Drug Resist Updat. 2011;14(6):280–96.
44. Chen Y, Yi C, Liu L, Li B, Wang Y, Wang X. Thymidylate synthase expression and prognosis in colorectal cancer: a meta-analysis of colorectal cancer survival data. Int J Biol Markers. 2012;27(3):e203–11.
45. Watson RG, Muhale F, Thorne LB, Yu J, O'Neil BH, Hoskins JM, et al. Amplification of thymidylate synthetase in metastatic colorectal cancer patients pretreated with 5-fluorouracil-based chemotherapy. Eur J Cancer. 2010;46(18):3358–64.
46. Calascibetta A, Contino F, Feo S, Gulotta G, Cajozzo M, Antona A, et al. Analysis of the thymidylate synthase gene structure in colorectal cancer patients and its possible relation with the 5-Fluorouracil drug response. J Nucleic Acids. 2010;2010:306754.
47. Humeniuk R, Menon LG, Mishra PJ, Gorlick R, Sowers R, Rode W, et al. Decreased levels of UMP kinase as a mechanism of fluoropyrimidine resistance. Mol Cancer Ther. 2009;8(5):1037–44.
48. Koopman M, Venderbosch S, van Tinteren H, Ligtenberg MJ, Nagtegaal I, Van Krieken JH, et al. Predictive and prognostic markers for the outcome of chemotherapy in advanced colorectal cancer, a retrospective analysis of the phase III randomised CAIRO study. Eur J Cancer. 2009;45(11):1999–2006.
49. Ramesh M, Ahlawat P, Srinivas NR. Irinotecan and its active metabolite, SN-38: review of bioanalytical methods and recent update from clinical pharmacology perspectives. Biomed Chromatogr. 2010;24(1):104–23.
50. Coppede F. Epigenetic biomarkers of colorectal cancer: focus on DNA methylation. Cancer Lett. 2014;342(2):238–47.
51. Crea F, Giovannetti E, Cortesi F, Mey V, Nannizzi S, Gallegos Ruiz MI, et al. Epigenetic mechanisms of irinotecan sensitivity in colorectal cancer cell lines. Mol Cancer Ther. 2009;8(7):1964–73.
52. Magrini R, Bhonde MR, Hanski ML, Notter M, Scherubl H, Boland CR, et al. Cellular effects of CPT-11 on colon carcinoma cells: dependence on p53 and hMLH1 status. Int J Cancer. 2002;101(1):23–31.
53. Faivre S, Chan D, Salinas R, Woynarowska B, Woynarowski JM. DNA strand breaks and apoptosis induced by oxaliplatin in cancer cells. Biochem Pharmacol. 2003;66(2):225–37.
54. Chaney SG, Campbell SL, Bassett E, Wu Y. Recognition and processing of cisplatin- and oxaliplatin-DNA adducts. Crit Rev Oncol Hematol. 2005;53(1):3–11.
55. Martin LP, Hamilton TC, Schilder RJ. Platinum resistance: the role of DNA repair pathways. Clin Cancer Res. 2008;14(5):1291–5. 14/5/1291 [pii].
56. Seetharam R, Sood A, Goel S. Oxaliplatin: preclinical perspectives on the mechanisms of action, response and resistance. Ecancermedicalscience. 2009;3:153.
57. Okugawa Y, Grady WM, Goel A. Epigenetic alterations in colorectal cancer: emerging biomarkers. Gastroenterology. 2015;149(5):1204–25 e12.
58. Nishida N, Yamashita S, Mimori K, Sudo T, Tanaka F, Shibata K, et al. MicroRNA-10b is a prognostic indicator in colorectal cancer and confers resistance to the chemotherapeutic agent 5-fluorouracil in colorectal cancer cells. Ann Surg Oncol. 2012;19(9):3065–71.
59. Kurokawa K, Tanahashi T, Iima T, Yamamoto Y, Akaike Y, Nishida K, et al. Role of miR-19b and its target mRNAs in 5-fluorouracil resistance in colon cancer cells. J Gastroenterol. 2012;47(8):883–95.
60. Rajesh C, Baker DK, Pierce AJ, Pittman DL. The splicing-factor related protein SFPQ/PSF interacts

with RAD51D and is necessary for homology-directed repair and sister chromatid cohesion. Nucleic Acids Res. 2011;39(1):132–45.
61. Salton M, Lerenthal Y, Wang SY, Chen DJ, Shiloh Y. Involvement of Matrin 3 and SFPQ/NONO in the DNA damage response. Cell Cycle. 2010;9(8):1568–76.
62. Joaquin M, Watson RJ. Cell cycle regulation by the B-Myb transcription factor. Cell Mol Life Sci. 2003;60(11):2389–401.
63. Rossi L, Bonmassar E, Faraoni I. Modification of miR gene expression pattern in human colon cancer cells following exposure to 5-fluorouracil in vitro. Pharmacol Res. 2007;56(3):248–53.
64. Chai H, Liu M, Tian R, Li X, Tang H. miR-20a targets BNIP2 and contributes chemotherapeutic resistance in colorectal adenocarcinoma SW480 and SW620 cell lines. Acta Biochim Biophys Sin (Shanghai). 2011;43(3):217–25.
65. Valencia CA, Cotten SW, Liu R. Cleavage of BNIP-2 and BNIP-XL by caspases. Biochem Biophys Res Commun. 2007;364(3):495–501.
66. Stiegelbauer V. MicroRNAs as novel predictive biomarkers and therapeutic targets in colorectal cancer. World J Gastroenterol. 2014;20(33):11727.
67. Faltejskova P, Besse A, Sevcikova S, Kubiczkova L, Svoboda M, Smarda J, et al. Clinical correlations of miR-21 expression in colorectal cancer patients and effects of its inhibition on DLD1 colon cancer cells. Int J Colorectal Dis. 2012;27(11):1401–8.
68. Valeri N, Gasparini P, Braconi C, Paone A, Lovat F, Fabbri M, et al. MicroRNA-21 induces resistance to 5-fluorouracil by down-regulating human DNA MutS homolog 2 (hMSH2). Proc Natl Acad Sci U S A. 2010;107(49):21098–103.
69. Deng J, Lei W, Fu JC, Zhang L, Li JH, Xiong JP. Targeting miR-21 enhances the sensitivity of human colon cancer HT-29 cells to chemoradiotherapy in vitro. Biochem Biophys Res Commun. 2014;443(3):789–95.
70. Yu Y, Sarkar FH, Majumdar AP. Down-regulation of miR-21 induces differentiation of chemoresistant colon cancer cells and enhances susceptibility to therapeutic regimens. Transl Oncol. 2013;6(2):180–6.
71. Li J, Hou N, Faried A, Tsutsumi S, Kuwano H. Inhibition of autophagy augments 5-fluorouracil chemotherapy in human colon cancer in vitro and in vivo model. Eur J Cancer. 2010;46(10):1900–9.
72. Zhang H, Tang J, Li C, Kong J, Wang J, Wu Y et al. MiR-22 regulates 5-FU sensitivity by inhibiting autophagy and promoting apoptosis in colorectal cancer cells. Cancer Lett. 2015;356(2 Pt B):781–90.
73. Shang J, Yang F, Wang Y, Xue G, Mei Q, Wang F, et al. MicroRNA-23a antisense enhances 5-fluorouracil chemosensitivity through APAF-1/caspase-9 apoptotic pathway in colorectal cancer cells. J Cell Biochem. 2014;115(4):772–84.
74. Wang CJ, Stratmann J, Zhou ZG, Sun XF. Suppression of microRNA-31 increases sensitivity to 5-FU at an early stage, and affects cell migration and invasion in HCT-116 colon cancer cells. BMC Cancer. 2010;10:616.
75. Akao Y, Noguchi S, Iio A, Kojima K, Takagi T, Naoe T. Dysregulation of microRNA-34a expression causes drug-resistance to 5-FU in human colon cancer DLD-1 cells. Cancer Lett. 2011;300(2):197–204.
76. Siemens H, Jackstadt R, Kaller M, Hermeking H. Repression of c-Kit by p53 is mediated by miR-34 and is associated with reduced chemoresistance, migration and stemness. Oncotarget. 2013;4(9):1399–415.
77. Li X, Zhao H, Zhou X, Song L. Inhibition of lactate dehydrogenase A by microRNA-34a resensitizes colon cancer cells to 5-fluorouracil. Mol Med Rep. 2015;11(1):577–82.
78. Akao Y, Khoo F, Kumazaki M, Shinohara H, Miki K, Yamada N. Extracellular disposal of tumor-suppressor miRs-145 and -34a via microvesicles and 5-FU resistance of human colon cancer cells. Int J Mol Sci. 2014;15(1):1392–401.
79. Zhang J, Guo H, Zhang H, Wang H, Qian G, Fan X, et al. Putative tumor suppressor miR-145 inhibits colon cancer cell growth by targeting oncogene Friend leukemia virus integration 1 gene. Cancer. 2011;117(1):86–95.
80. Liu RL, Dong Y, Deng YZ, Wang WJ, Li WD. Tumor suppressor miR-145 reverses drug resistance by directly targeting DNA damage-related gene RAD18 in colorectal cancer. Tumour Biol. 2015;36(7):5011–9.
81. Findlay VJ, Wang C, Nogueira LM, Hurst K, Quirk D, Ethier SP, et al. SNAI2 modulates colorectal cancer 5-fluorouracil sensitivity through miR145 repression. Mol Cancer Ther. 2014;13(11):2713–26.
82. He J, Xie G, Tong J, Peng Y, Huang H, Li J, et al. Overexpression of microRNA-122 re-sensitizes 5-FU-resistant colon cancer cells to 5-FU through the inhibition of PKM2 in vitro and in vivo. Cell Biochem Biophys. 2014;70(2):1343–50.
83. Karaayvaz M, Zhai H, Ju J. miR-129 promotes apoptosis and enhances chemosensitivity to 5-fluorouracil in colorectal cancer. Cell Death Dis. 2013;4:e659.
84. Song B, Wang Y, Xi Y, Kudo K, Bruheim S, Botchkina GI, et al. Mechanism of chemoresistance mediated by miR-140 in human osteosarcoma and colon cancer cells. Oncogene. 2009;28(46):4065–74.
85. Wilson AJ, Byun DS, Nasser S, Murray LB, Ayyanar K, Arango D, et al. HDAC4 promotes growth of colon cancer cells via repression of p21. Mol Biol Cell. 2008;19(10):4062–75.
86. Borralho PM, Kren BT, Castro RE, da Silva IB, Steer CJ, Rodrigues CM. MicroRNA-143 reduces viability and increases sensitivity to 5-fluorouracil in HCT116 human colorectal cancer cells. FEBS J. 2009;276(22):6689–700.

87. Qian X, Yu J, Yin Y, He J, Wang L, Li Q, et al. MicroRNA-143 inhibits tumor growth and angiogenesis and sensitizes chemosensitivity to oxaliplatin in colorectal cancers. Cell Cycle. 2013;12(9):1385–94.
88. Takahashi M, Cuatrecasas M, Balaguer F, Hur K, Toiyama Y, Castells A, et al. The clinical significance of MiR-148a as a predictive biomarker in patients with advanced colorectal cancer. PLoS One. 2012;7(10):e46684.
89. Boni V, Bitarte N, Cristobal I, Zarate R, Rodriguez J, Maiello E, et al. miR-192/miR-215 influence 5-fluorouracil resistance through cell cycle-mediated mechanisms complementary to its post-transcriptional thymidilate synthase regulation. Mol Cancer Ther. 2010;9(8):2265–75.
90. Song B, Wang Y, Kudo K, Gavin EJ, Xi Y, Ju J. miR-192 Regulates dihydrofolate reductase and cellular proliferation through the p53-microRNA circuit. Clin Cancer Res. 2008;14(24):8080–6.
91. Zhang JX, Song W, Chen ZH, Wei JH, Liao YJ, Lei J, et al. Prognostic and predictive value of a microRNA signature in stage II colon cancer: a microRNA expression analysis. Lancet Oncol. 2013;14(13):1295–306.
92. Song B, Wang Y, Titmus MA, Botchkina G, Formentini A, Kornmann M, et al. Molecular mechanism of chemoresistance by miR-215 in osteosarcoma and colon cancer cells. Mol Cancer. 2010;9:96.
93. Senfter D, Holzner S, Kalipciyan M, Staribacher A, Walzl A, Huttary N, et al. Loss of miR-200 family in 5-fluorouracil resistant colon cancer drives lymphendothelial invasiveness in vitro. Hum Mol Genet. 2015;24(13):3689–98.
94. Diaz T, Tejero R, Moreno I, Ferrer G, Cordeiro A, Artells R, et al. Role of miR-200 family members in survival of colorectal cancer patients treated with fluoropyrimidines. J Surg Oncol. 2014;109(7):676–83.
95. Toden S, Okugawa Y, Jascur T, Wodarz D, Komarova NL, Buhrmann C, et al. Curcumin mediates chemosensitization to 5-fluorouracil through miRNA-induced suppression of epithelial-to-mesenchymal transition in chemoresistant colorectal cancer. Carcinogenesis. 2015;36(3):355–67.
96. Li T, Gao F, Zhang XP. miR-203 enhances chemosensitivity to 5-fluorouracil by targeting thymidylate synthase in colorectal cancer. Oncol Rep. 2015;33(2):607–14. doi:10.3892/or.2014.3646.
97. Zhou Y, Wan G, Spizzo R, Ivan C, Mathur R, Hu X, et al. miR-203 induces oxaliplatin resistance in colorectal cancer cells by negatively regulating ATM kinase. Mol Oncol. 2014;8(1):83–92.
98. Amankwatia EB, Chakravarty P, Carey FA, Weidlich S, Steele RJ, Munro AJ, et al. MicroRNA-224 is associated with colorectal cancer progression and response to 5-fluorouracil-based chemotherapy by KRAS-dependent and -independent mechanisms. Br J Cancer. 2015;112(9):1480–90.
99. Guo ST, Jiang CC, Wang GP, Li YP, Wang CY, Guo XY, et al. MicroRNA-497 targets insulin-like growth factor 1 receptor and has a tumour suppressive role in human colorectal cancer. Oncogene. 2013;32(15):1910–20.
100. To KK, Leung WW, Ng SS. Exploiting a novel miR-519c-HuR-ABCG2 regulatory pathway to overcome chemoresistance in colorectal cancer. Exp Cell Res. 2015;338(2):222–31.
101. Zhang Y, Geng L, Talmon G, Wang J. MicroRNA-520g confers drug resistance by regulating p21 expression in colorectal cancer. J Biol Chem. 2015;290(10):6215–25.
102. Nakajima G, Hayashi K, Xi Y, Kudo K, Uchida K, Takasaki K, et al. Non-coding MicroRNAs hsa-let-7g and hsa-miR-181b are Associated with Chemoresponse to S-1 in Colon Cancer. Cancer Genomics Proteomics. 2006;3(5):317–24.
103. Bitarte N, Bandres E, Boni V, Zarate R, Rodriguez J, Gonzalez-Huarriz M, et al. MicroRNA-451 is involved in the self-renewal, tumorigenicity, and chemoresistance of colorectal cancer stem cells. Stem Cells. 2011;29(11):1661–71.
104. Dong Y, Zhao J, Wu CW, Zhang L, Liu X, Kang W, et al. Tumor suppressor functions of miR-133a in colorectal cancer. Mol Cancer Res. 2013;11(9):1051–60.
105. Zhang L, Pickard K, Jenei V, Bullock MD, Bruce A, Mitter R, et al. miR-153 supports colorectal cancer progression via pleiotropic effects that enhance invasion and chemotherapeutic resistance. Cancer Res. 2013;73(21):6435–47.
106. Fernandez de Mattos S, Villalonga P, Clardy J, Lam EW. FOXO3a mediates the cytotoxic effects of cisplatin in colon cancer cells. Mol Cancer Ther. 2008;7(10):3237–46.
107. Schimanski CC, Frerichs K, Rahman F, Berger M, Lang H, Galle PR, et al. High miR-196a levels promote the oncogenic phenotype of colorectal cancer cells. World J Gastroenterol. 2009;15(17):2089–96.
108. Tanaka S, Hosokawa M, Yonezawa T, Hayashi W, Ueda K, Iwakawa S. Induction of epithelial-mesenchymal transition and down-regulation of miR-200c and miR-141 in oxaliplatin-resistant colorectal cancer cells. Biol Pharm Bull. 2015;38(3):435–40.
109. Xu K, Liang X, Shen K, Sun L, Cui D, Zhao Y, et al. MiR-222 modulates multidrug resistance in human colorectal carcinoma by down-regulating ADAM-17. Exp Cell Res. 2012;318(17):2168–77.
110. Xu K, Liang X, Shen K, Cui D, Zheng Y, Xu J, et al. miR-297 modulates multidrug resistance in human colorectal carcinoma by down-regulating MRP-2. Biochem J. 2012;446(2):291–300.
111. Xu K, Liang X, Cui D, Wu Y, Shi W, Liu J. miR-1915 inhibits Bcl-2 to modulate multidrug resistance by increasing drug-sensitivity in human colorectal carcinoma cells. Mol Carcinog. 2013;52(1):70–8.

112. Rasmussen MH, Jensen NF, Tarpgaard LS, Qvortrup C, Romer MU, Stenvang J, et al. High expression of microRNA-625-3p is associated with poor response to first-line oxaliplatin based treatment of metastatic colorectal cancer. Mol Oncol. 2013;7(3):637–46.
113. Lee H, Kim C, Ku JL, Kim W, Yoon SK, Kuh HJ, et al. A long non-coding RNA snaR contributes to 5-fluorouracil resistance in human colon cancer cells. Mol Cells. 2014;37(7):540–6.
114. Heinemann V, Douillard JY, Ducreux M, Peeters M. Targeted therapy in metastatic colorectal cancer – an example of personalised medicine in action. Cancer Treat Rev. 2013;39(6):592–601.
115. Burotto M, Hartley ML, Marshall JL, Pishvaian MJ. Future of targeted agents in metastatic colorectal cancer. Colorectal Cancer. 2012;1(5):433–43. doi:10.2217/crc.12.52.
116. Grothey A, Galanis E. Targeting angiogenesis: progress with anti-VEGF treatment with large molecules. Nat Rev Clin Oncol. 2009;6(9):507–18.
117. Bronte G, Rolfo C, Peeters M, Russo A. How to find the Ariadne's thread in the labyrinth of salvage treatment options for metastatic colorectal cancer? Expert Opin Biol Ther. 2014;14(6):743–8.
118. Grothey A, Allegra C. Antiangiogenesis therapy in the treatment of metastatic colorectal cancer. Ther Adv Med Oncol. 2012;4(6):301–19.
119. Weickhardt AJ, Williams DS, Lee CK, Chionh F, Simes J, Murone C, et al. Vascular endothelial growth factor D expression is a potential biomarker of bevacizumab benefit in colorectal cancer. Br J Cancer. 2015;113(1):37–45.
120. Lieu CH, Tran H, Jiang ZQ, Mao M, Overman MJ, Lin E, et al. The association of alternate VEGF ligands with resistance to anti-VEGF therapy in metastatic colorectal cancer. PLoS One. 2013;8(10):e77117.
121. Bardelli A, Siena S. Molecular mechanisms of resistance to cetuximab and panitumumab in colorectal cancer. J Clin Oncol. 2010;28(7):1254–61.
122. De Mattia E, Cecchin E, Toffoli G. Pharmacogenomics of intrinsic and acquired pharmacoresistance in colorectal cancer: Toward targeted personalized therapy. Drug Resist Updat. 2015;20:39–70.
123. Stiegelbauer V, Perakis S, Deutsch A, Ling H, Gerger A, Pichler M. MicroRNAs as novel predictive biomarkers and therapeutic targets in colorectal cancer. World J Gastroenterol. 2014;20(33):11727–35.
124. Mlcochova J, Faltejskova P, Nemecek R, Svoboda M, Slaby O. MicroRNAs targeting EGFR signalling pathway in colorectal cancer. J Cancer Res Clin Oncol. 2013;139(10):1615–24.
125. Ragusa M, Majorana A, Statello L, Maugeri M, Salito L, Barbagallo D, et al. Specific alterations of microRNA transcriptome and global network structure in colorectal carcinoma after cetuximab treatment. Mol Cancer Ther. 2010;9(12):3396–409.
126. Ruzzo A, Graziano F, Vincenzi B, Canestrari E, Perrone G, Galluccio N, et al. High let-7a microRNA levels in KRAS-mutated colorectal carcinomas may rescue anti-EGFR therapy effects in patients with chemotherapy-refractory metastatic disease. Oncologist. 2012;17(6):823–9.
127. Cappuzzo F, Sacconi A, Landi L, Ludovini V, Biagioni F, D'Incecco A, et al. MicroRNA signature in metastatic colorectal cancer patients treated with anti-EGFR monoclonal antibodies. Clin Colorectal Cancer. 2014;13(1):37–45 e4.
128. Salendo J, Spitzner M, Kramer F, Zhang X, Jo P, Wolff HA, et al. Identification of a microRNA expression signature for chemoradiosensitivity of colorectal cancer cells, involving miRNAs-320a, -224, -132 and let7g. Radiother Oncol. 2013;108(3):451–7.
129. Saridaki Z, Weidhaas JB, Lenz HJ, Laurent-Puig P, Jacobs B, De Schutter J, et al. A let-7 microRNA-binding site polymorphism in KRAS predicts improved outcome in patients with metastatic colorectal cancer treated with salvage cetuximab/panitumumab monotherapy. Clin Cancer Res. 2014;20(17):4499–510.
130. Zhang W, Winder T, Ning Y, Pohl A, Yang D, Kahn M, et al. A let-7 microRNA-binding site polymorphism in 3′-untranslated region of KRAS gene predicts response in wild-type KRAS patients with metastatic colorectal cancer treated with cetuximab monotherapy. Ann Oncol. 2011;22(1):104–9.
131. Smits KM, Paranjape T, Nallur S, Wouters KA, Weijenberg MP, Schouten LJ, et al. A let-7 microRNA SNP in the KRAS 3′UTR is prognostic in early-stage colorectal cancer. Clin Cancer Res. 2011;17(24):7723–31.
132. Suto T, Yokobori T, Yajima R, Morita H, Fujii T, Yamaguchi S, et al. MicroRNA-7 expression in colorectal cancer is associated with poor prognosis and regulates cetuximab sensitivity via EGFR regulation. Carcinogenesis. 2015;36(3):338–45.
133. Mlcochova J, Faltejskova-Vychytilova P, Ferracin M, Zagatti B, Radova L, Svoboda M, et al. MicroRNA expression profiling identifies miR-31-5p/3p as associated with time to progression in wild-type RAS metastatic colorectal cancer treated with cetuximab. Oncotarget. 2015;6(36):38695–704.
134. Mussnich P, Rosa R, Bianco R, Fusco A, D'Angelo D. MiR-199a-5p and miR-375 affect colon cancer cell sensitivity to cetuximab by targeting PHLPP1. Expert Opin Ther Targets. 2015;19(8):1017–26.
135. Hansen TF, Sorensen FB, Lindebjerg J, Jakobsen A. The predictive value of microRNA-126 in relation to first line treatment with capecitabine and oxaliplatin in patients with metastatic colorectal cancer. BMC Cancer. 2012;12:83.
136. Hansen TF, Christensen R, Andersen RF, Sorensen FB, Johnsson A, Jakobsen A. MicroRNA-126 and epidermal growth factor-like domain 7-an angio-

genic couple of importance in metastatic colorectal cancer. Results from the Nordic ACT trial. Br J Cancer. 2013;109(5):1243–51.
137. Hansen TF, Carlsen AL, Heegaard NH, Sorensen FB, Jakobsen A. Changes in circulating microRNA-126 during treatment with chemotherapy and bevacizumab predicts treatment response in patients with metastatic colorectal cancer. Br J Cancer. 2015;112(4):624–9.
138. Kim JG, Park MT, Heo K, Yang KM, Yi JM. Epigenetics meets radiation biology as a new approach in cancer treatment. Int J Mol Sci. 2013;14(7):15059–73.
139. Burnet NG, Thomas SJ, Burton KE, Jefferies SJ. Defining the tumour and target volumes for radiotherapy. Cancer Imaging. 2004;4(2):153–61. doi:10.1102/1470-7330.2004.0054.
140. Seiwert TY, Salama JK, Vokes EE. The concurrent chemoradiation paradigm–general principles. Nat Clin Pract Oncol. 2007;4(2):86–100.
141. Begg AC, Stewart FA, Vens C. Strategies to improve radiotherapy with targeted drugs. Nat Rev Cancer. 2011;11(4):239–53.
142. Braendengen M, Tveit KM, Berglund A, Birkemeyer E, Frykholm G, Pahlman L, et al. Randomized phase III study comparing preoperative radiotherapy with chemoradiotherapy in nonresectable rectal cancer. J Clin Oncol. 2008;26(22):3687–94.
143. Yoon WH, Kim HJ, Kim CH, Joo JK, Kim YJ, Kim HR. Oncologic impact of pathologic response on clinical outcome after preoperative chemoradiotherapy in locally advanced rectal cancer. Ann Surg Treat Res. 2015;88(1):15–20.
144. Visvader JE, Lindeman GJ. Cancer stem cells in solid tumours: accumulating evidence and unresolved questions. Nat Rev Cancer. 2008;8(10):755–68.
145. Tichy ED, Stambrook PJ. DNA repair in murine embryonic stem cells and differentiated cells. Exp Cell Res. 2008;314(9):1929–36.
146. Kinsella TJ. Coordination of DNA mismatch repair and base excision repair processing of chemotherapy and radiation damage for targeting resistant cancers. Clin Cancer Res. 2009;15(6):1853–9.
147. Sengupta S, Harris CC. p53: traffic cop at the crossroads of DNA repair and recombination. Nat Rev Mol Cell Biol. 2005;6(1):44–55.
148. Dalhus B, Laerdahl JK, Backe PH, Bjoras M. DNA base repair–recognition and initiation of catalysis. FEMS Microbiol Rev. 2009;33(6):1044–78.
149. Tanei T, Morimoto K, Shimazu K, Kim SJ, Tanji Y, Taguchi T, et al. Association of breast cancer stem cells identified by aldehyde dehydrogenase 1 expression with resistance to sequential Paclitaxel and epirubicin-based chemotherapy for breast cancers. Clin Cancer Res. 2009;15(12):4234–41.
150. Ahsan A, Hiniker SM, Davis MA, Lawrence TS, Nyati MK. Role of cell cycle in epidermal growth factor receptor inhibitor-mediated radiosensitization. Cancer Res. 2009;69(12):5108–14.
151. Della Vittoria Scarpati G, Falcetta F, Carlomagno C, Ubezio P, Marchini S, De Stefano A, et al. A specific miRNA signature correlates with complete pathological response to neoadjuvant chemoradiotherapy in locally advanced rectal cancer. Int J Radiat Oncol Biol Phys. 2012;83(4):1113–9.
152. Ma W, Yu J, Qi X, Liang L, Zhang Y, Ding Y, et al. Radiation-induced microRNA-622 causes radioresistance in colorectal cancer cells by down-regulating Rb. Oncotarget. 2015;6(18):15984–94.
153. Zhang Y, Yu J, Liu H, Ma W, Yan L, Wang J, et al. Novel epigenetic CREB-miR-630 signaling axis regulates radiosensitivity in colorectal cancer. PLoS One. 2015;10(8):e0133870.
154. Yang XD, Xu XH, Zhang SY, Wu Y, Xing CG, Ru G, et al. Role of miR-100 in the radioresistance of colorectal cancer cells. Am J Cancer Res. 2015;5(2):545–59.
155. Zheng L, Zhang Y, Liu Y, Zhou M, Lu Y, Yuan L, et al. MiR-106b induces cell radioresistance via the PTEN/PI3K/AKT pathways and p21 in colorectal cancer. J Transl Med. 2015;13:252.
156. Wang G, Li Z, Zhao Q, Zhu Y, Zhao C, Li X, et al. LincRNA-p21 enhances the sensitivity of radiotherapy for human colorectal cancer by targeting the Wnt/beta-catenin signaling pathway. Oncol Rep. 2014;31(4):1839–45.

Non-coding RNAs: Therapeutic Strategies and Delivery Systems

12

Hui Ling

Abstract

The vast majority of the human genome is transcribed into RNA molecules that do not code for proteins, which could be small ones approximately 20 nucleotide in length, known as microRNAs, or transcripts longer than 200 bp, defined as long noncoding RNAs. The prevalent deregulation of microRNAs in human cancers prompted immediate interest on the therapeutic value of microRNAs as drugs and drug targets. Many features of microRNAs such as well-defined mechanisms, and straightforward oligonucleotide design further make them attractive candidates for therapeutic development. The intensive efforts of exploring microRNA therapeutics are reflected by the large body of preclinical studies using oligonucleotide-based mimicking and blocking, culminated by the recent entry of microRNA therapeutics in clinical trial for several human diseases including cancer. Meanwhile, microRNA therapeutics faces the challenge of effective and safe delivery of nucleic acid therapeutics into the target site. Various chemical modifications of nucleic acids and delivery systems have been developed to increase targeting specificity and efficacy, and reduce the associated side effects including activation of immune response. Recently, long noncoding RNAs become attractive targets for therapeutic intervention because of their association with complex and delicate phenotypes, and their unconventional pharmaceutical activities such as capacity of increasing output of proteins. Here I discuss the general therapeutic strategies targeting noncoding RNAs, review delivery systems developed to maximize noncoding RNA therapeutic efficacy, and offer perspectives on the future development of noncoding RNA targeting agents for colorectal cancer.

H. Ling (✉)
Department of Experimental Therapeutics, The University of Texas MD Anderson Cancer Center, Houston, TX, USA
e-mail: hling@mdanderson.org

O. Slaby, G.A. Calin (eds.), *Non-coding RNAs in Colorectal Cancer*, Advances in Experimental Medicine and Biology 937, DOI 10.1007/978-3-319-42059-2_12

Keywords
MicroRNA • Long noncoding RNA • Colorectal cancer • Delivery • Therapeutics • Mimics • anti-miR

12.1 Introduction

Noncoding RNAs are useful targets for therapeutic interventions of human cancer. Here we focus on the potential of microRNAs and long noncoding RNAs as therapeutic targets and tools in treatment of colorectal cancer. MicroRNAs have the advantages of targeting multiple protein-coding genes at once. In addition, microRNAs have their function in the physiological conditions, and thus restoring or reducing microRNA to their normal levels may lead to favorable consequences. Furthermore, microRNAs may be used to target proteins that are difficult to design small molecular chemical inhibitors. Another advantages of microRNAs are their relative simple structures, and their predictable mechanisms. These features made the design of mimics or anti-miRs easier than that of the conventional chemical drugs.

The levels of long noncoding RNAs are usually lower than those of the protein coding genes [1]. However, long noncoding RNAs tend to have more tissue specific expression pattern than the protein coding genes, and are thus possibly associated with certain cancer subtypes [1]. The fact that many long noncoding RNAs are identified from important cancer associated genomic locus suggests that they should be functionally important and relevant. Disruption of these transcripts, as demonstrated by previous studies, could lead to significant consequences in the biological activities and disease status. The theoretically specificity and efficacy of small interference RNA and antisense oligonucleotides in reducing levels of a long noncoding RNA, made readily available the means for manipulating such transcripts. Additionally, since long noncoding RNA could interact with proteins such as transcription factors and histone modifiers, targeting long noncoding RNA will lead to specific and delicate changes, which may be desirable in the cancer treatment [2].

12.2 Therapeutic Strategies Targeting Noncoding RNAs

Generally, there are two strategies for therapeutic targeting of noncoding RNAs in colorectal cancer (Fig. 12.1). The first is to restore the function of noncoding RNAs with tumor suppressor activities that are lost in colorectal cancer. The second is to block the actions of noncoding RNAs with oncogenic function that are aberrantly overexpressed in colorectal cancer. Both strategies could be applied to microRNAs. Since microRNAs are with small size, and often localized in the cytosolic subcellular fraction, it is possible to restore microRNA function with synthetic microRNA mimics. The function of microRNAs could also be blocked by a variety of strategies interfering with microRNA activities. For long noncoding RNAs, blocking their function is more plausible than restoring biological activities of such transcripts, because of several reasons. First, unlike microRNAs, long noncoding RNAs could fold into secondary and higher order structures, and its function is hard to predict with the sequence information [3]. This makes it difficult to synthesize long noncoding RNA to replace their original function. Second, many long noncoding RNAs are localized in the cell nucleus, thus the restoration of their function requires one more step of synthesized molecules entering the nucleus. Third, similar as mRNAs, long noncoding RNAs are more easily degraded than microRNAs, creating one more barrier for replacement strategy for long noncoding RNAs. Lastly, the mechanisms underlying long noncoding RNAs are not well elucidated. If a long noncoding RNA functions

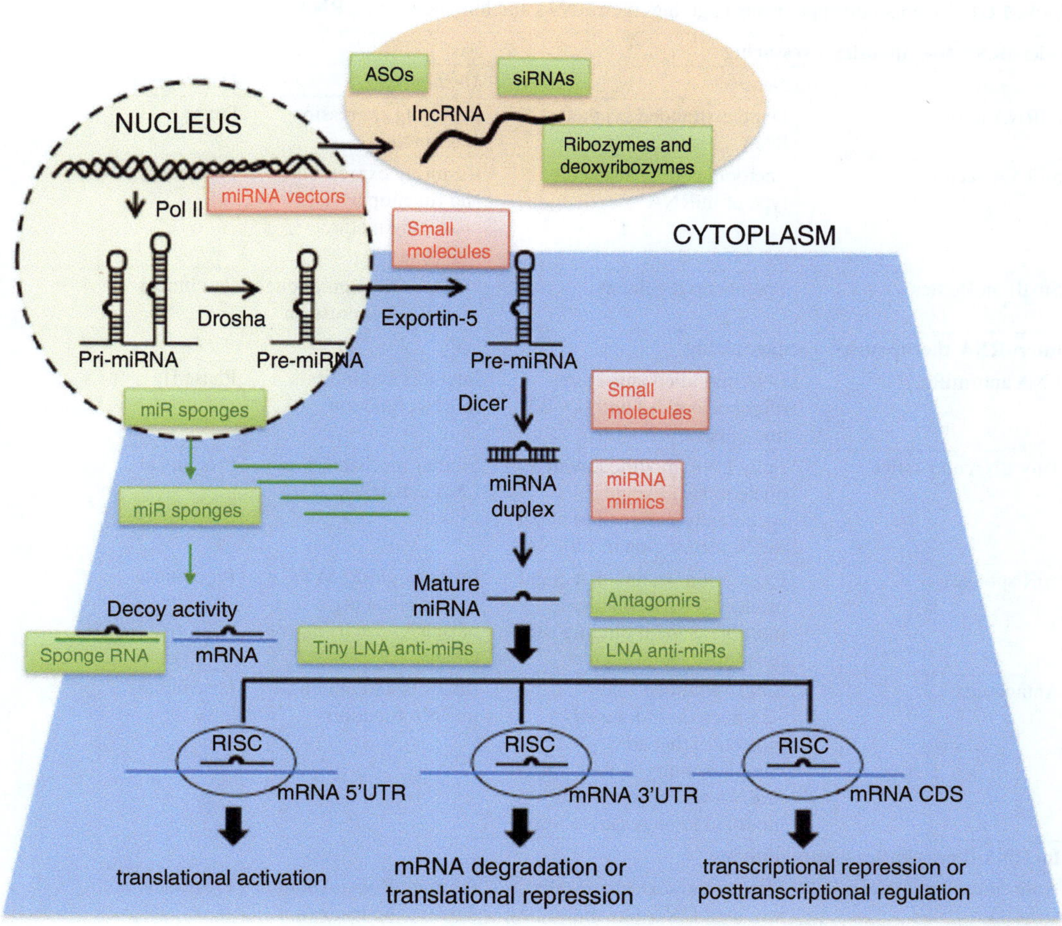

Fig. 12.1 Therapeutic strategies to activate or block noncoding RNA function

via *cis* regulatory mechanisms, it is impossible to restore its function without expressing at the specific genomic locus. However, blocking of the function of long noncoding RNAs could be easily achieved by several strategies, with the most straightforward approach of downregulation with RNA interference. The approaches discussed below were summarized in Table 12.1.

12.2.1 Restoring Noncoding RNA Function

To regain the function of a microRNA that is lost or downregulated in colorectal cancer, the simplest method is to supply with synthetic microRNA molecules with same function. This could be achieved with microRNA mimics or with microRNAs expression vectors [2]. microRNA mimics are chemically modified double stranded RNAs that mimic endogenous microRNAs [4]. When transfected into cells, microRNA mimics could be processed into single-strand microRNA molecule to target coding genes similar as the endogenous microRNA. An alternative way of replacing a microRNA function is to produce it in an expression vector. With the microRNA production machinery engineered, the designed vector could produce continuously the intended microRNA molecules for replacement. In addition, the microRNA expression vectors can be engineered with promoters to

Table 12.1 Therapeutic approaches targeting microRNAs and long noncoding RNAs

microRNA therapeutics – restoring			
Strategies	*Details*	*Mechanism*	*Development*
miRNA mimics	Double-stranded synthetic RNA	Restoring expression and function	Phase I
miRNA vectors	Vector encoding a specific type of miRNA	Restoring expression and function by producing mature miRNAs	Preclinical
Small molecules	Epigenetic regulators	Nonspecific regulation of miRNA expression	Preclinical
microRNA therapeutics – antagonizing			
LNA anti-miRs	LNA-modified antisense oligonucleotides with 13–22 nucleotides in length	Interacts and inhibits miRNA function	Phase IIa
Tiny LNA anti-miRs	Fully LNA-modified, with 8 nucleotides in length, specifically designed to target the 5′ seed region of miRNAs	Similar mechanism as LNA anti-miRs	Preclinical
miR sponges	Expression vectors that could produce RNAs containing multiple tandem binding sites to an miRNA of interest	Buffering miRNA to reduce silencing effect of miRNA on its native targets	Preclinical
Antagomirs	Single-stranded ~23-nucleotide-long RNA molecules that are complementary to the targeted miRNA and are chemically modified to increase stability	Interacts and inhibits miRNA function	Preclinical
lncRNA therapeutics – antagonizing			
Antisense oligonucleotides (ASOs)	Single-stranded, chemically modified DNA-like molecules, with 13–25 nucleotides in length, that are complementary to a selected RNA	Forming RNA–ASO duplex, and leads to RNase H-mediated cleavage of target RNA	Preclinical
Small interfering RNAs (siRNAs)	Double-stranded RNAs with perfect sequence homology to part of target RNA. The antisense strand is the functional one	Guide RNA-induced silencing complex to RNA target for endonucleolytic cleavage	Preclinical
Ribozymes and deoxyribozymes	An RNA (ribozyme) or DNA (deoxyribozyme) that catalyse specific biochemical reaction	Three repeated steps: base pairing, site-specific cleavage, and release of cleavage products	Preclinical

specifically express microRNA in a tumor- and tissue- specific manner, giving this expression vector method an advantage over microRNA mimics.

The loss of microRNA expression could be due to genomic deletion or epigenetic silencing. In the latter scenario, it is possible to recover a microRNA expression by reversing the epigenetic quenching. Decitabine and 5-azacytidine, two hypomethylating agents that have been approved for treatment of myelodysplastic syndromes, were shown to re-induce the expression of several miRNAs including miR-124a [5]. However, this regulation is non-specific to one microRNA. In addition, the spectrum of induced microRNAs is context dependent [5–7]. The antibacterial compound enoxacin has also been shown to boost the expression of a subset of microRNAs in colon cancer cell lines in vitro and in vivo by acting on TARBP2, a protein regulating microRNA processing [8].

Among these replacement strategies, microRNA mimics gain popularity in the development into therapeutic agents by biopharmaceutical companies. This can be reflected by the first miRNA replacement therapy entering clinical trial for treatment of human cancer - formulated miR-34a mimics for treatment of patients with advanced hepatocellular carcinoma [9].

12.2.2 Antagonizing Noncoding RNA Function

Current strategies to antagonize microRNA function include locked nucleic acids (LNA anti-miRs), tiny LNA anti-miRs, antagomirs, and miRNA sponges [2]. LNA anti-miRs are antisense oligonucleotides with several nucleotides substituted by bicyclic RNA analogues in a 'locked' conformation [10]. This LNA modification renders high affinity for the binding of the targeted microRNA by generating an ideal conformation for Watson–Crick binding, and allows for effective blockade of microRNA function with short sequences (13–22 nucleotides). Additionally, LNA anti-miRs are resistant to degradation, and efficient in uptake by many tissues. These features of LNA anti-miRs eliminate the need for sophisticated formulation and delivery, which is indispensible for most other antagonizing strategies. The exploration of LNA anti-miRs for clinical usages culminated with the entering of miravirsen (SPC3649; Santaris Pharma), an LNA anti-miR against miR-122, in Phase I and Phase IIa clinical trials for the treatment of hepatitis C virus (HCV) [11, 12]. Since 5′-seed region at positions 2–7 of a microRNA is essential for the binding of microRNA to its mRNA targets, tiny LNA anti-miRs specifically designed to target the microRNA seed region were explored [13]. Tiny LNA anti-miRs have the advantage of targeting multiple microRNAs within the same family; however, the specificity of microRNA targeting was compromised [13].

Antagomirs are synthetic, cholesterol-conjugated RNAs complementary to the targeted microRNA sequence, featured by a 2′-O-methyl linkage and phosphorothioate modification [14]. These added features help to increase cellular uptake and protect from degradation by nucleases [14]. While antagomirs have been shown to block microRNA function in mouse models, their uses are currently limited to experimental tools, probably due to high effective dosages associated with antagomirs [10].

Another strategy of blocking microRNA function is to generate microRNA sponges to competitively inhibit microRNA function [15]. These microRNA sponges contain multiple tandem bindings that are complementary to the microRNA seed sequence. By sequestering aimed microRNAs from their endogenous mRNA targets, this microRNA sponge method effectively blocks the microRNA function [15]. To achieve enough concentration of sponge RNAs, expression vectors with strong promoter are usually used to maintain high level of transcription. Several studies showed that microRNA sponges tend to have long-lasting effect [16]; however, because sponges are RNAs without chemical modification, the concentrations for effective inhibition of microRNA function may be much higher than other anti-miRs. Furthermore, whether the excess of sponge transcripts produce undesired effects remains to be determined by further studies.

The therapeutic exploration of long noncoding RNAs lags far behind the microRNA therapeutics. The function of long noncoding RNAs could be blocked by several strategies. First, the level of long noncoding RNA could be regulated by specifically designed siRNAs. The length of long noncoding RNA also makes the design of specific siRNAs not a difficult task. Previous studies have shown that siRNAs could successfully achieve knockdown of long noncoding RNAs, irrespective of their subcellular localization [17]. Considering the fact that many long noncoding RNA are upregulated in colorectal cancer, the use of siRNAs targeting such oncogenic long noncoding RNA could probably reverse the cancer malignancies. On the other hand, many protein-coding genes have corresponding natural antisense transcripts, which could negatively regulate expression of these protein-coding genes [18]. Therefore, targeting of natural antisense transcripts by single-stranded oligonucleotides represents a unique opportunity for therapeutic upregulation of tumor suppressor genes, which is difficult to realize with the conventional drug design of chemical compounds.

12.3 Delivery Systems

In almost all of the strategies of noncoding RNA therapeutics, safe and effective delivery of the oligonucleotides into the cancer tissue without causing deleterious side effects remains the premier challenges. Unmodified oligonucleotides are not stable in the circulation, can be attacked by immune system, and hardly penetrate into cells. Although modifications as discussed above could increase affinity to targets, and increase the stability, most of the oligonucleotide therapies need additional optimal delivery system to achieve the desired biological effects. Several aspects need to be considered when selecting a delivery system: stability against serum nucleases, evasion of the innate immune system, avoidance of non-specific interactions with serum proteins and non-target cells, prevention of renal clearance, release from blood vessels to target tissues, cell entry, incorporation into the RNA interference or other machinery [19].

Shielding the exterior of delivery vehicles with polyethylene glycol (PEG) is a common strategy to increase the circulation time for therapeutic oligonucleotides [20]. This strategy could prevent non-specific interaction of formulated particles with serum proteins, immune cells and other non-target tissues [20]. Particles with size of 8 nm to 20 nm in the circulation are subject to renal clearance, with the exception of dynamic polyconjugates (DPCs) and triantennary N-acetylgalactosamine (GalNAc) conjugates [19]. These two conjugates therefore could offer advantage of avoiding elimination of formulated particles by urine. To take effect on the target site, formulated particles need to release from the circulation into the aimed cancer tissues. Many solid tumors including colorectal cancer have discontinuous endothelia, and thus are more prone to permeation than the normal tissues [21]. Together with impaired lymphatic drainage in cancer, tumor tissues could accumulate more circulating particles.

Once reaching the tumor site, the delivery particles usually enter the cells via endocytosis. To facilitate such process, the delivery system could be engineered with targeting ligands that specifically recognize receptors on target cells. Alternatively, cell-penetrating peptides could increase the cellular uptake [22]. Tumors are characterized by acidic environment partially because of lack of nutrition and metabolic changes induced by Warburg effect [23]. The acidic environment of tumors offers opportunity to incorporate materials that can be released in the low pH environment. A recent study by the Slack group has developed a delivery system attaching antimiRs to a peptide with low pH-inducible transmembrane structure, and demonstrated the success of this system in blocking miR-155 function in a mouse model of lymphoma [23].

Lipid nanoparticles (LNPs) such as liposomes have been developed to protect oligonucleotides from nuclease degradation, avoid renal clearance, increase cellular uptake, and promote endosomal escape [24]. Several LNP RNAi drugs have passed the preclinical evaluation and entered clinical trials [25]. One example is the LNP drug ALN-VSP, a lipid delivery system developed by Alnylam Pharmaceuticals, which was recently evaluated in phase-I clinical trial for treatment of advanced solid tumors [26]. This study found that ALN-VSP successfully degraded target mRNA in tissue biopsies to exert antitumor activity at dosages well tolerated by patients [26]. As the first anticancer microRNA drug entering clinical trial, the miR-34 mimic MRX34 developed by Mirna Therapeutics is also liposome-based [9].

It should be noted that the delivery systems showing success in vivo vary largely in size, structure, and chemistry. For each specific case, unique designs of delivery system might be necessary to achieve best efficacy without causing deleterious side effect. LNPs are among the most effective formulations in the delivery of oligonucleotides for noncoding RNA therapy. Conjugate systems, which require minimal amounts of delivery material, have the advantage of defined molecular structures, and wide therapeutic window, also show promise as an effective delivery system [19].

12.4 Challenges of Noncoding RNA Therapeutics

Noncoding RNA therapeutics is a new concept that differs from the conventional chemical drug design. Numerous challenges exist for the therapeutic use of noncoding RNAs in the treatment of colorectal cancer. For instance, while the fact that microRNAs target multiple mRNAs can be an advantage itself, this also cause ambiguity as to the scope of genes that are exactly controlled by microRNAs. Making this even more complicated, studies show that microRNA functions are fine-tuned and context-dependent [27]. The microRNA targets identified by the cell model system or animal models may not be applicable to the clinical scenarios. To serve as a candidate for clinical evaluations, the functional phenotype and mechanisms of a microRNA need to be well elucidated and validated in the most stringent way. For long noncoding RNAs, the challenges are even bigger. The functioning mechanisms of long noncoding RNAs are not well understood, and general principles governing the functioning mechanisms are missing. In addition, long noncoding RNAs are more tissue-specific than protein-coding genes. This adds further challenges in targeting noncoding RNAs in the specific tissue or subcellular compartments. Detailed understanding of the biology and functioning mechanisms holds the key for translation of such knowledge into clinical usages.

For the noncoding RNAs with well-defined activity and functioning mechanisms, the biggest challenges lie in the delivery system. Even the most advanced formulations do not solve the technical requirement for a clinically useful drug. For instance, the manufacturing production of nanoparticles needs a better controlled mixing processes to achieve consistent quality [28]. In addition, the mechanisms underlying the delivery process are not well elucidated, and the established formulation guidelines may not always lead to expected biological phenomenon. Most of the oligonucleotide delivery systems are for well-perfused tissues such as liver, which physiologically allows for the distribution of therapeutic particles into target tissues. Novel delivery systems need to be developed for targeting colorectal cancer. Considering the importance of cancer stem cells in the initial and progression of colorectal cancer, it can be conceived that conjugated ligand specifically recognizing colon cancer stem cells could be used for delivery of therapeutic materials to destroy cancer stem cells. Recent studies showed that microRNAs could be packaged into multivesicular bodies and released into the extracellular environment as exosomes [29]. This represents a natural delivery system and may offer more advantages than the synthetic delivery systems. The detailed understanding of exosome microRNAs in colorectal cancer progression, metastasis, and drug response might offer novel strategies for cancer treatment, and aid the design of more efficient tumor specific delivery systems.

12.5 Conclusions and Summary

Noncoding RNA therapeutics for colorectal cancer is still in its infancy. Nonetheless, the field of noncoding RNA therapeutics is developing fast. Just two decades after the initial discovery of microRNA link with human cancer in 2002, MRX34 entered clinical trials for treating advanced hepatocarcinoma. Both academia and pharmaceutical companies have been enthusiastically pursing the therapeutic value of noncoding RNAs. Companies such as Regulus

Therapeutics and Mirna Therapeutics have developed pipelines for microRNA therapeutics in treating diseases including cancer. In addition, companies such as RaNA Therapeutics are exploring the therapeutic potential of long noncoding RNAs. With the experience gained from developing oligonucleotides-based therapeutics, many obstacles that noncoding RNA therapeutics face might be cleared. Colorectal cancer is characterized by genetic alterations; noncoding RNAs including microRNAs and long noncoding RNAs have pivotal role in the regulation of these genetic events. We believe that with improved understanding of noncoding RNA biology and delivery system innovation, we will see in the near future the utility of noncoding RNA in the treatment of patients with colorectal cancer, in combination with chemotherapy and radiotherapy.

References

1. Rosenbloom KR, Dreszer TR, Long JC, Malladi VS, Sloan CA, Raney BJ, et al. ENCODE whole-genome data in the UCSC Genome Browser: update 2012. Nucleic Acids Res. 2012;40(Database issue):D912–7. Epub 2011/11/15.
2. Ling H, Fabbri M, Calin GA. MicroRNAs and other non-coding RNAs as targets for anticancer drug development. Nat Rev Drug Discov. 2013;12(11):847–65. Epub 2013/11/01.
3. Wan Y, Qu K, Zhang QC, Flynn RA, Manor O, Ouyang Z, et al. Landscape and variation of RNA secondary structure across the human transcriptome. Nature. 2014;505(7485):706–9. Epub 2014/01/31.
4. Bader AG, Brown D, Stoudemire J, Lammers P. Developing therapeutic microRNAs for cancer. Gene Ther. 2011;18(12):1121–6. Epub 2011/06/03.
5. Lujambio A, Ropero S, Ballestar E, Fraga MF, Cerrato C, Setien F, et al. Genetic unmasking of an epigenetically silenced microRNA in human cancer cells. Cancer Res. 2007;67(4):1424–9. Epub 2007/02/20.
6. Bandres E, Agirre X, Bitarte N, Ramirez N, Zarate R, Roman-Gomez J, et al. Epigenetic regulation of microRNA expression in colorectal cancer. Int J Cancer. 2009;125(11):2737–43. Epub 2009/06/13.
7. Saito Y, Jones PA. Epigenetic activation of tumor suppressor microRNAs in human cancer cells. Cell Cycle. 2006;5(19):2220–2. Epub 2006/10/03.
8. Melo S, Villanueva A, Moutinho C, Davalos V, Spizzo R, Ivan C, et al. Small molecule enoxacin is a cancer-specific growth inhibitor that acts by enhancing TAR RNA-binding protein 2-mediated microRNA processing. Proc Natl Acad Sci U S A. 2011;108(11):4394–9. Epub 2011/03/04.
9. Bouchie A. First microRNA mimic enters clinic. Nat Biotechnol. 2013;31(7):577. Epub 2013/07/11.
10. Lennox KA, Behlke MA. Chemical modification and design of anti-miRNA oligonucleotides. Gene Ther. 2011;18(12):1111–20. Epub 2011/07/15.
11. Janssen HL, Reesink HW, Lawitz EJ, Zeuzem S, Rodriguez-Torres M, Patel K, et al. Treatment of HCV infection by targeting microRNA. N Engl J Med. 2013;368(18):1685–94. Epub 2013/03/29.
12. Lieberman J, Sarnow P. Micromanaging hepatitis C virus. N Engl J Med. 2013;368(18):1741–3. Epub 2013/03/29.
13. Obad S, dos Santos CO, Petri A, Heidenblad M, Broom O, Ruse C, et al. Silencing of microRNA families by seed-targeting tiny LNAs. Nat Genet. 2011;43(4):371–8. Epub 2011/03/23.
14. Krutzfeldt J, Rajewsky N, Braich R, Rajeev KG, Tuschl T, Manoharan M, et al. Silencing of microRNAs in vivo with 'antagomirs'. Nature. 2005;438(7068):685–9. Epub 2005/11/01.
15. Ebert MS, Neilson JR, Sharp PA. MicroRNA sponges: competitive inhibitors of small RNAs in mammalian cells. Nat Methods. 2007;4(9):721–6. Epub 2007/08/19.
16. Xie J, Ameres SL, Friedline R, Hung JH, Zhang Y, Xie Q, et al. Long-term, efficient inhibition of microRNA function in mice using rAAV vectors. Nat Methods. 2012;9(4):403–9. Epub 2012/03/06.
17. Khalil AM, Guttman M, Huarte M, Garber M, Raj A, Rivea Morales D, et al. Many human large intergenic noncoding RNAs associate with chromatin-modifying complexes and affect gene expression. Proc Natl Acad Sci U S A. 2009;106(28):11667–72. Epub 2009/07/03.
18. Wahlestedt C. Targeting long non-coding RNA to therapeutically upregulate gene expression. Nat Rev Drug Discov. 2013;12(6):433–46. Epub 2013/06/01.
19. Kanasty R, Dorkin JR, Vegas A, Anderson D. Delivery materials for siRNA therapeutics. Nat Mater. 2013;12(11):967–77. Epub 2013/10/24.
20. Alexis F, Pridgen E, Molnar LK, Farokhzad OC. Factors affecting the clearance and biodistribution of polymeric nanoparticles. Mol Pharm. 2008;5(4):505–15. Epub 2008/08/05.
21. Kanasty RL, Whitehead KA, Vegas AJ, Anderson DG. Action and reaction: the biological response to siRNA and its delivery vehicles. Mol Ther. 2012;20(3):513–24. Epub 2012/01/19.
22. Bolhassani A. Potential efficacy of cell-penetrating peptides for nucleic acid and drug delivery in cancer. Biochim Biophys Acta. 2011;1816(2):232–46. Epub 2011/08/16.
23. Cheng CJ, Bahal R, Babar IA, Pincus Z, Barrera F, Liu C, et al. MicroRNA silencing for cancer therapy targeted to the tumour microenvironment. Nature. 2015;518(7537):107–10. Epub 2014/11/20.

24. Alabi C, Vegas A, Anderson D. Attacking the genome: emerging siRNA nanocarriers from concept to clinic. Curr Opin Pharmacol. 2012;12(4):427–33. Epub 2012/06/26.
25. Burnett JC, Rossi JJ, Tiemann K. Current progress of siRNA/shRNA therapeutics in clinical trials. Biotechnol J. 2011;6(9):1130–46. Epub 2011/07/12.
26. Tabernero J, Shapiro GI, LoRusso PM, Cervantes A, Schwartz GK, Weiss GJ, et al. First-in-humans trial of an RNA interference therapeutic targeting VEGF and KSP in cancer patients with liver involvement. Cancer Discov. 2013;3(4):406–17. Epub 2013/01/30.
27. Dvinge H, Git A, Graf S, Salmon-Divon M, Curtis C, Sottoriva A, et al. The shaping and functional consequences of the microRNA landscape in breast cancer. Nature. 2013;497(7449):378–82. Epub 2013/05/07.
28. Gindy ME, Leone AM, Cunningham JJ. Challenges in the pharmaceutical development of lipid-based short interfering ribonucleic acid therapeutics. Expert Opin Drug Deliv. 2012;9(2):171–82. Epub 2012/01/19.
29. Melo SA, Sugimoto H, O'Connell JT, Kato N, Villanueva A, Vidal A, et al. Cancer exosomes perform cell-independent microRNA biogenesis and promote tumorigenesis. Cancer Cell. 2014;26(5):707–21. Epub 2014/12/03.

MicroRNAs as Therapeutic Targets and Colorectal Cancer Therapeutics

13

Hirofumi Yamamoto and Masaki Mori

Abstract

The diagnosis and treatment of colorectal cancer (CRC) have improved greatly over recent years; however, CRC is still one of the most common cancers and a major cause of cancer death worldwide. Several recently developed drugs and treatment strategies are currently in clinical trials; however, there is still a compelling need for novel, highly efficacious therapies. MicroRNAs (miRNAs) are short non-coding RNAs consisting of 20–25 nucleotides that regulate post-transcriptional gene expression by binding to the 3′-untranslated region of mRNAs. miRNAs are known to regulate cancer pathways and to be expressed aberrantly in cancer. Since their initial discovery, a large number of miRNAs have been identified as oncogenes, whereas others function as tumor suppressors. Furthermore, signaling pathways that are important in CRC (e.g. the WNT, MAPK, TGF-β, TP53 and PI3K pathways) are regulated by miRNAs. A single miRNA can simultaneously regulate several target genes and pathways, indicating the therapeutic potential of miRNAs in CRC. However, significant obstacles remain to be overcome, such as an efficient miRNA delivery system, and the assessment of safety and side effects. Thus, miRNA therapy is still developing and possesses great potential for the treatment of CRC. In this chapter, we focus on miRNAs related to CRC and summarize previous studies that emphasize the therapeutic aspects of miRNAs in CRC.

H. Yamamoto (✉)
Department of Surgery, Gastroenterological Surgery, Graduate School of Medicine, Osaka University, Yamadaoka 2-2, Suita City, Osaka 565-0871, Japan

Department of Molecular Pathology, Division of Health Sciences, Graduate School of Medicine, Osaka University, Yamadaoka 1-7, Suita City, Osaka 565-0871, Japan
e-mail: hyamamoto@gesurg.med.osaka-u.ac.jp

M. Mori
Department of Surgery, Gastroenterological Surgery, Graduate School of Medicine, Osaka University, Yamadaoka 2-2, Suita City, Osaka 565-0871, Japan

O. Slaby, G.A. Calin (eds.), *Non-coding RNAs in Colorectal Cancer*, Advances in Experimental Medicine and Biology 937, DOI 10.1007/978-3-319-42059-2_13

Keywords
MicroRNAs • Colorectal cancer • Therapeutics

13.1 Introduction

The alterations of miRNA expressions can influence global gene expression networks, leading to drastic changes of cell fates including cancer initiation and progression. The aberrant miRNA expressions are observed in a wide variety of human malignancies, indicating a potential use of miRNAs as diagnostic markers and therapeutic targets. The natural endogenous expression and its remarkable stability make miRNAs a safe and efficient treatment option in cancer treatment. Now the global pharmaceutical market of miRNA-related therapy is huge and rapidly growing. It is predicted to reach 6 hundred million US dollars in 2014 and 10 hundred million in 2019. In this decade miRNA-targeting drugs have been developed all over the world, and some of them are already under investigation in preclinical randomized controlled trials. For examples, MRX34, a double-stranded RNA mimic of miR-34a encapsulated in a liposomal nanoparticle formulation, has already been in clinical trials in patients with primary liver cancer or other selected solid tumors or hematologic malignancies [1]. Moreover, miravirsen and RG-101, effective inhibitors of liver specific miR-122 that the hepatitis C virus requires for replication, have also been in clinical trials. Miravirsen is a Locked Nucleic Acid (LNA)-modified oligonucleotide complementary to miR-122, and RG-101 is Regulus' wholly-owned GalNAc-conjugated anti-miR-122 for the treatment of HCV.

However, systemic delivery technology of miRNAs as therapeutic targets/therapeutics for solid tumors has been obstructed by many limitations [2], including drug delivery systems, low specificity, adverse effects and miRNA instability.

This chapter focused on the molecular background and its clinical application of candidate miRNAs in colorectal cancer (CRC) (Table 13.1).

13.2 MicroRNAs Studied as Therapeutic Targets in Colorectal Cancer

13.2.1 miR-34a

Mutation of tumor suppressor p53 is observed in 50–75 % of CRCs [3]. Some miRNAs are known to be transcriptionally activated by p53 and exert its tumor suppressive effect through regulating a various kinds of targets [4]. miR-34a is one of the representative downstream molecules of p53. Target genes of miR-34a are associated with almost all kinds of biological processes including cell-cycle progression, apoptosis, DNA repair and angiogenesis. Upon DNA damage p53 directly activates miR-34a, and subsequent inhibition of miR-34a targets leads to a global cell protective response including cell cycle arrest and induction of apoptosis [5]. These antiproliferative effect are disadvantage for cancers, therefore the pathway should be inactivated in tumors. Indeed, downregulation of miR-34a is a common feature of human malignancies including CRC.

Recent evidence suggests that p53-dependent expression of miR-34a blocks IL-6R/STAT3/miR-34 feedback loop and consequently inhibit tumor progression in CRC [6]. As STAT3 and IL-6R play a central role in cancer proliferation, the restoration of miR-34a could be a useful treatment strategy for CRC. Nugent et al. have shown that the expression levels of miR-34a significantly decreased in CRC patients compared with healthy individuals, suggesting that miR-34a could be a useful biomarker as well as a therapeutic target in CRC [7, 8].

Notch signaling pathway is a critical regulator of asymmetric cell division, in which stem cells simultaneously generate both a daughter stem cell for self-renewal and a differentiated daughter cell to create cellular diversity [9–11]. Interestingly, recent study demonstrated that

Table 13.1 Overview of in vivo studies as potential miRNAs therapeutic targets/therapeutics in CRC

miRNA	Animal models	Oligonucleotides format	DDS	Results	References
miR-34a	Transgenic mice	–	–	Anti-tumor effect	[6]
miR-135b	Xenotransplantation of tumor-derived organoids to mice	Antisense	–	Anti-tumor effect	[13]
miR-143	Xenograft mice	3′-BP modified	–	Anti-tumor effect	[17]
miR-145	Xenograft mice	3′-BP modified	–	Negative	[17]
miR-4689	Xenograft mice	Mimic	sCA	Anti-tumor effect	[39]

miRNA, *miR* microRNA, *DDS* Drug Delivery System, *BP* benzen-pyridine, *sCA* Super carbonate apatite

expression levels of miR-34a might define a cell division as symmetric or asymmetric [12]. High expression levels of miR-34a inhibit Notch signaling pathway and promote daughter cells to create non-CCSCs, whereas its low expression levels facilitate Notch signaling and promote daughter cells to remain CCSC. Because non-CCSCs are likely to susceptible to chemotherapy and irradiation, induction of miR-34a could be a useful therapeutic strategy through promoting asymmetric division rather than maintaining CSCs.

13.2.2 miR-135b

MiR-135b plays an important role as a key downstream effector of oncogenic pathways and could be a crucial therapeutic target in CRC [13]. Furthermore, anti-miR-135b therapy shows a promise because miR-135b expression in normal colorectal tissue and other organs is very low, in contrast to other miRNAs (e.g., miR-21). Another research showed that miR-135a/b target the 3′ untranslated region of APC, suppress its expression, and induce downstream Wnt pathway activity. This study showed a considerable up-regulation of miR-135a/b expressions in colorectal adenomas and carcinomas, which correlated with low APC mRNA levels [14]. Moreover, a recent study showed that miR-135b overexpression was triggered in mice and humans by APC loss, PTEN/PI3K pathway deregulation, and SRC overexpression and promoted tumor transformation and progression [13]. This study also demonstrated that miR-135b up-regulation was common in sporadic and inflammatory bowel disease-associated human CRCs and correlates with tumor stage and poor clinical outcome. Inhibition of miR-135b in CRC mouse models reduced tumor growth by controlling genes involved in proliferation, invasion, and apoptosis. These observations suggest that miR-135b is a key downstream effector of oncogenic pathways and a potential target for CRC treatment.

13.2.3 miR-143, 145

Michael et al. first studied microRNAs changed in the adenomatous and cancer stages of colorectal neoplasia and identified that miR-143 and miR-145 act as potential tumor suppressors [15]. Consistent with this notions, the upregulation of the tumor suppressor miR-143 and miR-145 in post-therapeutic tumor tissue stand in line with the antitumor properties of the chemotherapy. This suggests that the expression levels of these miRs may be associated with prognosis or therapeutic outcome in CRC [16].

Both miR-143 and -145 have been shown to inhibit cell proliferation *in vitro* [17]. Moreover, it was reported that miR-143 directly binds to and suppresses KRAS, DNMT3A, and ERK5 and that miR-145 targets IRS-1, c-Myc, YES1,

STAT1 and FLI1 [18]. In particular, administration of miR-143 potently inhibits colorectal tumor growth in xenograft mice models. miR-143 may be a promising option as potential miRNA therapeutics for colorectal tumors [17].

13.2.4 miR-101

The Wnt/β-catenin pathway is known to play a central role in an early colorectal carcinogenesis, where inactivation of the adenomatous polyposis coli (APC) gene is one of the major tumor initiating events. More than 60 % of colorectal adenomas and carcinomas, carries inactivating mutation in APC gene, which results in a stimulation of the Wnt/β-catenin pathway [3]. Recent evidence suggests that miRNAs represent a novel mechanism for WNT regulation in CRC. For example, miR-93 suppresses colorectal cancer development via downregulating Wnt/β-catenin pathway by partially targeting Smad7. It has been reported that activation of the Wnt/β-catenin pathway significantly induced miR-101 repression, which was reverted by blocking β-catenin activity [19]. Interestingly, miR-101 overexpression in CRC cells impaired β-catenin nuclear localization and inhibited the expression of stem/EMT-related genes, while miR-101 silencing exerted opposite effects in normal colon epithelial cells. These findings suggest that pharmacological restoration of miR-101 may inhibit the aggressive behavior of CRC.

13.2.5 miR-21

miR-21 is overexpressed in a wide variety of cancers including CRC [20, 21]. Recent meta-analysis revealed that circulating miR-21 is a useful diagnostic marker for CRC with adequate sensitivity and specificity [22]. Importantly, the expression levels of miR-21 in serum is elevated even in early diseases, indicating the possible use of miR-21 in early diagnosis [23, 24]. Mechanistically, miR-21 negatively regulates *PDCD4*, which inhibits transformation and invasion in cancer. Asangani et al. identified a specific binding site for miR-21 in the *PDCD4* 3′-UTR at nucleotide position 228–249. Indeed, antisense oligonucleotides against miR-21 (Anti-miR-21) restored the expression levels of PDCD4 protein, leading to a remarkable inhibition of cancer migration, whereas overexpression of miR-21 promotes the invasive behavior of CRC cell lines [25]. A recent study also demonstrated that miR-21 is associated with invasive capacity of colorectal cancer cells through promoting nuclear translocation of β-catenin. Interestingly, this was only observed in *adenomatous polyposis coli* (*APC*)-mutated cells but not in *APC*-wild-type cells. CRC patients with high expression levels of serum miR-21 exhibit poorer prognosis in APC mutated cases, while this correlation was not observed in *APC*-wild type CRC patients [26]. Furthermore, Valeri et al. revealed that miR-21 confers resistance to 5-fluorouracil (5-FU) through downregulation of human MutS homolog 2 (*MSH2*). They also performed cell-cycle analysis and showed that G2/M arrest and apoptosis induced by 5-FU was decreased by overexpression of miR-21 [27]. miR-21 inhibitor (2′-F and 2′-MOE bicyclic sugar-modified antisense inhibitor) against hepatocellular carcinoma is currently being developed by Regulus Therapeutics [28]. Although the possible adverse effects of systemic induction of antisense oligonucleotides need to be overcome [29], anti-miR-21 therapy could be a promising therapeutic option in many types of cancers including CRC.

13.2.6 miRNAs Related to EGFR Signaling Pathway (KRAS and PI3K Pathways)

The epidermal growth factor receptor (EGFR) pathways including KRAS and PI3K contribute to promotion and progression of broad spectrum of solid tumors and it is a promising target for anticancer therapy [30]. The emerging role of EGFR signaling in cancers has led the development of anti-EGFR agents, including tyrosine kinase inhibitors (TKIs) and monoclonal antibodies against EGFR. Previously, it was considered that only patients with KRAS mutations in codons 12 and 13 of exon 2 did not have a response to anti-EGFR therapy. However, recent clinical studies revealed that other mutations in genes of the RAS family (KRAS exon 3 and 4 and NRAS exon 2, 3 and 4) are also associated with reduced response to anti-EGFR agents [31, 32]. In addition, it is estimated that 19.9 % of KRAS exon 2 wild-type tumors harbor at least one of these new RAS mutations [33]. Therefore, novel therapeutic strategies are urgently needed to treat CRC patients with RAS mutation. In this context, increasing numbers of evidence indicates that miRNAs are correlated with the drug resistance to anti-EGFR agents and regulate the EGFR signaling. For example, let-7 miRNA family has been reported to directly target KRAS oncogene [34]. Let-7 miRNA post-transcriptionally downregulates KRAS, and let-7 administration reduced tumor formation in animal cancer models expressing activating KRAS mutations. Higher let-7a expression was significantly associated with better survival outcomes in patients with mutant KRAS CRC who received salvage cetuximab (an anti-EGFR monoclonal antibody) plus irinotecan. These findings suggest that high let-7a microRNA levels in KRAS-mutated CRCs may rescue anti-EGFR therapy effects in patients with chemotherapy-refractory metastatic CRC [35].

Another central signaling pathway downstream from EGFR and important in CRC development is the phosphatidylinositol-3-kinase (PI3K)-AKT pathway. Recent study revealed that KRAS, PIK3CD and BCL2 were identified as direct and functional targets of miR-30b. Moreover, miR-30b promoted G1 arrest and induced apoptosis, suppressing CRC cell proliferation *in vitro* and tumor growth in vivo. Expression analyses using CRC clinical samples showed that a low expression level of miR-30b was closely related to poor differentiation, advanced TNM stage and poor prognosis of CRC [36]. According to other recent studies, the p85β regulatory subunit involved in stabilizing and propagating the PI3K signal was demonstrated to be a direct target of miR-126 [37]. Furthermore, this p85β reduction mediated by miR-126 was accompanied by a substantial reduction in phosphorylated AKT levels in the cancer cells, suggesting a suppression of PI3K signaling. MiR-612 was also identified to directly target AKT2, which in turn inhibited the downstream epithelial-mesenchymal transition-related signaling pathway [38]. Comprehensive microarray profiled analysis identified miR-4689 as one of the significantly down-regulated miRNAs in mutated KRAS (G12V)- overexpressing cells [39]. MiR-4689 was found to exhibit potent growth-inhibitory and pro-apoptotic effects both *in vitro* and in vivo. Further analysis revealed that miR-4689 expression was significantly down-regulated in cancer tissues compared to normal mucosa, and it was particularly decreased in mutant KRAS CRC tissues. MiR-4689 directly targets both KRAS and AKT1, suggesting KRAS overdrives this signaling pathway through inhibition of miR-4689. These observations suggested that miR-4689 might be a promising therapeutic agent in mutant KRAS CRC (Fig. 13.1). Another important regulatory component of PI3K signaling pathway is a tumor suppressor gene PTEN (phosphatase and tensin homologue). Recent study revealed that PTEN was a direct target of miR-17-5p in CRC cells [40]. Overexpression of miR-17-5p promoted chemo-resistance and tumor metastasis of CRC by repressing PTEN expression. Gain and loss -of-function studies revealed that miR-32 directly target PTEN, suggesting that miR-32 was crucially involved in tumorigenesis of CRC at least in part by suppressing PTEN [41].

13.2.7 MiRNAs in TGF-β/Smad Signaling Pathway

The epithelial to mesenchymal transition (EMT) is a critical process in tumor invasion, metastasis, and tumorigenesis. Various signaling pathways can induce EMT and include key molecules such as transforming growth factor beta (TGF-β), platelet-derived growth factor (PDGF), and the proteins nuclear factor kappa-light-chain-enhancer of activated B cells (NF-kB), Wnt, Notch and hedgehog proteins [42]. Among them,

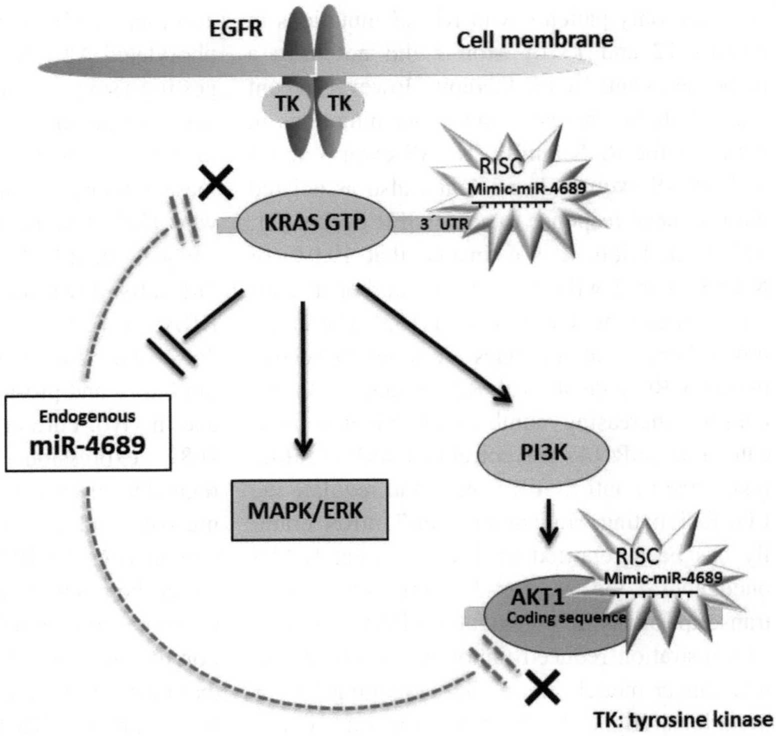

Fig. 13.1 miR-4689 regulates EGFR signaling pathway

TGF-β is one of the major inducers of EMT. TGF-β binds to its receptors (TGF-βR), leading to the activation through phosphorylation of Smad. The complex is translocated into the nucleus where it regulates the expression of DNA binding factors, such as Snail, ZEB, and Twist. miRNAs are important regulators in controlling the TGF-β/Smad signaling pathway. Recently, miRNAs have been suggested to be involved in the acquisition of stem-cell-like properties for cancer cells by regulating EMT signaling. It is reported that TGF-β2 is a predominant target of the miR-200 family. Further study has demonstrated that miR-200c aberrantly expressed in metastatic colon tumor tissues and colon cancer cells [43]. This upregulated miR-200c was correlated with a reduction of the expression of its target genes: zinc finger E-box binding homeobox 1 (ZEB1), which resulted in increased E-cadherin and reduced vimentin expression, sequentially led to an inhibition of EMT signaling pathway. In CRC cell lines, transfection of miR-200c precursors resulted in increased cell proliferation but reduced invasion and migration. Therefore, TGF-β/ZEB/miR-200 signaling regulatory network controls the plasticity between the epithelial and mesenchymal states of the CRC cells [42, 43]. Recent clinical cohort study revealed that miR-1269a expression was upregulated in late-stage CRC and was associated with relapse and metastasis of disease-free 100 stage II CRC patients [44]. In vivo and *in vitro* experiments, SW480 cells treated with miR-1269a promoted CRC cells to undergo EMT and to metastasize. Furthermore, miR-1269a directly targeted Smad7 and HOXD10 to enhance TGF-β signaling, which in turn caused TGF-β mediated up-regulation of miR-1269a via Sox4. These indicate that TGF-β and miR-1269a constitute a positive feedback loop. Taken together, miR-1269a could be a potential marker for CRC patients as well as a potential therapeutic target to suppress metastasis.

Other kinds of upregulated miRNAs in CRCs, miR-130a/301a/454 family is also shown to regulate TGF-β signaling pathway through inhibiting SMAD4. Overexpression of these miRNAs enhanced cell proliferation and migration in

HCT116 and SW480 colon cancer cells, while an inhibition decreased cell survival [45].

Another study demonstrated that miR-21 is involved in the maintenance of cancer stem cells by modulating transforming growth factor beta receptor 2 (TGFβR2) expression in colorectal cancer cells. Cell lines with increased fraction of cancer stem cells exhibit a relatively high expression of miR-21 [46].

13.3 Future Perspectives

Since the first study of miRNAs, a huge number of miRNAs have been studied as biomarkers and prognostic factors. However, only a small number of miRNAs are available as therapeutic tools. Against this background, a clinical trial of miR-34 mimics (MRX34) against hepatocellular carcinoma and metastatic liver cancer is now in phase I (ClinicalTrials.gov identifier: NCT01829971). The limited number of miRNAs available as therapeutic tools might be due to several factors. First, since miRNAs are short non-coding RNAs of 20–25 nucleotides, one miRNA could regulate several target genes transcriptionally, indicating the difficulty of targeting specific genes. At the same time, this nonspecificity leads to the possibility that one miRNA could regulate several targets and pathways simultaneously. To overcome this issue, further studies are necessary to elucidate the real therapeutic target miRNAs, which might avoid side effects of this therapy. Second, the optimal system for delivering miRNAs has not been established yet. In some in vivo studies, nanomolecules were used and their efficacy was reported (e.g. polymer nanoparticles, lipid nanoparticles, and liposomes). Recently, a new anti-miR delivery system was reported, which showed that anti-miRNAs with a low-pH-induced transmembrane structure (pHLIP) were efficiently delivered to the tumor in lymphoma cases [47]. This method could transport anti-miRNAs through the plasma membrane under acidic conditions and then deliver miRNAs specifically to tumors. Additionally, two clinical trials using Dicer substrate short-interfering RNA (DsiRNA™) are ongoing (ClinicalTrials.gov identifiers: NCT02110563 and NCT02314052). DsiRNAs are synthesized 27mer RNA duplexes that are processed by Dicer into 21mer siRNAs. This new treatment related to microRNA biogenesis is also thought to improve the delivery of miRNAs to specific targets. Thus, the systems for delivering miRNAs are continuing to advance, but further investigations are necessary for their actual use in clinical practice.

On the other hand, as mentioned previously, several target miRNAs for the therapy of CRC were elucidated and directly used for anti-miRNA therapy in vivo. Furthermore, some miRNAs (e.g. miR-17-5p, miR-140, and miR-192) have also been reported to be associated with chemotherapy resistance, which indicates the possibility of combination therapy with miRNAs and anticancer drugs. Thus, miRNA therapy has great potential to expand the therapeutic options for CRC. Although several obstacles to this still remain, miRNA therapy should lead to novel discoveries relevant to the diagnosis and treatment of CRC.

References

1. Misso G, Di Martino MT, De Rosa G, Farooqi AA, Lombardi A, Campani V, et al. Mir-34: a new weapon against cancer? Mol Ther Nucleic Acids. 2014;3:e194.
2. Decuzzi P, Causa F, Ferrari M, Netti PA. The effective dispersion of nanovectors within the tumor microvasculature. Ann Biomed Eng. 2006;34:633–41.
3. Fearon ER, Vogelstein B. A genetic model for colorectal tumorigenesis. Cell. 1990;61:759–67.
4. Hermeking H. MicroRNAs in the p53 network: micromanagement of tumour suppression. Nat Rev Cancer. 2012;12:613–26.
5. Chang TC, Wentzel EA, Kent OA, Ramachandran K, Mullendore M, Lee KH, et al. Transactivation of miR-34a by p53 broadly influences gene expression and promotes apoptosis. Mol Cell. 2007;26:745–52.
6. Rokavec M, Oner MG, Li H, Jackstadt R, Jiang L, Lodygin D, et al. IL-6R/STAT3/miR-34a feedback loop promotes EMT-mediated colorectal cancer invasion and metastasis. J Clin Invest. 2014;124:1853–67.
7. Nugent M, Miller N, Kerin MJ. Circulating miR-34a levels are reduced in colorectal cancer. J Surg Oncol. 2012;106:947–52.
8. Hahn S, Jackstadt R, Siemens H, Hunten S, Hermeking H. SNAIL and miR-34a feed-forward regulation of

ZNF281/ZBP99 promotes epithelial-mesenchymal transition. EMBO J. 2013;32:3079–95.
9. Neumuller RA, Knoblich JA. Dividing cellular asymmetry: asymmetric cell division and its implications for stem cells and cancer. Genes Dev. 2009;23:2675–99.
10. Sugiarto S, Persson AI, Munoz EG, Waldhuber M, Lamagna C, Andor N, et al. Asymmetry-defective oligodendrocyte progenitors are glioma precursors. Cancer Cell. 2011;20:328–40.
11. Clevers H. The cancer stem cell: premises, promises and challenges. Nat Med. 2011;17:313–9.
12. Bu P, Chen KY, Chen JH, Wang L, Walters J, Shin YJ, et al. A microRNA miR-34a-regulated bimodal switch targets Notch in colon cancer stem cells. Cell Stem Cell. 2013;12:602–15.
13. Valeri N, Braconi C, Gasparini P, Murgia C, Lampis A, Paulus-Hock V, et al. MicroRNA-135b promotes cancer progression by acting as a downstream effector of oncogenic pathways in colon cancer. Cancer Cell. 2014;25:469–83.
14. Nagel R, le Sage C, Diosdado Bvan der Waal M, Oude Vrielink JA, Bolijn A, et al. Regulation of the adenomatous polyposis coli gene by the miR-135 family in colorectal cancer. Cancer Res. 2008; 68:5795–802.
15. Michael MZ, O' Connor SM, van Holst Pellekaan NG, Young GP, James RJ. Reduced accumulation of specific microRNAs in colorectal neoplasia. Mol Cancer Res. 2003;1:882–91.
16. Drebber U, Lay M, Wedemeyer I, Vallbohmer D, Bollschweiler E, Brabender J, et al. Altered levels of the onco-microRNA 21 and the tumor-supressor microRNAs 143 and 145 in advanced rectal cancer indicate successful neoadjuvant chemoradiotherapy. Int J Oncol. 2011;39:409–15.
17. Akao Y, Nakagawa Y, Hirata I, Iio A, Itoh T, Kojima K, et al. Role of anti-oncomirs miR-143 and -145 in human colorectal tumors. Cancer Gene Ther. 2010;17:398–408.
18. Schetter AJ, Okayama H, Harris CC. The role of microRNAs in colorectal cancer. Cancer J. 2012;18:244–52.
19. Strillacci A, Valerii MC, Sansone P, Caggiano C, Sgromo A, Vittori L, et al. Loss of miR-101 expression promotes Wnt/beta-catenin signalling pathway activation and malignancy in colon cancer cells. J Pathol. 2013;229:379–89.
20. Kanaan Z, Rai SN, Eichenberger MR, Roberts H, Keskey B, Pan J, et al. Plasma miR-21: a potential diagnostic marker of colorectal cancer. Ann Surg. 2012;256:544–51.
21. Pan X, Wang ZX, Wang R. MicroRNA-21. A novel therapeutic target in human cancer. Cancer Biol Ther. 2010;10:1224–32.
22. Xu F, Xu L, Wang M, An G, Feng G. The accuracy of circulating microRNA-21 in the diagnosis of colorectal cancer: a systematic review and meta-analysis. Color Dis. 2015;17:O100–7.
23. Yamada A, Horimatsu T, Okugawa Y, Nishida N, Honjo H, Ida H, et al. Serum miR-21, miR-29a, and miR-125b are promising biomarkers for the early detection of colorectal neoplasia. Clin Cancer Res. 2015;21:4234–42.
24. Toiyama Y, Takahashi M, Hur K, Nagasaka T, Tanaka K, Inoue Y, et al. Serum miR-21 as a diagnostic and prognostic biomarker in colorectal cancer. J Natl Cancer Inst. 2013;105:849–59.
25. Asangani IA, Rasheed SA, Nikolova DA, Leupold JH, Colburn NH, Post S, et al. MicroRNA-21 (miR-21) post-transcriptionally downregulates tumor suppressor Pdcd4 and stimulates invasion, intravasation and metastasis in colorectal cancer. Oncogene. 2008;27:2128–36.
26. Lin PL, Wu DW, Huang CC, He TY, Chou MC, Sheu GT, et al. MicroRNA-21 promotes tumour malignancy via increased nuclear translocation of beta-catenin and predicts poor outcome in APC-mutated but not in APC-wild-type colorectal cancer. Carcinogenesis. 2014;35:2175–82.
27. Valeri N, Gasparini P, Braconi C, Paone A, Lovat F, Fabbri M, et al. MicroRNA-21 induces resistance to 5-fluorouracil by down-regulating human DNA MutS homolog 2 (hMSH2). Proc Natl Acad Sci U S A. 2010;107:21098–103.
28. Wagenaar TR, Zabludoff S, Ahn SM, Allerson C, Arlt H, Baffa R, et al. Anti-miR-21 suppresses hepatocellular carcinoma growth via broad transcriptional network deregulation. Mol Cancer Res. 2015;13:1009–21.
29. Huang L, Wang X, Wen C, Yang X, Song M, Chen J, et al. Hsa-miR-19a is associated with lymph metastasis and mediates the TNF-alpha induced epithelial-to-mesenchymal transition in colorectal cancer. Sci Rep. 2015;5:13350.
30. Ciardiello F, Tortora G. EGFR antagonists in cancer treatment. N Engl J Med. 2008;358:1160–74.
31. Douillard JY, Oliner KS, Siena S, Tabernero J, Burkes R, Barugel M, et al. Panitumumab-FOLFOX4 treatment and RAS mutations in colorectal cancer. N Engl J Med. 2013;369:1023–34.
32. Schwartzberg LS, Rivera F, Karthaus MF, Asola G, Canon JL, Hecht JR, et al. PEAK: a randomized, multicenter phase II study of panitumumab plus modified fluorouracil, leucovorin, and oxaliplatin (mFOLFOX6) or bevacizumab plus mFOLFOX6 in patients with previously untreated, unresectable, wild-type KRAS exon 2 metastatic colorectal cancer. J Clin Oncol. 2014;32:2240–7.
33. Sorich MJ, Wiese MD, Rowland A, Kichenadasse G, McKinnon RA, Karapetis CS. Extended RAS mutations and anti-EGFR monoclonal antibody survival benefit in metastatic colorectal cancer: a meta-analysis of randomized, controlled trials. Ann Oncol. 2015;26:13–21.
34. Johnson SM, Grosshans H, Shingara J, Byrom M, Jarvis R, Cheng A, et al. RAS is regulated by the let-7 microRNA family. Cell. 2005;120:635–47.

35. Ruzzo A, Graziano F, Vincenzi B, Canestrari E, Perrone G, Galluccio N, et al. High let-7a microRNA levels in KRAS-mutated colorectal carcinomas may rescue anti-EGFR therapy effects in patients with chemotherapy-refractory metastatic disease. Oncologist. 2012;17:823–9.
36. Liao WT, Ye YP, Zhang NJ, Li TT, Wang SY, Cui YM, et al. MicroRNA-30b functions as a tumour suppressor in human colorectal cancer by targeting KRAS, PIK3CD and BCL2. J Pathol. 2014;232:415–27.
37. Guo C, Sah JF, Beard L, Willson JK, Markowitz SD, Guda K. The noncoding RNA, miR-126, suppresses the growth of neoplastic cells by targeting phosphatidylinositol 3-kinase signaling and is frequently lost in colon cancers. Genes Chromosomes Cancer. 2008;47:939–46.
38. Sheng L, He P, Yang X, Zhou M, Feng Q. miR-612 negatively regulates colorectal cancer growth and metastasis by targeting AKT2. Cell Death Dis. 2015;6:e1808.
39. Hiraki M, Nishimura J, Takahashi H, Wu X, Takahashi Y, Miyo M, et al. Concurrent targeting of KRAS and AKT by MiR-4689 is a novel treatment against mutant KRAS colorectal cancer. Mol Ther Nucleic Acids. 2015;4:e231.
40. Fang L, Li H, Wang L, Hu J, Jin T, Wang J, et al. MicroRNA-17-5p promotes chemotherapeutic drug resistance and tumour metastasis of colorectal cancer by repressing PTEN expression. Oncotarget. 2014;5:2974–87.
41. Wu W, Yang J, Feng X, Wang H, Ye S, Yang P, et al. MicroRNA-32 (miR-32) regulates phosphatase and tensin homologue (PTEN) expression and promotes growth, migration, and invasion in colorectal carcinoma cells. Mol Cancer. 2013;12:30.
42. Zaravinos A. The regulatory role of MicroRNAs in EMT and cancer. J Oncol. 2015;2015:865816.
43. Hur K, Toiyama Y, Takahashi M, Balaguer F, Nagasaka T, Koike J, et al. MicroRNA-200c modulates epithelial-to-mesenchymal transition (EMT) in human colorectal cancer metastasis. Gut. 2013;62:1315–26.
44. Bu P, Wang L, Chen KY, Rakhilin N, Sun J, Closa A, et al. miR-1269 promotes metastasis and forms a positive feedback loop with TGF-beta. Nat Commun. 2015;6:6879.
45. Wang J, Du Y, Liu X, Cho WC, Yang Y. MicroRNAs as regulator of signaling networks in metastatic colon cancer. Biomed Res Int. 2015;2015:823620.
46. Yu Y, Kanwar SS, Patel BB, Oh PS, Nautiyal J, Sarkar FH, et al. MicroRNA-21 induces stemness by downregulating transforming growth factor beta receptor 2 (TGFbetaR2) in colon cancer cells. Carcinogenesis. 2012;33:68–76.
47. Cheng CJ, Bahal R, Babar IA, Pincus Z, Barrera F, Liu C, et al. MicroRNA silencing for cancer therapy targeted to the tumour microenvironment. Nature. 2015;518(7537):107–10.

Index

A

Adenomatous polyp, 154, 160
Adenomatous polyposis coli (APC), 26, 57, 110, 124, 155, 191, 241, 242
Aldehyde-dehydrogenase 1 (ALDH-1), 96
Alternative polyadenylation (APA), 125, 126, 133–134, 142
Alu repeats, 10, 11
Angiogenesis, 21, 22, 24–26, 28, 29, 54–57, 220, 240
Antagomirs, 104, 232, 233
Anti-EGFR therapy, 27, 28, 185, 218, 219, 243
AOM/DSS induced tumours, 58
Argonaute protein, 5, 7, 40, 158

B

B-cell lymphoma 6 (*BCL6*), 112, 116
Bcl2, 22, 56, 59, 62, 83, 100, 211–215, 217, 219, 243
Bevacizumab, 218–220
Bisulfite sequencing PCR (BSP), 117
BMI-1, 98, 99, 101
Bone morphogenic protein (BMP), 72, 96, 97
BRAF, 30, 216, 218, 220
BRAF-activated non-protein coding RNA (BANCR), 32, 36, 37
Brain-derived neurotrophic factor (BDNF), 112, 114, 133, 134

C

CA19-9, 38, 155, 165, 167, 187
Cancer stem cells (CSCs), 21, 30, 98, 213, 216, 221, 222, 235, 241, 245
Carcino-embryonic antigen (CEA), 155, 156, 165, 167, 187
β-Catenin, 21–26, 28, 29, 32, 37, 57–58, 63, 72, 75, 77, 78, 96, 98–100, 102, 221, 222, 242
CD133 (promonin-1), 96, 100, 101
CD166, 96
CD34, 95
CD38, 95
CD44, 96, 98, 99, 101
CDK4, 83, 112, 113
CDR1as, 40
CDX1, 101
Cell cycle, 10, 22–25, 35, 37, 56–58, 83, 101, 110, 126, 143, 191, 193, 209–215, 217, 219, 221, 240
Cetuximab, 27, 98, 100, 185, 193, 218–220, 243
Chemoresistance, 21, 24, 30–34, 98, 99, 101, 102, 194, 209, 211–217
Circular RNAs (circRNAs), 9, 12, 20, 39–40, 159
Circulating miRNAs, 163–165, 172–174, 178, 179
Circulating tumor cells (CTCs), 72, 74
CiRS-7, 40, 159
CLMAT3, 31
Colitis associated colorectal cancer (CAC), 54–64
Colon cancer associated transcript 1 (CCAT1), 32, 38, 160
Colon cancer associated transcript 2 (CCAT2), 32, 38, 143, 160, 196
Colorectal cancer screening, 171
Colorectal cancer stem cells (CRCSCs), 94–104
Colorectal neoplasia differentially expressed (CRNDE), 32, 36, 160, 166
Competing endogenous RNAs (ceRNAs), 9, 12, 13, 39, 159
CpG island hypermethylation, 115, 125, 159
CpG island methylator phenotype (CIMP), 110, 113
Crohn's disease (CD), 54, 60
CXCL3, 61
Cyclin D1, 56, 57, 60, 62, 96
Cyclooxygenase 2 (COX2), 22, 56, 102, 216
Cytochrome C, 212, 216

D

Delivery systems, 240, 245
DGCR8, 5, 7
Differentiation, 4–5, 7, 9–12, 21, 30, 34, 36, 55, 59, 64, 73, 74, 77, 94–99, 101–103, 110, 111, 129, 143, 185, 190, 196, 213, 217, 243
Dihydrofolate reductase (DHFR), 8, 9, 23, 128, 212, 215
Dihydropyrimidine dehydrogenase (DPD), 210, 213
DLEU2, 83
DNA damage, 13, 28, 56–58, 61, 63, 64, 98, 100, 101, 126, 209–211, 214, 216, 217, 220–222, 240
DNA hypomethylation, 113

O. Slaby, G.A. Calin (eds.), *Non-coding RNAs in Colorectal Cancer*, Advances in Experimental Medicine and Biology 937, DOI 10.1007/978-3-319-42059-2

DNA methylation, 22, 23, 31, 80, 103, 109, 111, 112, 114–118, 124, 155, 162
DNA methyltransferase (DNMT) enzymes, 110, 115, 116
DNA methyltransferase 1 (DNMT1), 110, 112, 113, 115, 116
DNMT3A, 22, 110, 112, 113, 241
DNMT3B, 110, 112, 113, 116
Drosha, 5, 7, 111
Dynamic polyconjugates (DPCs), 234

E
Early detection of colorectal cancer, 172, 175–177, 184, 208
E-cadherin, 25, 28, 29, 36, 37, 63, 73–75, 77, 97, 114, 215, 244
ELF3, 73, 74, 77
EMT-inducing transcription factors (EMT-TFs), 72–84
Endogenous siRNAs (endo-siRNAs), 7
Enhancer RNAs (eRNAs), 8, 9
Epidermal growth factor (EGF), 26, 72, 97
Epidermal growth factor receptor (EGFR) signaling, 22–24, 26–28, 30, 63, 110, 185, 218–220, 242–244
Epithelial to mesenchymal transition (EMT), 22–30, 36–38, 56, 57, 63, 64, 76–77, 79, 82, 94, 96–98, 100, 101, 110, 114, 195, 214, 215, 217, 243, 244
ErbB family, 26
ERCC1, 211
EREG, 101
Exosomes, 163, 164, 172, 177, 194, 235
EZH2, 36, 112–114

F
Familial adenomatous polyposis (FAP), 154
Fecal occult blood testing (FOBT), 154, 155, 161, 167, 171
5-Fluorouracil (5-FU), 30, 31, 77, 99, 128, 130, 185, 197, 209–217, 242
Fms-related tyrosine kinase 1 (FLT1), 23, 29
FOLFOX treatment, 99, 179, 217
FOSL1, 76, 78, 83
FOXO3A, 24, 212, 217

G
GAS5, 33, 189, 196
1000 Genomes Project, 124
GRHL2, 74, 77

H
H19, 12, 33, 36, 38, 160
H3K27, 36, 113
H3K27me3, 11, 36
Hepatocyte growth factor (HGF), 26, 97
Hereditary non-polyposis colon cancer (HNPCC), 154
HIF1α, 72, 78, 84
Hippo tumor suppressor pathway, 77
Histone deacetylases (HDACs), 112, 116
Homeobox C (HOXC) gene, 10, 36
Housekeeping ncRNAs, 7–8, 12
HOX transcript antisense RNA (HOTAIR), 10, 31, 33, 36, 160, 166, 189, 195, 197
HOXD gene cluster, 10, 36

I
ICAM, 56
IGF/insulin pathway, 55, 60–61
IGF-1, 23, 55, 216
IGF-IR, 55
IL-22, 55
IL-23, 55, 60, 61
IL-23/IL-17 axis, 55
IL-6, 55–57, 60–63
Immune evasion, 54, 95
Induced pluripotent stem cells (iPSCs), 97, 102, 103
Inducible nitric oxide synthase (iNOS), 58, 64
Inflammation, 57, 59–61, 72, 192
Inflammatory bowel disease (IBD), 54–64, 154, 164
Inflammatory microenvironment, 54
Interleukin-1 (IL-1), 55, 56
Invasion, 21–25, 28, 29, 32, 34, 36–38, 54, 55, 63, 76, 78, 80–84, 96, 98, 99, 101, 113, 114, 159, 176, 178, 185, 186, 190–192, 194–196, 213, 214, 216, 241–244
Irinotecan (CPT-11), 209

K
KRAS mutations, 22, 23, 26–28, 30, 77, 99, 112, 136, 137, 139, 155, 185, 187, 191, 194, 216, 218–220, 241–243
Krüppel-like factor 4 (KLF4), 24, 25, 30, 36, 61, 62, 77, 98, 100

L
LCS6 polymorphism, 219
LEF1, 73, 75, 76, 83
Let-7 family, 27, 38, 139, 191, 219, 243
Lgr5, 96, 99, 101
LIN28, 62
Linc-MD1, 12
LincRNA-p21, 12, 36, 102, 221, 222
LINE-1, 112
Lipid nanoparticles (LNPs), 234, 235, 245
Liposomes, 234, 245
LNA anti-miRs, 232, 233
LNA modification, 233
LncRNA-422, 31, 33
Long non-coding RNAs (lncRNAs), 4, 8–13, 20, 30–40, 94, 97, 102–103, 115, 140, 143, 159–160, 166, 172, 175, 189, 190, 194–197, 208, 209, 217, 220, 222, 230, 232–236
LSD1, 10, 33, 36, 112–114

M

MDM2, 34, 73
MDR transporter ABCG2, 216
MEG3, 34, 160, 189, 196
Mesenchymal to epithelial transition (MET), 22, 28, 30, 74, 84, 112, 113
Metastasis, 8, 10, 20–34, 36, 37, 39, 40, 54, 55, 75–78, 80–84, 96–102, 110, 113, 118, 130, 143, 162, 176–178, 190–193, 195–197, 235, 243, 244
Metastasis-Associated Lung Adenocarcinoma transcript 1 (MALAT1), 34, 36, 37, 160, 189, 195, 197
Microprocessor complex, 5
MicroRNA mimics, 54, 104, 222, 230–233
MicroRNAs (miRNAs), 4–5, 7, 9, 12, 20–30, 38–39, 73, 79, 81, 82, 94, 124, 140, 154, 157, 158, 161–165, 172–179, 190, 194, 208, 230–235
MicroRNAs expression vectors, 231
Microsatellite instability (MSI), 110, 114, 155, 157, 191
MiR-106b, 24, 98, 100, 103, 158, 174, 221, 222
MiR-122, 24, 61, 211, 212, 214, 233, 240
MiR-126, 22, 29, 161, 220, 243
MiR-135, 21, 26, 110, 161
MiR-143, 21, 22, 29, 60, 61, 63, 112, 157, 158, 161, 162, 174, 187, 191, 194, 211, 212, 215, 241–242
MiR-145, 21, 22, 26, 30, 60, 61, 63, 81, 83, 98, 99, 103, 157–159, 161, 162, 164, 165, 174, 211, 212, 214, 241–242
MiR-146, 59–62
MiR-150, 59–61, 64, 158
MiR-155, 21, 24, 54, 59–62, 82, 102, 110, 116, 133, 134, 164, 165, 188, 194, 234
MiR-15a/16-1, 24, 28, 83
MiR-17~92 gene cluster, 5, 7, 21, 24, 28, 59–61, 63, 81, 97, 162, 178, 187, 191, 193, 212
MiR-181a, 24, 28
MiR-200 family, 21, 24, 29, 73, 74, 80, 81, 100, 112, 114, 211, 215, 244
MiR-21, 21, 24, 26, 28, 29, 60–63, 82, 97–99, 103, 110, 157, 158, 161, 162, 164, 165, 174, 178, 186, 190, 192, 193, 211–213, 242, 245
MiR-215, 98, 100, 101, 103, 158, 187, 193, 215
MiR-221, 21, 25, 102, 158, 161, 162, 164, 165
MiR-222, 21, 137, 211, 212, 217
MiR-223, 23, 28, 60, 61, 63, 64, 158, 161, 162, 164, 165
Mir-29 family, 60, 61, 112, 113
MiR-31, 21, 24–27, 102, 116, 157, 158, 164, 165, 174, 186, 190, 192, 211, 212, 214, 219
MiR-34, 60, 61, 63, 64, 73, 80, 83, 84, 99, 112, 113, 118, 134, 137, 214, 234, 245
MiR-34a, 21, 22, 28–30, 63, 80–84, 98–100, 103, 158, 162, 211, 212, 214, 233, 240–241
MiR-7, 28, 40, 159, 165, 219, 220
Miravirsen, 233, 240
MiRNA "seed" sequence, 125, 126, 132, 133, 233
MiRNA precursor (pre-miRNAs), 7, 110, 111, 128, 131, 140
MiRNA Sponge, 12, 38, 39, 159, 160, 233
MiRNA targets, 5, 40, 112, 125–128, 132–133, 140–142, 222
MiRSNP, 126, 128, 132–143
Mitogen activated protein kinase (MAPK) pathways, 8, 22, 23, 25, 55, 56
MLH1, 110, 124, 211
MMR deficiency, 211
MRX34, 234, 235, 240, 245
MS-MLPA, 116, 117
MYB, 59, 213
MYC, 22, 28, 32, 36, 38, 61, 63, 83, 84, 99, 113, 143, 196, 219
MYH-associated polyposis, 154

N

NANOG, 12, 30, 36, 98, 99, 102
ncRAN, 34, 189, 196
Neuropilin-1, 25, 29
Nitric oxide synthase (NOS2), 56, 58
NOD2, 55, 60, 61
NOD-like receptors, 55
Notch, 22, 23, 72, 96–100, 240, 243
NRAS, 185, 243
Nuclear factor-kB (NF-kB), 56, 57, 62, 97
Nuclear-enriched abundant transcript 1 (NEAT1), 10, 34, 166, 189, 195
Nucleotide excision-repair (NER) pathway, 210, 211, 221

O

Oct4, 12, 36, 97–102
Oncomirs, 21, 27, 28, 34, 209, 213
Oxaliplatin (L-OHP), 209, 211
Oxidative stress, 55, 56, 58, 64, 133

P

p53, 8, 22–24, 26–28, 30, 33, 34, 38, 39, 56–58, 61, 63, 64, 73, 80–84, 99, 102, 103, 110, 113, 155, 176, 214–217, 240
Panitumumab, 185, 218–220
Piwi-associated RNAs (piRNAs), 4–7, 20, 114, 125, 143, 158, 161, 190
Plasmacytoma Variant Translocation 1 (PVT1), 34, 38, 160
Platelet-derived growth factor (PDGF), 26, 72, 243
Platelet-derived growth factor receptor (PDGFR), 97
Polycomb repressive complex 2 (PRC2), 9–11, 33, 36, 113, 195
polyethylene glycol (PEG), 234
Primary miRNA transcripts (pri-miRNAs), 5, 7, 111, 125
Progenitor miRNA (pro-miRNA), 5, 7
Programmed cell death protein 4 (PDCD4), 21, 24, 25, 27, 62, 98, 99, 190, 192, 212, 213, 242
Proliferation, 10, 21–26, 28–30, 32–34, 36–38, 54–60, 62–64, 72, 95–101, 110, 111, 143, 159, 185, 190, 193–196, 208, 209, 213–218, 220, 222, 240, 241, 243, 244

Prostate cancer-associated non-coding RNA 1 (PRNCR1), 31, 34, 38, 143, 160
Proteome, 156
PTEN pseudogene (PTENP), 12, 21, 24–28, 56, 60–62, 98, 100, 213, 218, 222, 241, 243
PTEN pseudogene 1(PTENP1), 12, 34, 38
PTEN/PI3K signaling, 24, 28–29, 100

R
Radiation, 33, 72, 209, 220–222
Radioresistance, 98, 100, 103, 221
Ribosomal RNAs (rRNAs), 4, 8, 158
RNA-induced silencing complex (RISC), 5, 126, 232
RNA Polymerase II, 5, 7, 158, 159
ROCK, 22, 29

S
Single-nucleotide polymorphisms (SNPs), 38, 61, 124–144
Sirtuin 1 (SIRT1)-p53 pathway, 22, 30
SLUG, 22, 72–77, 97, 214
SMAD, 25, 28, 72, 78, 243–245
Small interfering RNAs (siRNAs), 4, 5, 7, 20, 125, 232, 233, 245
Small nuclear RNAs (snRNAs), 4, 166
Small nucleolar RNAs (snoRNAs), 4, 7, 12, 114, 125, 158–159, 190
SNAIL, 22, 56, 63, 72–77, 80, 83, 84, 97, 100, 101, 244
SNORD50A/B, 8
SNPs in miRNA target genes, 126, 133
Sonic Hedgehog signaling, 96
Sox2, 12, 23, 30, 36, 76, 97–102
STAT3, 24, 56, 57, 60–63, 72, 76–78, 83, 240
Staufen-1 (STAU1), 9, 11
Stem-cell self-renewal, 21, 30, 57, 72, 77, 95–97, 99, 103, 158, 240
Stool-based screening tests, 160

T
TAZ, 77
TGF-β/Smad signaling, 243–245
TH17 differentiation, 59
Thymidylate synthase (TYMS), 209, 210, 212, 215, 216
TINCR, 9, 11
TLR4 signaling pathway, 59
TNM staging, 82, 185, 197
Toll-like receptors, 55–56, 61, 62
Total colonoscopy (TC), 154
TP53 status, 23, 95, 134, 191
Transfer RNAs (tRNAs), 4, 8, 12, 125
Transforming growth factor beta receptor 2 (TGFβR2), 99, 245
Transforming growth factor-beta (TGF-β), 26, 38, 97, 110, 243–245
TRBP, 5, 7
Triantennary N-acetylgalactosamine (GalNAc) conjugates, 234, 240
TRNA-derived fragments (tRFs), 8
Tumor necrosis factor-α (TNF-α), 30, 55
Tumor suppressor candidate 7 (TUSC7), 34, 38, 160
Tumor-associated antigens (TAAs), 155, 156
Tumor-initiating colorectal cancer cells, 95
Twist, 56, 72, 74, 76, 77, 84, 97, 244

U
Ulcerative colitis (UC), 54, 57, 81, 161
3' Untranslated region (3' UTR), 4, 25, 61, 111, 125, 216, 241
Uridine diphosphogluronysltransferase 1A1 (UGT1A1), 211
3'UTR associated SNPs, 125
3'UTR heterogeneity, 125, 142

V
Vincristine (VCN), 31
Volatile organic compounds (VOCs), 156

W
Wnt/β-catenin signaling, 21–23, 26, 32, 37, 57–58, 63, 77, 96, 99, 102, 242

X
X-chromosome inactivation, 4, 11, 110
XELOX treatment, 217, 220

Y
YAP1, 61, 63, 77

Z
ZFAS1, 35, 159, 160
Zinc finger E-box binding homeobox 1 (ZEB1), 22, 24, 26, 29, 37, 38, 61, 72, 74, 76, 77, 80, 84, 97, 98, 100, 101, 112, 114, 212, 215, 217, 244
Zinc finger E-box binding homeobox 2 (ZEB2), 22–24, 26, 29, 37, 38, 72, 74, 76, 80, 98, 101, 112, 114, 215
ZNF281, 22, 73, 75, 80, 83

MIX
Papier aus verantwortungsvollen Quellen
Paper from responsible sources
FSC® C105338

If you have any concerns about our products,
you can contact us on
ProductSafety@springernature.com

In case Publisher is established outside the EU,
the EU authorized representative is:
Springer Nature Customer Service Center GmbH
Europaplatz 3, 69115 Heidelberg, Germany

Printed by Libri Plureos GmbH
in Hamburg, Germany